D1243425

Thermally Stable Polymers

Thermally Stable Polymers

Syntheses and Properties

Patrick E. Cassidy

Department of Chemistry
Southwest Texas State University
San Marcos, Texas

Marcel Dekker, Inc. New York and Basel

Library of Congress Cataloging in Publication Data

Cassidy, Patrick E [date]
Thermally stable polymers.

Bibliography: p.
Includes indexes.
1. Polymers and polymerization--Thermal properties.
I. Title.
QD382.H4C37 547.8'4 80-18571
ISBN 0-8247-6969-4

Most of the experimental procedures in this volume were adapted from *Journal of Polymer Science, Journal of Applied Polymer Science,* and *Macromolecular Syntheses* as noted in the text, with permission of John Wiley and Sons, Inc.

MARCEL DEKKER, INC.
270 Madison Avenue, New York, New York 10016

Current printing (last digit):
10 9 8 7 6 5 4 3 2 1

PRINTED IN THE UNITED STATES OF AMERICA

To my son, Andrew

Foreword

Since the mid-1950s there has been much activity in the study of thermally stable polymers, much of which has been activated and supported by the Materials Laboratory of the U.S. Air Force at Wright-Patterson Air Force Base, Ohio. The field was reviewed by Frazer in 1968. Since then many advances have been made. The new developments are well covered in this new review of the field by Dr. Cassidy and these are dovetailed with the discussions of the older work.

Dr. Cassidy has brought the literature references up to mid-1979. As new features in the book he has collected good preparative methods for the various types of thermally stable polymers. He has treated the problems of their fabrication. He has also included an appendix of commercial products, their names, formulas, and producers. This review will be welcomed by the host of people working in this still very active field.

Carl S. Marvel

Preface

It is generally recognized that the science of thermally stable materials entered a new phase in the mid-1970s. For over 15 years the emphasis in this area was synthesis of new types of backbones. Numerous functional groups were incorporated into polymers in an effort to find an organic material significantly more resistant to thermooxidation than those known. Recently, however, synthesis research has been directed toward the improvement of processability of the polymers and fabrication of hardware. Primary uses have been as adhesives, coatings, fibers, and ablatives.

Other applications have also been sought for these new materials. In many cases the new areas of use were not dependent on the thermal stability of the polymer. For example, it appears that high-modulus fibers, reverse-osmosis membranes, flame-resistant cloth, and corrosion-preventive coatings may be important areas in which these new polymers serve very well.

There is increased interest also in the development of inorganic backbones because chemists have generally agreed that the limit of thermooxidative stability for organic backbones has been attained.

For the applications-oriented scientists and engineers in materials science, the next few years may prove to be even more important than the recent past. In order for this to happen, however, the synthetic chemist must take a new look at polymers and their synthetic methods. Easier, less expensive preparations will be needed for polymers which can then be processed or fabricated into an end-use item.

The purpose of this book is to provide the research scientist or advanced student with a single source for syntheses and properties of all polymers which may be considered to be thermally stable. It is also meant to provide an updated and expanded source beyond those by A. H. Frazer and V. V. Korshak, published in 1968 and 1969, respectively. The literature has been thoroughly surveyed through mid-1979, but not all references to a particular topic were necessarily included. The references which were chosen are those that give the most recent or best synthetic procedure and properties. Detailed preparative methods and properties of new polymers are included with few exceptions. Some data on commercially available, thermally stable polymers are also given. The reader will see that some of the commercial materials (aromatic polyamides, silicones, and polyesters) are not discussed in great detail. However, some of the less well-known polymers in these classes are presented. Monomer synthesis has not been included in detail in most cases; for this the reader is referred to T. W. Campbell, *Condensation Monomers* (J. K. Stille, Ed.), Wiley-Interscience, New York, 1972.

One of the very real dangers in preparing a compilation such as this one is the error of omission. There is much that could be included but is not—work that is important and well done. To those who have published such research which is not cited here, I can only apologize and ask for their understanding of this immense task.

The book begins with introductory material such as definitions, history, and analytic methods. The following chapters present polymers according to backbone composition or functional group in the order: nonheterocyclic, heterocyclic, ladder, and organometallic and inorganic. Since there is an inordinate amount of information available on nonheterocyclic and heterocyclic systems, these two subjects were broken into smaller units according to backbone composition for the former and ring size for the latter.

Each chapter begins with a table of structures, names, and some properties of the polymers which are discussed in that chapter. For each polymer the historical development is given briefly, followed by a discussion of the chemistry, properties, and uses. Each section on a particular polymer terminates with the detailed description of the synthesis.

Three appendixes have been prepared. The first is a cross-reference of nomenclature and structure for each functional group included in the book. This appendix facilitates the use of this book by those whose background in organic nomenclature (particularly heterocycles) is not extensive. This first appendix also points out some of the different names which have been given to the same structure.

The second appendix is a list of abbreviations and acronyms which are used frequently in this field and other work.

The third appendix is a list of commercially available, thermally stable polymers. This listing includes commercial name or designation, composition and structure (as much as it is known), and supplier or manufacturer.

A bibliography of useful books and reviews is included at the end of the present volume.

In the preparation of this manuscript, the author has recognized the untold value of other persons—not only those who did the original research, but those who assisted more directly. Southwest Texas State University and the faculty of the Department of Chemistry, Dr. B. J. Yager, Chairman, contributed significantly in allowing the author to experience the luxury of working on one major project without interruption. This was made possible by a development leave for the Spring semester of 1976. Dr. Newton C. Fawcett, a postdoctoral fellow at Southwest Texas, was a key individual in this process because he helped to direct the group of students involved in my research group at Southwest Texas State during the preparation of this manuscript. Further, Dr. Fawcett was coauthor of part of Chapter 5 on polyoxadiazoles and polythiazoles, a more detailed version of which appeared in the *Journal of Macromolecular Science — Reviews in Macromolecular Chemistry* [*C17*(2), 209-266 (1979)].

The ongoing laboratory research was possible only through the generous support of the Robert A. Welch Foundation, Dr. W. O. Mulligan, Director of Research. Mr. Frank Lee was diligent in helping to collect data from numerous references. Both Texas Research Institute, Dr. J. Scott Thornton, President, and the Department of Chemistry at the University of Arizona, Dr. Lee B. Jones, Chairman, were extremely helpful in providing space and materials. Of course the typists, Sandra Posey, Derra Raymor, and Roxie Smeal, were essential in the completion of the work. Dr. Jack Preston provided several useful suggestions. Mr. Richard W. Thomas drew the many and complex structures herein, an act which earned my undying gratitude, and he assisted in many other ways. W. R. Leverich was very helpful in preparing the index and converting units to SI, and Miss Debra Harper aided greatly in the final proofreading. Others, especially Carl Eaton and Carter King, provided the inspiration and encouragement so necessary at times. Finally, my heartfelt appreciation goes to Professor C. S. Marvel who has been an extraordinary influence on my professional life. As a man and as a teacher, he has shown me and hundreds of others the true meaning of greatness in chemistry.

Patrick E. Cassidy

Contents

Thermally Stable Polymers

Chapter 1

Introduction

One of the newest specialty areas within polymer science is that of high-temperature or thermally stable materials. It has been recognized as a separate area within polymer chemistry for nearly 20 years. Like many scientific disciplines, this one received most of its momentum from the space program owing to the need for ablative systems, high-temperature adhesives and coatings, and heat- and flame-resistant fibers. Some of the more recently recognized benefits derived from this research were serendipitous, i.e., not related to thermal properties. Two of these are high-modulus fibers and reverse osmosis membranes.

By far the largest amount of research has dealt with organic polymers, especially aromatic heterocycles. Inorganic and organometallic systems are gaining notoriety and are covered in the last chapter. Of course, organic compounds are not conducive to high-temperature applications. When one addresses high-temperature uses for polymers, one must define the thermal stress in terms of time and temperature. An increase in either of these factors shortens the expected lifetime; and if both are increased the use time is shortened logarithmically. In general terms it has been said that for a polymer to be considered thermally stable, it must retain its physical properties at 250°C for extended periods, at 500°C for intermediate periods, or up to 1000°C for a very short time (seconds). More specifically, the total stress has been given as 177°C (350°F) for 30,000 hr, 260°C (500°F) for 1000 hr, 538°C (1000°F) for 1 hr, or 816°C (1500°F) for 5 min [1]. Whether a usual application (such as a wire coating) or a special application (such as an ablative shield) is to be encountered, test conditions must be designed to be similar to anticipated use. This is so because different classes of polymers do not show parallel behavior over different tests. The temperature/time requirements mean that organic polymers have an inherent disadvantage. In fact, most polymer scientists now agree that the ultimate temperature stability has been observed and further work toward this single goal is futile. However, this inherent problem notwithstanding, there are some reasons that organic systems are still favored. These advantages include low density (more specifically, a high strength-to-weight ratio), ease of processing (compared to metals and ceramics), availability of wide structure variations, flexibility of properties, and the ability to have properties tailored to a specific use by structural changes.

The requirements for practical, thermally stable polymers dictate high melting (softening) temperatures, resistance to oxidative degradation at elevated temperatures, resistance to other (nonoxidative) thermolytic processes, and stability to radiation and chemical reagents. The problem is even more challenging when the requirement for processibility is added.

There are three ways to improve the thermal stability of a polymer: increase crystallinity, crosslink, and/or remove thermooxidative "weak links." Crystallinity development has limited application for truly high temperatures and results in lowered solubility and more rigorous processing conditions. Crosslinking of oligomers is certainly useful and does make a real, albeit irreversible, change in properties. This approach is being developed for some of the linear backbones to allow their use as adhesives because no gaseous byproducts are formed. The weak links to be avoided are mostly alkylene, alicyclic, unsaturated, and nonaromatic hydrocarbons and NH. Those functions which have proven to be desirable are aromatic (benzenoid or heterocyclic), ether, sulfone, and some carboxylic acid derivatives (imide, amide, etc.). Further, these stable functions can be incorporated into bridged or ladder configurations to further improve their degradation resistance.

For practical applications it is desirable to have use temperatures (stable physical properties) above 550°C but processing temperatures below 400°C. Regarding processing, one must also consider the intended application vis-a-vis the reaction (if any) during final cure and the possibility of off-gas. For example, an adhesive which generates a gas on curing would be unacceptable in a closed, laminated system since numerous voids would be created (1 mol of gas at 205°C is 40 liters in volume).

This chapter is designed to provide an historical background on the beginning of the field and some recent trends. The types of thermal data available to the engineer and chemist and the techniques commonly used to test thermally stable polymers are also discussed briefly, including advantages, disadvantages, and interpretation limitations. Finally, a bibliography of some excellent books and review articles can be found at the end of the book.

The nomenclature used throughout this work is that used by the authors of journal articles. However, the reader should be made aware of "Structure Based Nomenclature" [*Macromolecules, 1*(3), 193 (1968)] as a rigorous but somewhat cumbersome system to apply to polymers.

1.1 Historical Background

The longtime goal of the high-temperature polymer chemist has been to achieve a material which can exist for a long period in air at 500°C. Unfortunately, for the purposes of historical comparison, most thermal stability data are obtained by thermogravimetric analysis (TGA) rather than by long-term, isothermal aging. (See Sections 1.2.2.2 and 1.2.2.3 for detailed discussions and comparisons of these experimental techniques.) It is unfortunate because the two types of data cannot be compared for different polymer types. Generally, if a material is to withstand use at 500°C, then its TGA should show a stability of 780-800°C.

In Figure 1.1 TGA data for polymers are arranged over the year of their discovery, permitting one to see graphically any progress toward higher stabilities. The graph, although approximate, reveals some interesting trends, or the lack of same.

First, even from the very beginning of the era, stabilities of 400-500°C in air (by TGA) were realized and within 6 years reached 600°C in air; and from 1966 on these values ranged from 500 to 600°C. For nitrogen atmospheres, 800°C stability was possible early in the period with polyphenylenes and again with polyquinoxalines, but no material has surpassed 800°C in a TGA test. Not only is there no progression in polymer stabilities with time, but also all stabilities in air fall between 400 and 600°C.

It seems then that we have seen the temperature limits of organic and even some inorganic and organometallic polymeric backbones. It is unfortunate that one cannot graph solubility or tractibility (i.e., solubility and heat softening) or processability, for then a more significant trend would be obvious. By the mid-1960s inherent limitations in stability were suspected and the problem of tractibility was readily apparent in that most polymers up to that point were "brick dust"-like in physical properties. The research at that point began to turn to the modification of known systems to make them more soluble, processable materials. These structural modifications included reduction of crystallinity, phenylation of the

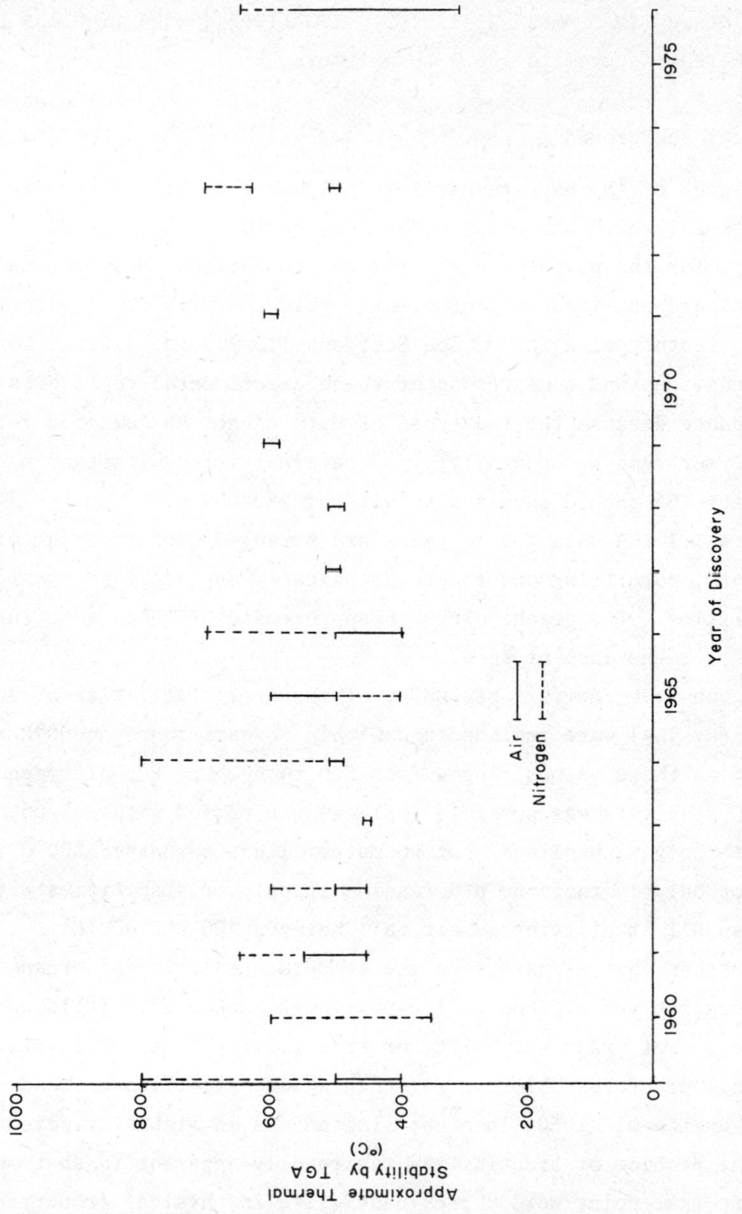

Figure 1.1 Thermal stabilities of newly discovered polymers: in air (——), in inert gas (usually nitrogen) (----).

backbone, introduction of flexibilizing units (ether, sulfone, alkylene, etc.), copolymerization, and constructing an ordered backbone structure. Almost without exception these changes led to lower stability, but were nonetheless necessary if applications were to be developed.

This problem of thermal stability vs tractibility points out nicely the opposing forces facing most polymer chemists. On one hand, organic molecules are not heat-resistant and are destined to be oxidized below 250-350°C in air. On the other hand, nearly any approach to improve stability is detrimental to tractibility. In commercial materials one frequently sees the sacrifice of thermal stability to improve the ease of processing. Hence, the past 10 years' research has been directed not toward higher stability in polymers but toward retention of as much stability as possible while introducing solubility or moldability.

The above phenomena can also be seen perhaps in the following table of significant dates. A flurry of activity in the synthesis of new polymers is seen in the mid-1960s due mostly to the success of polybenzimidazoles. After that time, however, the frequency of appearance of new polymers declined as known polymeric materials were moved into the product, or end-use, development phase. (The activity level just might be connected also to the amount of funding available for research! As money becomes less available people are more interested in justifying expenditures by more directly applicable developments.)

1.2 Thermal Analysis

Perhaps the most important and yet least reliable factor in the study of heat-stable polymers is the measurement or evaluation of thermal stability. There are numerous methods used, and no two can be directly compared in all cases. Sometimes even the same technique is not comparable in different reports or for different polymer compositions.

Basically, the goal is to monitor some physical property in regard to temperature. The unreliability of measurements is due to (1) the complexity of polymer change or degradation and (2) the effect of this decomposition on the measured properties. A further complication is the environment of the decomposition. One must be concerned with the removal or retention of volatile products of decomposition, i.e., is a gas to be flowing over the sample and, if so, at what rate? If the atmosphere around the sample is static, what effect do the decomposition products have on further degradation? A reaction can also ensue between polymer and gaseous environment

Table 1.1 Significant Dates and Occurrences in High-Temperature Polymers

1944	Polythiazoles isolated but not characterized
1950-1956	Polyacrylonitrile ladder polymers isolated but not fully cyclized
1955	Poly(p-xylylene)
1957	Low molecular weight polyphenylenes
1958	(a) Polyoxadiazoles first recognized
	(b) Aromatic polyamides patented
1959	Polyphenylenes of high molecular weight (through 1967)
1960	(a) Regular ladder polyquinizarine attempted from polyacrylonitrile
	(b) Polythiazoles
	(c) Polyperfluoroalkyltriazine elastomers
1961	(a) Polybenzimidazole, first significant aromatic polyheterocycle
	(b) Polyoxadiazoles
	(c) Polybisthiazoles
	(d) Polytriazoles
	(e) Aromatic polyamides
1962	Polyborobenzimidazoline
1963	(a) Fibers produced from polyoxadiazoles and polytriazoles
	(b) Processable polytriazoles
1964	(a) Polytetraazopyrene
	(b) Polyquinoxalines
	(c) Ladder polyaromatic heterocycles suggested
	(d) Ordered aromatic copolyamides
1965	(a) Polythiazoles improved
	(b) Polypyrrazoles
	(c) Polybenzimidazoles with no hydrogen present on N
	(d) Poly(ether quinoxalines)
	(e) Copoly(benzimidazole oxadiazole)
	(f) Polybenzimidazole fibers produced
	(g) Pyrrolone ladders, first good ladder system
	(h) Polybenzthiazoles
	(i) Polybenzoxazoles
1966	(a) Regular cyclized ladder from polyacrylonitrile
	(b) Polyisoxazoles

Table 1.1 (Continued)

	(c) BBB and BBL (polybisbenzimidazophenanthroline ladders)
	(d) Poly(pyrazine quinoxaline) ladders
	(e) Poly(pyrenequinoxaline) ladders
	(f) Ordered aramid fibers (aromatic polyamides)
1967	(a) Polythiadiazoles
	(b) Polyetherpyrrones and biphenylpyrrones
	(c) Linear polyphenylquinoxaline
1968	Experimental polyimide fibers revealed
1969	Pyrazine polyimides (no hydrogen in structure)
1971	(a) Poly(carborane siloxanes)
	(b) Ladder polymers produced from aromatic dianhydrides and tetraaminoanthraquinone
1972	Reverse osmosis properties published
1973	(a) Poly(pyrazine pyrrolones)
	(b) Adhesive developments
1976	Polyimidines and poly(benzodipyrolediones)

at extreme temperatures. In an inert atmosphere, usually nitrogen, helium, or argon, the weight of property loss is less severe and it begins at a higher temperature than with air. Further, the loss does not go to 100%, even at extreme temperatures in the absence of a reactive gas (air). The last phenomenon is due to the formation of a carbonaceous char- or graphite-like residue. In air at elevated temperature a polymer will, of course, eventually oxidize with subsequent chain scission and crosslinking. Ultimately, oxidation and bond cleavage will yield all gaseous products and a 100% weight loss. There is only one polymer backbone functional group, carborane, which shows anomalous results. The change of weight data for these polymers are not valid when the experiment is done in air due to oxidation of the carborane nucleus to a stable oxide, which results in a weight gain. The situation is somewhat more complicated than this, in that weight is being lost by virtue of oxidation of the organic radical; and simultaneously weight is gained by oxide formation. The net result is a very slight loss.

Therefore, weight loss is used to assess stability. Usually a precipitous loss (over a 10°C range) is observed as a polymer is heated to higher temperatures while continuously monitoring weight. The weight change can be monitored against a steadily increasing temperature (dynamic) or against increasing time spans at constant temperature (static). The former method has been used more in the past because it is faster and the data are fewer and are easier to analyze. However, when one considers the relationship between weight loss and change in physical properties the picture again becomes cloudy.

Polymer decomposition is further complicated by factors which are often unknown, such as initiator residues, sensitive terminal groups, impurities (especially inorganic types), residual solvent, etc. Few investigators concern themselves with these parameters, even though significant differences can be caused by them.

The above problems notwithstanding, one must know the meaning, advantages, and shortcomings of each type of result of the tests applied to heat-stable polymers. The following two subsections discuss first the types of thermal analytic data usually found in the polymer literature; second, the techniques by which these data are generated are outlined. The discussion includes advantages, disadvantages, and factors which affect the outcome of thermal analyses. This information, however, is presented as an overview or introduction to the methods and not as a comprehensive source. For more detail the reader is referred to the Bibliography.

Finally, it should be noted that the terminology used by the polymer scientists may not conform to the standards recommended by ICTA (International Committee for Thermal Analysis). When differences occur the ICTA nomenclature is noted in parentheses.

1.2-1 Thermal Analytic Data

Table 1.2 is a list of five types of thermal analytic determinations with a brief description of each. After some inspection it will be obvious that the order of these on a temperature scale is Tm > softening point > PMT > HDT > Tg. This is due to either (or both) the type of transition involved, i.e., primary or secondary, and physical conditions of the test, i.e., load or no load. As a material experiences these transitions, its mechanical properties can change by an order of magnitude.

1.2.1.1 Glass Transition Temperature The glass transition temperature, Tg, is the temperature at which the amorphous phase of the polymer is converted between rubbery and glassy states. Since cooling causes the polymer to freeze into an unordered solid, this process is also referred

Table 1.2 Five Types of Thermal Analytic Determinations

Determination	Type of thermodynamic transition and load conditions
(1) Tg: glass transition temperature, the rubbery to glassy phase transition	Secondary/no load
(2) Tm: crystalline transition temperature, melting point of polymer crystalline regions (crystallites)	Primary/no load
(3) PMT: polymer melt temperature or melting point, temperature of adhesion when a slight force is exerted on a polymer which is resting on a hot surface	Secondary load
(4) Softening point: qualitative observation of temperature at which polymer softens under no load	Secondary/no load
(5) HDT: heat deflection (or distortion) temperature, an ASTM test for rigidity vs temperature with a constant load applied (stress)	Secondary/load

to as "vitrification." Tg is a secondary thermodynamic transition related to the flexibility of a polymer backbone, the secondary interchain forces acting on it, and the free volume of the system. These same parameters determine Tm; therefore, it is reasonable that Tm and Tg vary similarly from polymer to polymer. Generally, Tm exceeds Tg by a factor of 1.4-1.7 on an absolute temperature scale. The more able a chain or chain segment is to move, the less energy is required and therefore the lower the temperature at which movement occurs. This motion is now thought to be a segmental "jump-rope" rotation.

Numerous contributing factors to Tg can be listed: bond type; bond length; bond polarity; rotational freedom; type of functional group(s); types of secondary bonds; regularity (tacticity) of backbone; type, size, number, and placement of pendant groups; etc. These all regulate the freedom of motion of a polymer chain and are, of course, interrelated.

It should be noted here that the Tg being discussed is only one secondary transition of which there may be many. The Tg is the most thermodynamically significant of these and is designated as the alpha transition. Others which occur at lower temperatures are labeled beta, gamma, etc., and are due to rotational motions of side groups or end groups or small portions of the backbone. For example, PTFE (polytetrafluoroethylene) has indicated the presence of six secondary transitions.

One can measure the Tg in a number of ways either mechanically or calorimetrically. Naturally the value of Tg obtained is dependent on the method used for detection and even on the heating rate used. It is quite possible to pass through a transition region either too slowly or too rapidly and thereby never detect the actual Tg. The calorimetric methods will be discussed in Section 1.2.2.1. The mechanical determination can be made by monitoring almost any physical property while gradually changing the temperature because the properties change so drastically from the rubbery to the glassy state. Some of these properties are hardness, volume, strength, percent elongation-to-break, modulus, and permeability. The temperature at which a relaxation occurs produces a change in the effect of the property with temperature. It is a secondary change and not a discontinuity as in the case with Tm. This process is expressed graphically in Figure 1.2. The value of Tg is obtained by extrapolating the linear values of the property above and below this Tg to their intersection. Above this transition area the amorphous portion (disregarding now any crystallinity which, if present, is unaffected) is in the rubbery state while below this point it is rigid or frozen into an irregular backbone conformation.

Two other properties which are either important or of interest in relation to Tg are modulus and gas solubility. Each of these plotted against temperature gives the response shown in Figure 1.3. The modulus response (Figure 1.3A) indicates an extreme softening or increase in flexibility as the temperature increases up to and beyond the Tg. The gas solubility curve (Figure 1.3B) is more difficult to explain. It appears that

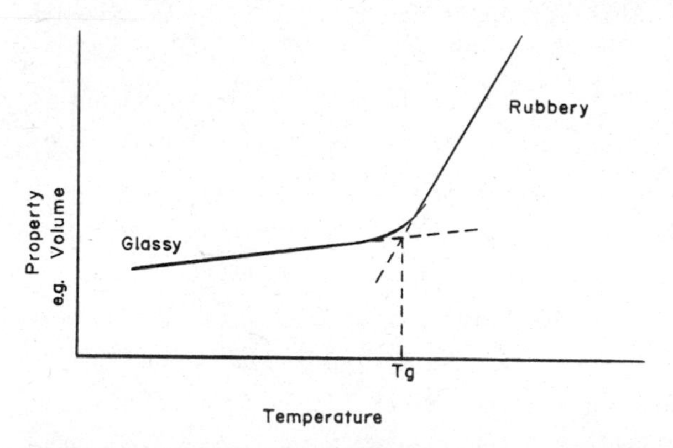

Figure 1.2 The relationship of properties to temperature through the glass transition region.

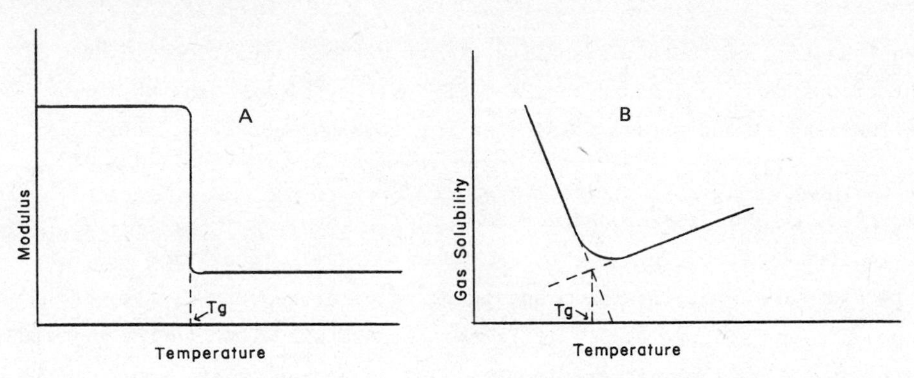

Figure 1.3 The relationship of (A) modulus and (B) gas solubility to temperature through the glass transition region.

above the Tg, gas solubility is dependent upon free volume in the rubbery, amorphous (disordered) phase, while below the Tg, it is inversely proportional to the kinetic energy of the amorphous region, much like the solubility of a gas in a liquid. Of course any crystallinity is disregarded since gas solubility in a crystallite (i.e., crystalline region) is negligible compared to that in the amorphous phase.

The Tg is important for end users of a polymeric product in that most applications demand that no significant property changes occur over the temperature range of its use. As seen later, the Tg is, as one would expect, closely related to other mechanical property changes with temperature (HDT) which are important for use by engineers in design criteria.

The Tg can be affected externally or internally on a molecular level as regards the chain backbone. It can be lowered by the addition of plasticizers which allow a greater molecular freedom in the polymeric matrix. This could be considered an external modification which almost always lowers the Tg. However, some fillers or additives can serve to form secondary bonds to flexible portions of the backbone and thereby raise the Tg. Internal processes can also affect the Tg in either direction. "Internal" refers now to a chemical modification rather than a physical one. Anything which will decrease the flexibility (stiffer functional groups in the backbone, crosslinking, etc.) will increase Tg, and those factors which increase backbone flexibility (primarily bonding) will lower Tg. In the former category are aromatics (para catenation gives a higher Tg than ortho or meta), ladder structures, highly polar substituents (-CN), and multiple bonding. In the latter area of flexibilizing functions are alkylene, ether, sulfide, sulfone, and m-phenylene. Unfortunately the flexibilizing groups also cause a lessening of thermal stability almost without exception. Even pendant groups have been found to affect Tg, alkyl, alkoxy, and phenoxy groups lowering it and phenyl raising it for some heterocyclic, thermally stable systems [2].

1.2.1.2 Crystalline Transition (Melting) Point The crystalline transition temperature, Tm, is a primary thermodynamic transition which exhibits a discontinuity in the property vs temperature curve rather than a change in slope (Figure 1.4). It is the temperature at which the crystallites (if any) lose their intermolecular bonding and go to the amorphous, rubbery state. Like the melting of ordinary organic molecules, this transition can be observed and accurately determined by use of a polarizing microscope and hot stage. Since these types of intermolecular bond forces are much stronger than those in the amorphous glass, the Tm always occurs at a value higher than Tg. The ratio of Tm/Tg on an absolute temperature scale is generally found to be in the region of 1.4-1.7. For this reason little change in property is seen at the Tm for polymers with low crystallinity; i.e., the polymer is largely already rubbery (the continuous phase of the system is rubbery), and the bulk properties are determined primarily by the continuous phase which does not experience the Tm. Figure 1.5 shows the Tg and Tm relationships on one graph.

1.2.1.3 Polymer Melt Temperature The PMT test takes on something of a less scientific character than DTA or DSC (differential thermal analysis or differential scanning colorimetry). The experimental procedure involves a heated platform, and across this with a spatula is pushed a sample of the polymer at ever-increasing temperatures until the sample sticks and can be caused to spread under stress. When comparing PMTs one must be

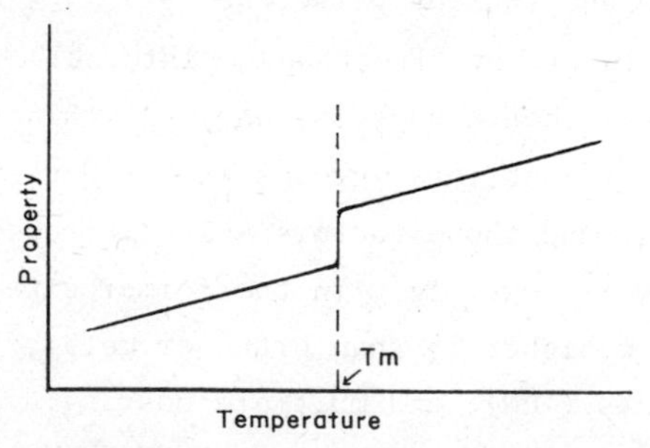

Figure 1.4 The relationship of property to temperature through the crystalline transition region.

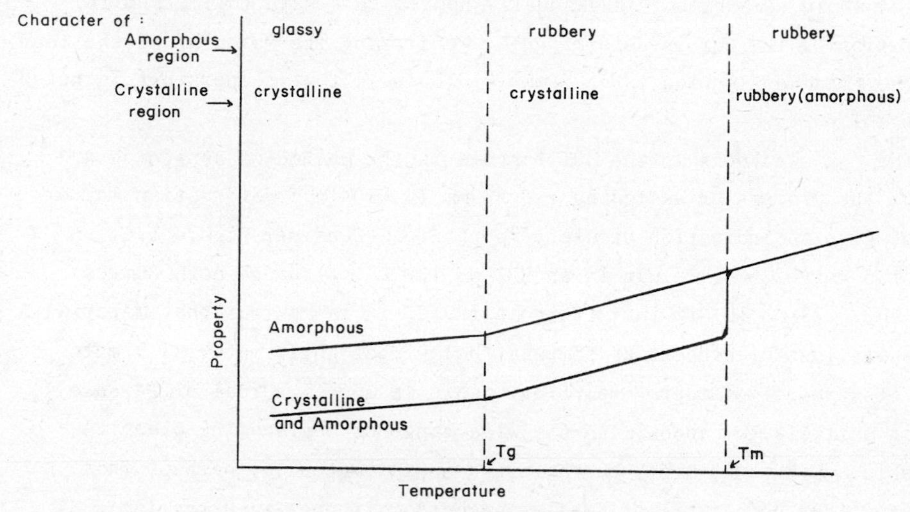

Figure 1.5 A composite of the relationship of the properties of both amorphous and crystalline materials to temperature.

aware of the variable techniques, observations, and polymer behavior; that is to say the data reported may be considered to cover a range of temperatures. Small differences in PMT are not significant.

1.2.1.4 Softening Point (or Temperature) Softening temperature data are frequently reported but infrequently detailed with respect to experimental procedure. They have been reported as being taken in a capillary melting point device or on a hot plate. Each instance is unique and must be judged on the particular technique used (if delineated). Generally, the softening point is the temperature at which deformation of the polymer can be observed with no applied load. The qualitative nature of this process usually dictates that a range of temperatures be reported.

1.2.1.5 Heat Distortion (or Deflection) Temperature The HDT test is used by those who are interested in structural application of a polymeric material because this technique measures the temperature at which a certain deformation occurs under load (ASTM D-648). Specifically, a bar is prepared to be 1.27 x 1.27 x 12.7 cm (1/2 x 1/2 x 5 in.) and is rested on fulcrums 10.2 cm (4 in.) apart. On the top center of the bar a static load is placed to generate either 455 kpa (66 psi) (or 185,000 stones per square rod for the purist) or 182 kpa (264 psi) stress, depending on the desired severity of the test. (The lower stress will, of course, give a higher HDT value.) Then the whole arrangement is gradually heated in a bath (oil), and the deformation of the bar is determined by monitoring the movement of the load. When the total deflection is 0.25 mm (0.010 in.), that temperature is noted as the HDT.

The main weakness in the HDT test is in the method of assigning a value. The process of assigning a 0.25-mm (0.010-in.) deformation may or may not give an indication of use temperature. Consider Figure 1.6, on which two curves both result in an HDT of 100°C. Although both curves cross the 0.25-mm (0.010-in.) limit at 100°C, it is obvious that material A begins its loss of properties somewhat below that while material B maintains its property-temperature relationship to 150°C, a real difference and not unlikely to happen with the wide range of engineering plastics available. Since the break in the curve approximates Tg, perhaps this indicates that both types of data are necessary to obtain a realistic picture of material expectations.

Now it may seem that HDT and Tg are at least related since both types of data are based on the same phenomena. The expected correlation has been shown to exist in at least one system, a crosslinked epoxy resin [3]. Figure 1.7 shows the rise in both HDT and Tg with an increase in the amount of crosslinking agent epoxytetrahydrophthalic anhydride (ETHPA). (A higher crosslink density leads, of course, to a more rigid material and a higher

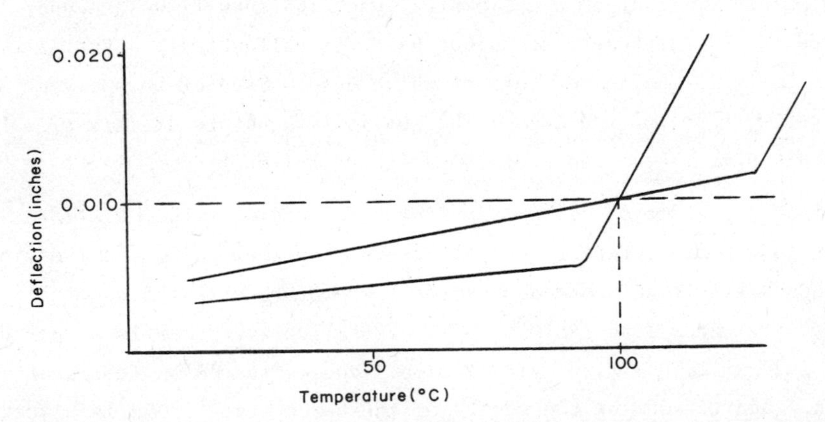

Figure 1.6 Two possible representations of deflection vs temperature.

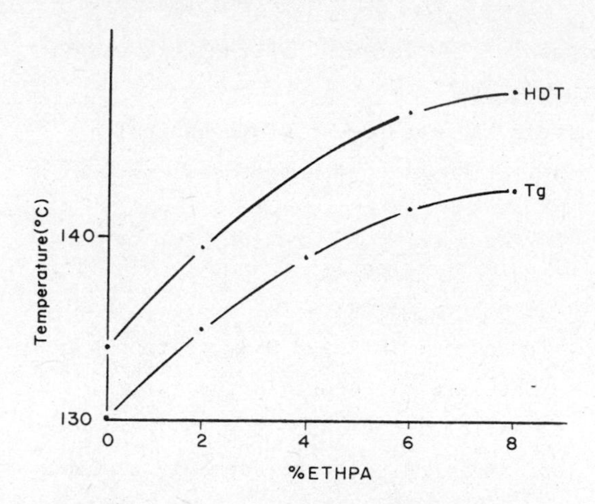

Figure 1.7 A comparison of HDT and Tg values for an epoxy resin with varying amounts of crosslinking.

Tg and HDT.) The two criteria parallel one another nicely, with the Tg being below HDT by 4-7°C. It follows that Tg < HDT for any composition, because before a deflection can occur, the material must be heated through its Tg to allow some flexibility. This relationship is more useful than it appears since it serves to reduce the conflict between the processor or user of a polymeric material and the laboratory research chemist.

1.2-2 Thermal Analytic Techniques

Many techniques and instruments are known to those involved in thermal analysis. The six listed below, however, are the ones found most applicable to polymeric materials.

1. DTA and DSC: differential thermal analysis and differential scanning calorimetry; methods to determine Tg and Tm
2. TGA thermogravimetric analysis, a measurement of weight vs temperature. (ICTA terminology: TG --dynamic thermogravimetry).
3. Isothermal thermogravimetric analysis: a measurement of weight vs time at constant temperature. (ICTA terminology: isothermal weight-change determination)
4. EGA: effluent (or evolved) gas analysis, a monitoring of volatile byproducts of thermal or oxidative decomposition processes (also called TEA --thermal evolution analysis. (ICTA terminology: EGD --evolved gas detection)

5. TBA: torsional braid analysis: a measurement of complex mechanical properties vs temperature
6. TMA: thermomechanical analysis, a measurement of mechanical response (commonly probe penetration at constant stress) with temperature

1.2.2.1 Differential Thermal Analysis/Differential Scanning Calorimetry There are two common laboratory methods of which the polymer chemist avails himself to determine both the Tg and Tm: DTA and DSC. Although both give similar-looking graphic results, they are different in the thermal techniques used. Both methods utilize an insulated chamber with thermocouples in contact with the sample, and an inert ballast, commonly aluminum oxide. With DTA a third thermocouple is present to monitor the temperature of the chamber, all of which is heated. In DSC, however, there are two small heaters each in contact with sample or reference so that the materials are heated individually rather than by the chamber. The output is traced by an x-y recorder. The abscissa is temperature and the ordinate is the thermal character of a process; a positive peak is exothermic and negative is endothermic.

Briefly, then, what is the difference in principle of operation of DTA and DSC? In DTA the observed thermogram is a measure of the difference in temperature between the sample and the inert reference material as the temperature of the common chamber is raised. So, if a material experiences a Tg or Tm, energy is consumed and its temperature lags behind that of the reference, hence a negative slope.

In DSC the measured datum is not temperature differential, but rather a measure of power input required to maintain equal temperatures between sample and reference as both are raised in temperature according to a predetermined program. So, in this situation the current required for the individual heaters is measured and this differential is plotted. As the name indicates, DSC can be used to conduct quantitative experiments regarding a thermal process with 1-2% accuracy.

An ideal (i.e., one which is never seen in practice) curve demonstrating both Tg and Tm is shown in Figure 1.8. The first endothermic break in the curve represents Tg and the sharp endothermic peak the Tm. Frequently the base line is seen to come back gradually toward the original location after experiencing Tg.

There is another thermodynamic change which is seen occasionally, and that is an exotherm due to crystallization. This occurs between the Tg and

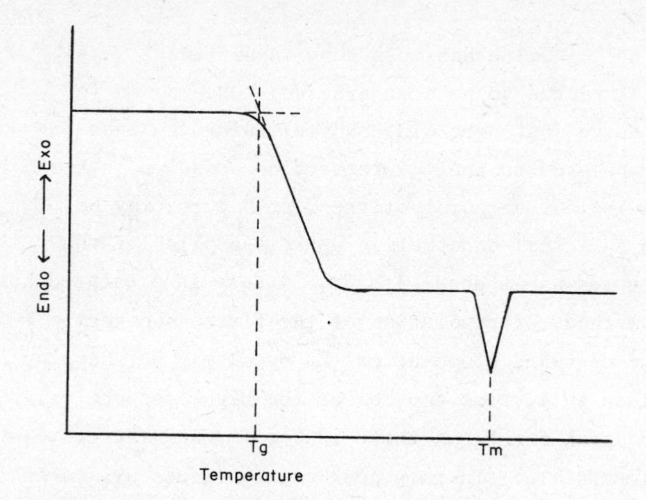

Figure 1.8 An ideal representation of Tg and Tm determination by DSC or DTA.

Tm, and is due to the fact that the amorphous region may develop crystalline regions after it is in its mobile rubbery state. Therefore a slightly more complex figure may present itself with three transitions, the crystallization exotherm just preceding the Tm (Figure 1.9). If one traces the cooling curve of the sample, the crystallization exotherm is readily obtained (only if, of course, crystallization occurs). It occurs at a slightly lower temperature than Tm due to some supercooling effect. If the sample is rapidly quenched, then no crystallites form and no exotherm is observed on cooling and no endotherm on reheating the sample.

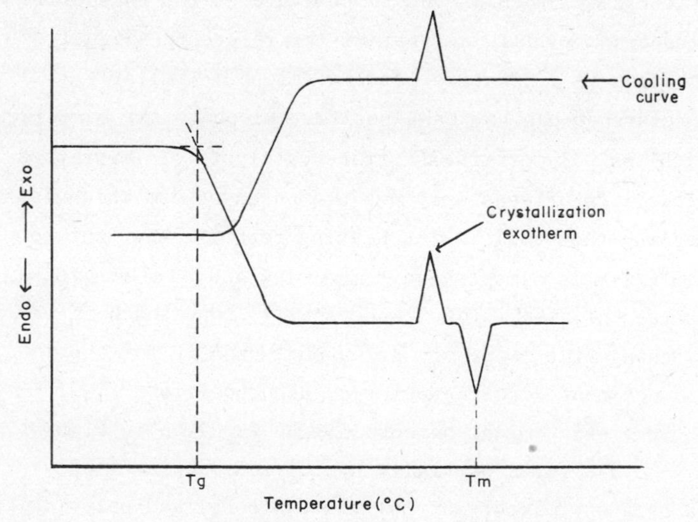

Figure 1.9 Ideal DSC or DTA heating and cooling curves showing a crystallization exotherm.

The interpretation of the Tg value has been done in different ways. There is a need for these techniques to be standardized. One way is for the linear portions of the curve just preceding and following the endothermic Tg break to be extrapolated to their intersection as shown in the figures above. Second, the inflection point of the flat S curve may be determined; and third, when an actual endothermic peak is available, the minimum of the peak may be noted as is done with Tm. A case is presented here to make standard the first method, extrapolation of the first endothermic break, for two reasons. The downward slope after the break may be sharp or gradual and therefore may lead to a value too low or too high, respectively. Also, there is frequently no peak for Tg as there is for Tm, and therefore the third method is often impossible. In this book all Tg values are taken as the extrapolated endothermic break where possible. This method has been shown to be reliable and reproducible [4].

As mentioned above, the graphic data are ideal to the extreme. Frequently one witnesses a sloping baseline, gradual breaks, broadly rounded curves, and nonlinear portions of the trace, all of which make interpretation difficult at best. These are due to imbalance between thermocouples, broad molecular weight distributions, thermocouple placement, heat flow problems, sample contact with holder, etc. Interpretation then becomes one of recognition based on experience alone. For example, an imbalance between heat capacities of the reference (ballast) and sample can cause a baseline slope for either method. If the two materials are similar in weight but the sample has a high specific heat, then as the temperature of the chamber is increased the sample temperature will lag that of the reference slightly. The result of this situation is a negative baseline slope. Similarly, excess ballast in the reference pan causes a positive slope to the baseline. Reactions may occur with the polymer itself, trace catalysts or impurities to create anomolous peaks. An off-gas will produce an endotherm which is irreproducible on recycling the sample. The heating rate is important to visibility of the Tg; 5°C/min is almost always too slow and 30°C/min too fast to be able to observe the transition. And finally, repeated heat cycles will frequently change the shape of the output, usually for the better if the sample is not heated to the point of decomposition.

A more realistic curve which might be encountered is given in Figure 1.10, showing three DTA curves for a partially hydrolyzed sample of polyvinyl acetate, i.e., a backbone mixture of polyvinyl alcohol and polyvinyl acetate. The first heating curve usually gives the poor results shown and

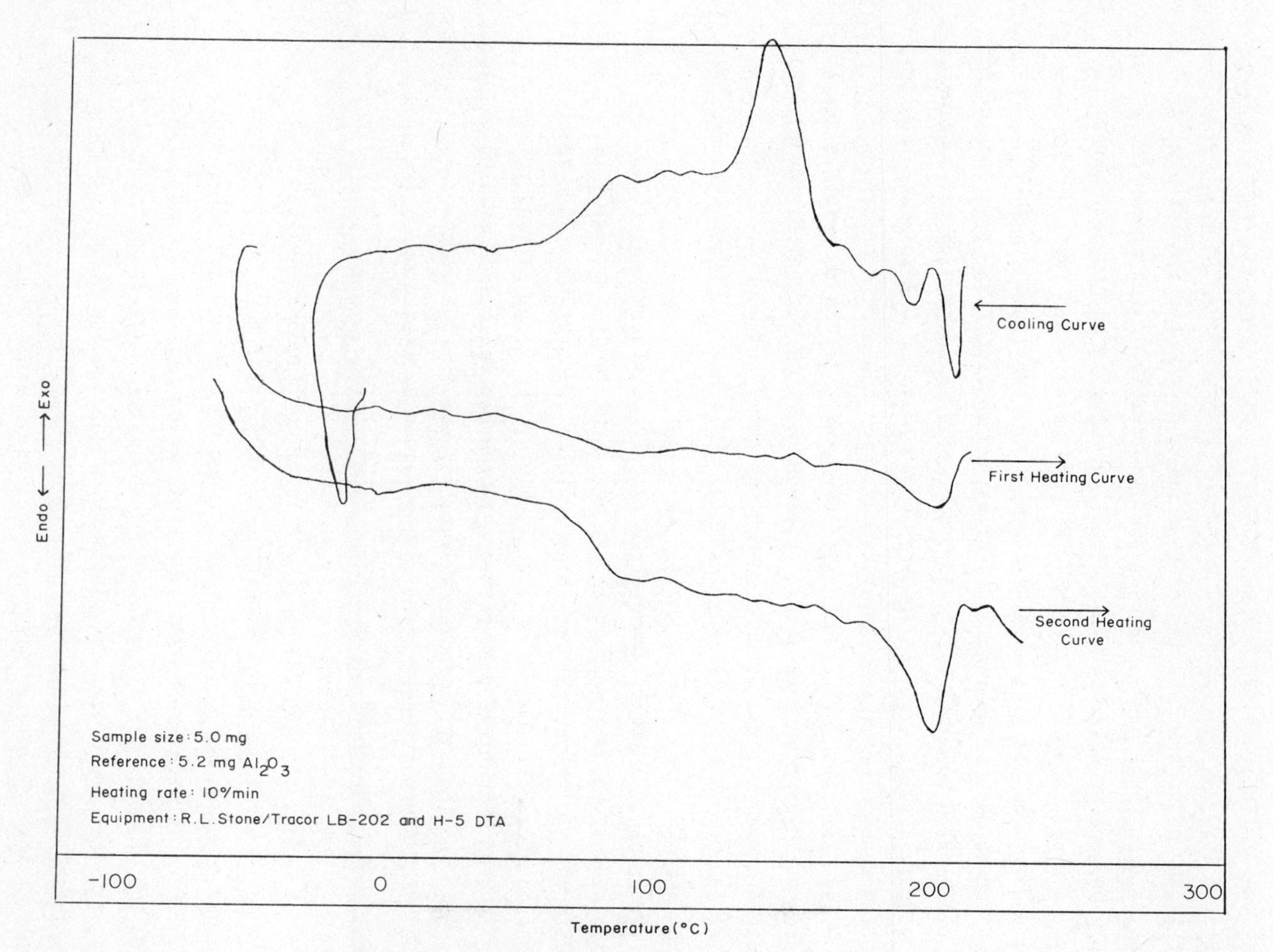

Figure 1.10 Actual DTA curves for partially hydrolyzed polyvinyl acetate. (Courtesy of Mr. Fred Weber, Texas Research Institute, Austin, Texas.)

therefore should be repeated. This insensitive curve may be due to the fact that the sample merely is not in good contact with the pan. Furthermore, the first trace is dependent on the thermal history of the sample; so the first heating provides a known history. The second heating process affords a more reliable curve on which the desired data are more evident. On this curve one observes (by extrapolation of the first endothermic break) a Tg of 60°C and a sharp crystalline melting point at 200°C. These same phenomena are observed in the cooling curve in an opposite manner. First, however, one must learn to disregard certain artifacts of the equipment such as the "ringing" at the start of the cooling curve (~200°C) due to overcorrection of heating and cooling processes. With controlled cooling one experiences an exotherm at 135-140°C which is due to crystallization of the rubbery material. It is lower than the Tm value due to a supercooling effect. Near 70°C the endothermic slope is encountered due to formation of the glass. If, however, one quenches the sample from a temperature above its Tm by flooding it with liquid nitrogen, it will not be able to crystallize. A thermogram following a quench then will show no Tm.

1.2.2.2 Thermogravimetric Analysis Thermal stability is a property sought by both the synthetic chemist and the design engineer; unfortunately, the former needs a facile comparative test for different materials, while the latter needs test data which tell him something about a material in real conditions. TGA gives the research chemist a readily comparable indication of thermal stability but is of little value to the applications people. This is due to the time-temperature relationship for polymers which complicates the problems of assigning a stability value. TGA is generally thought of as being somewhat insensitive in comparison to isothermal aging because weight loss detected by TGA can occur either prior to or following the loss of mechanical properties.

TGA data are collected by subjecting a continuously weighed sample (5-10 mg) to an increasing temperature (5-20°C/min) in a controlled, flowing atmosphere (air, nitrogen, or argon). Some of the complicating factors for TGA are sample size and packing, particle size, heating rate, gas flow rate, diffusion of pyrolysis products out of the sample, weight gain due to reaction with the atmosphere, and placement of the sensing thermocouple.

The methods of interpreting TGA results are numerous and also lack standardization. Some of these note the temperature of a 7% weight loss, or a 10% loss (the latter done for this book), or by the temperature inflection of the downward slope, or by the first detectable break in the curve (termed the *incipient weight loss* by some workers). Still another method requires integration of the area under the first plateau. It is suggested here that the most reliable interpretation is by extrapolation of the break in the curve, much like that done for Tg. By investigation of the extreme examples of TGA curves in Figure 1.11, one can see the pitfalls of various interpretive methods. By assuming that a 10% loss is the thermal stability limit by TGA, one can assign a value of 400°C to samples 1, 2, and 3 although they have widely different behavior. Furthermore, sample 4 would be assigned a stability of 100°C even though it is quite similar to sample 3 and in fact is better than sample 3 at elevated temperatures. Samples 3 and 4 are experiencing a loss of a volatile residue and should be preheated to 100°C before testing. Water, either physically absorbed or chemically bound, can be lost from numerous polymers anywhere below 300°C, and care must be taken not to confuse water loss with thermal or oxidative degradation.

A differential curve of thermogravimetric data is seen infrequently but serves to obviate some of the interpretation problems. This method, DTG (differential thermogravimetric) analysis, provides a first derivative of the weight retention curve (see Figure 1.12). The stability of the polymer is then taken as the maximum of the first derivative, i.e., the maximum rate of degradation.

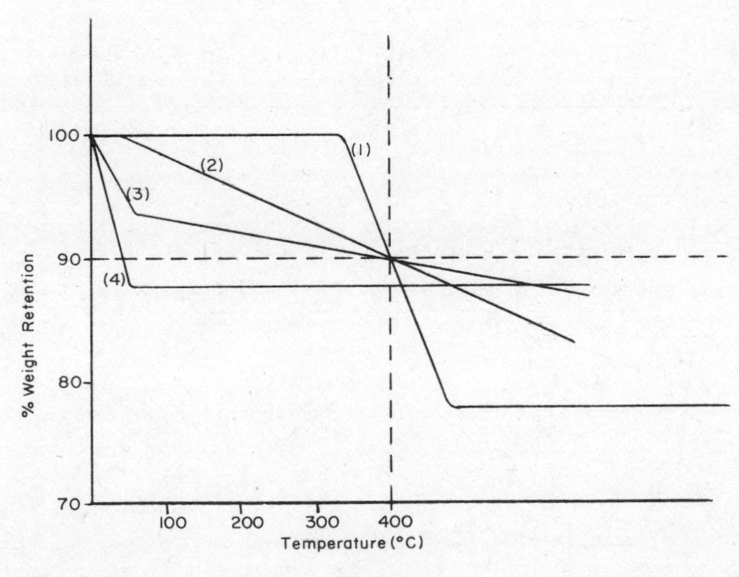

Figure 1.11 Possible routes to a 10% weight loss value by TGA.

In air one usually observes a single break in the curve whereupon the weight retention proceeds to zero. In an inert atmosphere (argon or nitrogen), however, the initial weight loss occurs at a higher temperature and generally proceeds through a plateau at 40-60% to a quite high temperature (800-1000°C), commonly to the limit of the equipment. This plateau is caused by the formation of an extremely stable carbonaceous char. These two behaviors are demonstrated in Figure 1.12.

TGA and DTA or DSC data can be useful when compared; and they can be directly compared since they are taken under like time-temperature conditions. For example, an endotherm detected by DTA/DSC may be Tg or Tm or merely a loss of a volatile byproduct. If the endotherm is caused by a material loss, it is readily observable by TGA.

1.2.2.3 Isothermal Weight Loss Analysis The isothermal weight loss analysis method is analogous to TGA except that weight loss is monitored at a constant temperature with time. The advantage is that a much more real situation exists in testing. It is much more severe than TGA in that for a long-term exposure distinctly lower temperatures can be tolerated. Unfortunately, there is no correlation between, or even a way to compare, TGA and isothermal data. Examples of TGA and isothermal data from two polymeric systems are given in the following table. Note the time variable in the last column.

Polymer	Weight loss (%)	TGA stability (°C)	Isothermal stability
Polyimide	5	520	450°C/10 hr
Polybenzoxazole polyimide	10	580	350°C/9 days

It is even difficult to compare isothermal weight loss data from two different experimental sources owing to the variables and complexity of the

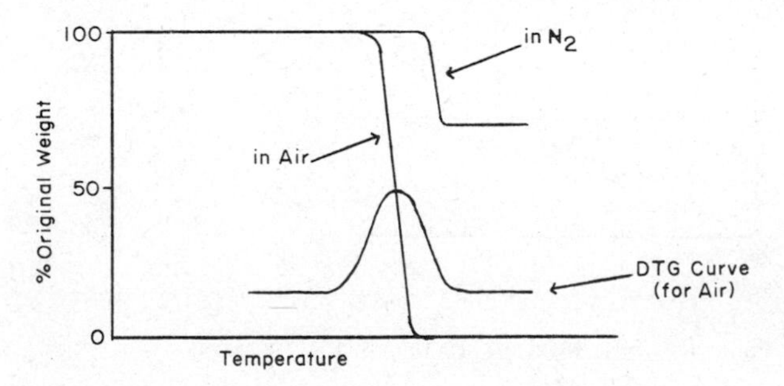

Figure 1.12 TGA curves with a differential (DTG) curve superimposed.

data. In some instances a percent weight or property loss is reported for a certain time period and temperature; and in other cases the time or temperature necessary to reach a loss value is reported. This is all to say that three variables exist in an isothermal evaluation: weight (or property) loss, time, and temperature. One must be careful in comparing time data to ascertain that the temperature and losses are identical in all cases. TGA experiments, on the other hand, are much more readily comparable since time is not a variable.

The effect of test temperature in an isothermal situation on weight loss can be seen in the above table and in Figure 1.13. The isothermal stability of a polymer is seriously shortened by increasing temperature.

The following table shows the drastic effect of time exposure for different materials by comparing TGA data to use temperatures, another type of isothermal test. Of course, use temperature implies more than merely weight retention. It also means that no phase change occurs (i.e., the material does not become thermoplastic), discoloration is minimal, and mechanical properties are not significantly affected (modulus, strength, creasibility, adhesion, etc.). Use temperature, then, is a very severe criterion, perhaps the most stringent set of conditions one can use to qualify a polymer.

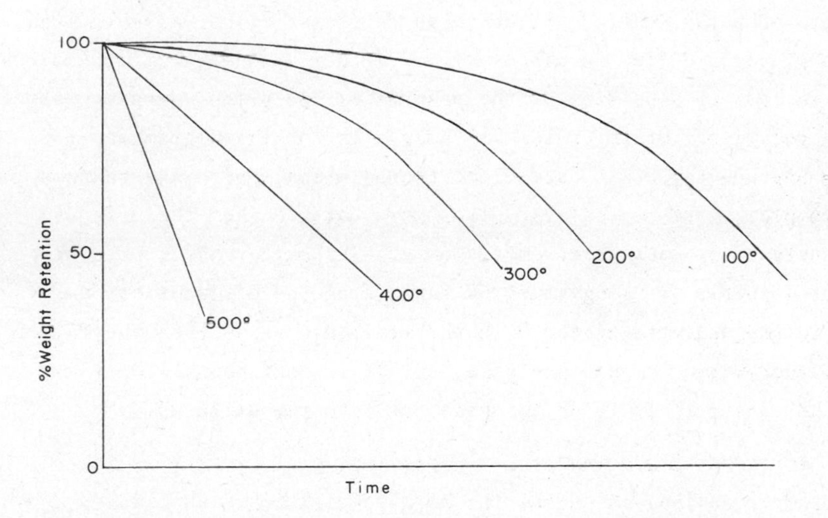

Figure 1.13 A family of isothermal gravimetric analysis curves showing time of stability vs test temperature.

Polymer	TGA stability in air/N_2 (10% wt loss) (°C)	Max. use temp. (°C) for 10 min	200 hr
Polyimide	554/582	377	358
Polybenzimidazole	454/660	>650	320
Polybenzthiazole	632/-	538	330-343
Polyphenylquinoxaline	460-520/550	-	>316

It is interesting that although TGA stabilities vary by nearly 200°C, the 200-hr use temperatures are all within about 40°C of one another.

Another example of the difference between TGA and isothermal analysis has been demonstrated using polystyrene and a substituted polyphosphazene [5]. The TGA curves of these materials are quite similar, showing 10% weight losses in air at 340° and 400°C, respectively. However, under aging at 165°C, styrene degrades by random scission 10 times faster than the polyphosphazene.

1.2.2.4 Effluent Gas Analysis EGA is used in connection with DTA, DSC, and TGA to monitor the gases given off during a heating cycle. The simplest way to accomplish this is to have a thermoconductivity detector in the effluent gas stream through the sample chamber. Of course, this does not afford qualitative data since only the presence of a contaminant is determined and not its identity. For this reason, then, the method is sometimes, and more properly, called effluent gas detection. The value in such analyses with DTA or DSC lies in discovery of whether an endotherm or exotherm is due to loss of a portion of the pendant or end groups or a contaminant which is volatile. Other causes or changes in the thermogram may not have volatile byproducts, e.g., oxidation, crosslinking, and phase changes.

A more complex system which also gives more data is when the off-gas is actually analyzed by mass spectrometry or gas chromatography. Of course in these cases a sample or "slug" must be taken from the effluent stream and admitted to the analytic system. In the strictest sense then the former system is only detection and not analysis; but it is continuous. On the other hand, the latter is truly an analysis but is not continuous.

1.2.2.5 Torsional Braid Analysis Torsional braid analysis, infrequently referred to as TBA, is one of the more complex tests applied to high-temperature polymers, but is important because it is a dynamic process, all others discussed being static or passive types, i.e., with no applied variable load. By this technique a polymer which cannot support its own weight can be studied mechanically vs temperature. Investigations such as resin curing or environmental degradation can be made.

In this procedure one uses a thin piece of polymeric film or a glass fiber braid coated with the polymer (~100 mg). To one end a torsional stress is applied, and the strain is measured at the other end, a sort of torsional pendulum stress-strain determination. The strain lags behind the applied stress in an amount measured by the internal friction. This internal friction is perhaps better thought of as a damping mechanism, i.e., an energy absorption process. As a material is heated through its Tg, the energy absorption realizes a maximum at the Tg owing to the molecular process of secondary bond breakage.

The Tg (or other secondary transition) can be determined, obviously, by a change in modulus or damping. In TBA a sample is slowly heated while being monitored for modulus and internal friction to result in the idealized graphs shown in Figure 1.14. The damping phenomenon also occurs at the Tm and at the temperature of weight loss by TGA.

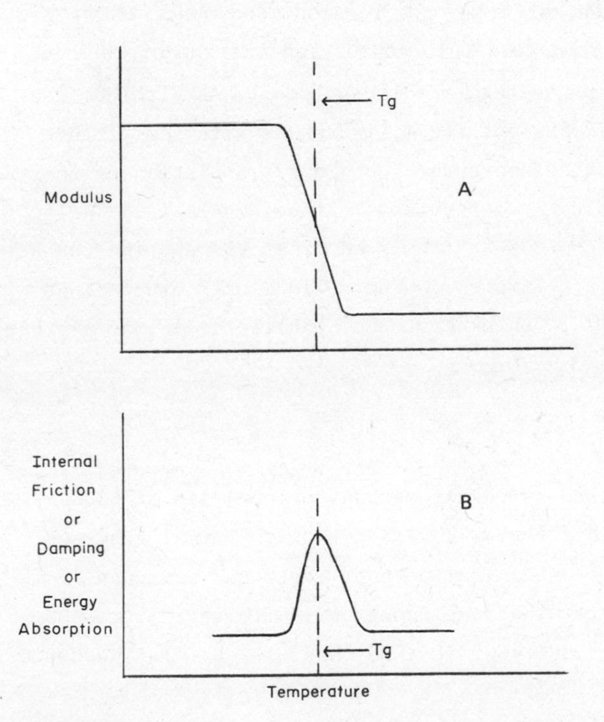

Figure 1.14 The relationship of (A) modulus and (B) damping to temperature through the glass transition region.

At low test frequencies (~1 cycle/sec), by far more common, one observes the decay of torsional oscillations upon termination of the driving stress. The internal friction is related to the ratio of successive amplitudes of the decay, while the modulus of the material is proportional to the frequency. At high-stress frequencies (10^3 cycles/sec), an uncommon case, the drive is maintained at a constant amplitude and varying frequency, and one measures the displacement (strain) amplitude.

Frequently, thermomechanical loss occurs at temperatures below the Tg. These damping maxima are attributed to relaxation mechanisms which are available at low temperature. For example, a polybenzimidazole shows a significant decrease in rigidity near 400°C, which is accompanied by a damping maximum [6]. However, two other damping or loss peaks occur at -70° and 310°C. This means that the material has a toughness or impact resistance far below the Tg.

1.2.2.6 Thermal Mechanical Analysis TMA encompasses a number of types of data, all pertaining to mechanical response with a change in temperature. One type already discussed is HDT. Others are stress-strain relationships including torsion, dilatometry, tension, and penetration. These phenomena have been incorporated into commercial instruments which monitor the penetration of a probe into a solid polymer. As with HDT the effect, penetration (volume expansion, etc.), increases with temperature (~5°/min) much more significantly above the Tg. So extrapolation of the two slopes can yield a mechanically determined Tg (see Figure 1.15). The configuration of a penetration probe and the load or stress applied can be changed to produce useful data. Samples can be solid blocks or coatings less than 0.025 mm (0.001 in.) in thickness. For tensile tests one can use films or fibers.

1.3 Recent Developments

An extensive and well-documented survey of methods of analysis of high polymers has appeared [7]. This review covers not only thermal analyses (DSC, DTA, and TMA) but also a mechanical method [torsional braid analysis (TBA)] and molecular level determinations (neutron scattering, microscopy, x-ray, molecular weight, GPC, ESCA, GC, TLC, MS, NMR, IR, UV, luminescence and fluorescence spectroscopy).

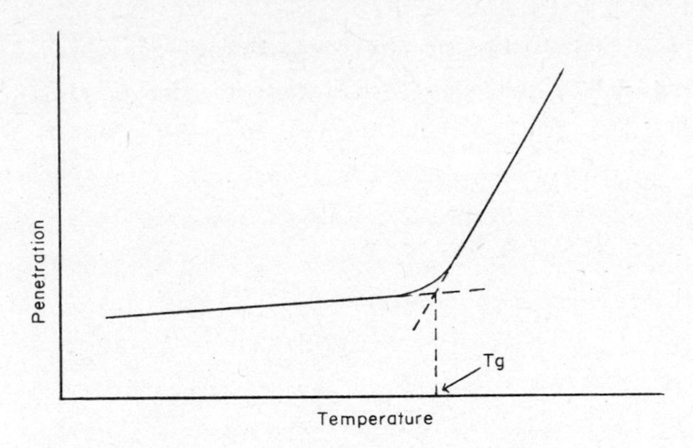

Figure 1.15 Determination of Tg by the measurement of mechanical penetration with increasing temperature.

The thermal degradation of a series of polymers has been studied [8]. By the investigation of polyquinazolones, poly(quinazolone-diones), and polybenzoxazinones, two interesting observations were made. First, the order of thermal stability (by TGA in nitrogen) of some linkages are -O- > $-CH_2-$ > $-SO_2$ > -S-S-. Second, caution was expressed toward the comparison of thermal stabilities of polymers, even those with similar structure. This is because of different degradation mechanisms, residual groups which may lead to chain scission or crosslinking, possible incomplete cyclization, or functions which show an increase in weight due to oxygen pickup.

The proceedings of a symposium on rigid chain polymers, their synthesis and properties have been published [9].

References

1. H. H. Levine, *Ind. Eng. Chem.*, *54*, 22 (1962).
2. P. M. Hergenrother, *Macromolecules*, *7*(5), 575 (1975).
3. P. E. Cassidy and D. K. McCarthy, *J. Appl. Polym. Sci.*, *12*, 1239 (1968).
4. P. D. Garn and O. Menis, *Polym. Preprints*, *17*(2), 151 (1976).
5. G. L. Hagnauer and B. R. LaLiberte, *J. Appl. Polym. Sci.*, *20*, 3073 (1976).
6. H. H. Levine, *AFML Technical Report 64-365*, Part 1, Vol. 1 (November 1964); and J. R. Hall and D. W. Levi, *Pastec Report*, *28* (July 1966).
7. J. G. Cobler and C. D. Chow, *Anal. Chem. Applic. Rev.*, *49*(5), 159R (1977).

8. A. Ghafoor and R. H. Still, *J. Appl. Polym. Sci.*, *21*, 2905 (1977).

9. G. C. Berry and C. E. Sroog, Eds., *J. Polym. Sci.: Polym. Symp.*, *65*, 1-224 (1978).

Chapter 2

Nonheterocyclic Polymers: Backbones Containing Carbon Only

There is a large amount of information available on thermally stable polymers which could be classified as linear or, more precisely, nonheterocyclic functions. Perhaps this large category contains more commercial polymers than any other. For this reason the linear backbones have been divided among four chapters based on composition of the backbone.

The three polymers discussed in this chapter are polyphenylene, polybenzyl, and poly(p-xylylene), which are summarized in Table 2.1.

2.1 Polyphenylenes

The most obvious choice perhaps for a stable backbone is the phenylene unit. Although it could be expected to provide a rigid rod- or ribbon-like backbone, and therefore a brittle, crystalline product, its thermal stability should be among the best.

Table 2.1 Polyphenylene, Polybenzyl, and Poly(p-xylylene)

Polymer	Structure	Comments
Polyphenylene	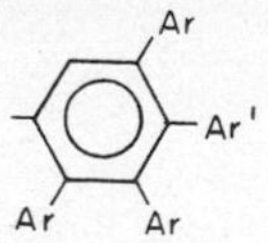 Ar, Ar', Ar, Ar	Intractible when unsubstituted. Lubricants, ablatives, graphite precursors. Stable to 500°C/air. Crystalline. Phenylated version soluble
Polybenzyl	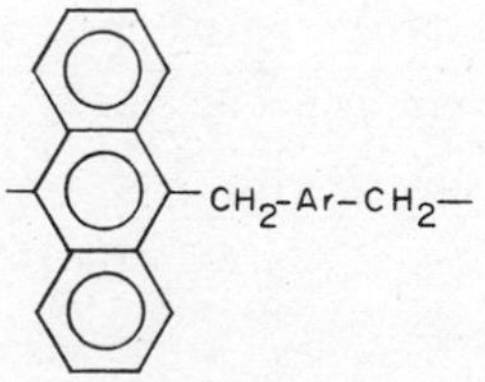 $-CH_2-Ar-CH_2-$	Softens at 400°C. Fluorescent. Intractible.
Poly(p-xylylene)	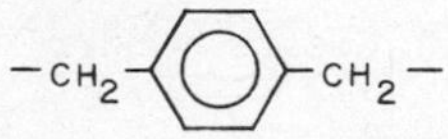 $-CH_2-$ $-CH_2-$	Insoluble/crystalline. mp 300-400°C. Forms coating by vacuum vapor deposition. Oxidizes above 100°C.

Shortly after the discovery and synthesis of polyphenylene, numerous applications were investigated and/or patented. The preparations and uses were discussed in a truly monumental and infinitely detailed review [19] with 410 references which will not be repeated here. This earlier compilation includes topics from 1964 to 1969 encompassing use of polyphenylenes as semiconductors, photoconductors, solid lubricants, lubricant additives, insulators, pigments, ablatives, stabilizers to heat and light, laminate binders, graphite fiber precursors, and ion exchange resins.

A large number of preparative methods have been found to yield polyphenylene. These have led to products with wide ranges of molecular weights, isomeric forms, colors, solubilities, etc. All, however, have significant thermal stabilities, commonly above 400°C in air and 500°C in N_2 (by TGA). Total weight losses in nitrogen have been reported at only 7-18% at 800-900°C.

Prior to the synthesis of phenylated polyphenylene solubility and processing of these materials were serious limitations. Interestingly, it was assumed that polyphenylene was black and insoluble by virtue of its rigid conjugated backbone and that this condition could not be altered. The phenylated version when synthesized was off-white to yellow and soluble, however. It has been proposed that these pendant phenyl groups serve to prevent crystallinity thereby restoring solubility. This concept has been repeated on other backbone types with similar results. Excellent comprehensive reviews are readily available [9,15,19].

Although numerous methods have been used successfully to produce polyphenylene and substituted versions of it, only the major processes will be covered here, i.e., those which produce it in significant yield and molecular weight.

2.1-1 By Direct Coupling

Perhaps the most obvious approach is that of condensing aromatic nuclei. Polyphenylenes have been obtained by electrolysis of benzene with catalysts present [17] and by electrical discharge in benzene vapor [18]. Numerous other methods of aromatic coupling give low yields and low molecular weights (<2000), such as Ullman and Fittig condensations of dihalobenzene and Grignard coupling [19].

The first significant amount of material was obtained by the oxidative, cationic coupling of benzene in a Friedel-Crafts manner in which metal halides

$$C_6H_6 \xrightarrow[\text{or } MoCl_5 \text{ or } FeCl_3]{AlCl_3\text{-}CuCl_2} \left[C_6H_4 \right]_n$$

act as Lewis acids and oxidants [11]. This process resulted in a brown or black, "brick dust" product with a thermal stability of 400°C (10% weight loss) which was determined by heating a sample in an open crucible, a rather more severe test than TGA. One of the most active catalyst systems was ferric chloride with water cocatalyst in a 1:1 molar ratio.

Method [10] Benzene and anhydrous ferric chloride (2:1 mole ratio) are rapidly weighed and added to a flask with stirrer, condenser, and nitrogen blanket. Water (equimolar with $FeCl_3$) is added dropwise with stirring while the temperature is maintained below 25°C. Then the mixture is heated to reflux for 2 hr, quickly cooled, and filtered, whereupon the residue is washed with benzene, stirred with dilute HCl and ice, filtered, and triturated repeatedly with boiling HCl to yield a colorless filtrate. The polymeric residue was then boiled in 2 M sodium hydroxide and again triturated in boiling hydrochloric acid until the filtrate was colorless and finally washed with water and dried at 150°C. The yield is approximately only 8 g from 2 mol of benzene and 1 mol of ferric chloride.

A more recent approach is by coupling benzidine-tetrazonium chloride with copper or iron chlorides as catalysts [14]. A copper-ammonia complex yields an insoluble product. The polymer is 65% soluble in benzene and has a molecular weight of 20,000. Its TGA shows a 10-15% weight loss at 500°C in nitrogen, which is not as good as most polyphenylenes. The soluble portion showed few side groups, while the insoluble material was 25% substituted (one phenyl substituent per four backbone phenylene groups). This is somewhat anomolous behavior since phenylation of a backbone generally increases the solubility. Heating to 500°C increased the number of side groups.

A film of the soluble fraction demonstrated an electrical conductivity of 10^{-2}-10^{-3} ohm^{-1} cm^{-1} at 25°C which decreased considerably on compression of the sample.

Method: Monomer preparation [14] To a mixture of 13 g of benzidine hydrochloride, 30 ml of water, and 15 ml of 35% HCl a solution of 15 g of sodium nitrite in 30 ml of water is added slowly with stirring while the temperature is kept at 0-10°C. After the mixture is stirred overnight at 0°C, yellow crystals are collected, dissolved in water, recrystallized by the addition of acetone, and stored in a refrigerator. The yield of benzidinetetrazonium chloride (BTC) is 70%.

Complexation [14] To 35 ml of a 20 wt % aq. solution of BTC at 5°C is slowly added 5 g of powdered cuprous chloride. The brown precipitate of 1:1 complex is isolated by filtration, washed repeatedly with water, and dried in vacuo for a yield of 85-90%.

Polymerization [8] The complex is heated in water at 80°C for 1 hr whereupon the solid is isolated, heated at reflux for 2 hr in dilute HCl, washed with hot water, and dried. The solid is then heated at reflux in methanol for 10 hr, and the dark brown product is extracted with benzene. The filtrate is treated with petroleum ether to precipitate the polymer in a yield of 65%.

2.1-2 By Poly(cyclohexadiene) Aromatization

The second approach was developed in order to obtain an intermediate which could be characterized prior to aromatization and consequent insolubility. Cyclohexadiene was anionically polymerized as a 1,3-diene to 17,000 molecular weight as a white, soluble powder. Aromatization of this material again afforded a black, insoluble, infusible solid. Aromatization took place either in one step by heating with chloranil or in a two-step process consisting of allylic bromination followed by dehydrohalogenation [2].

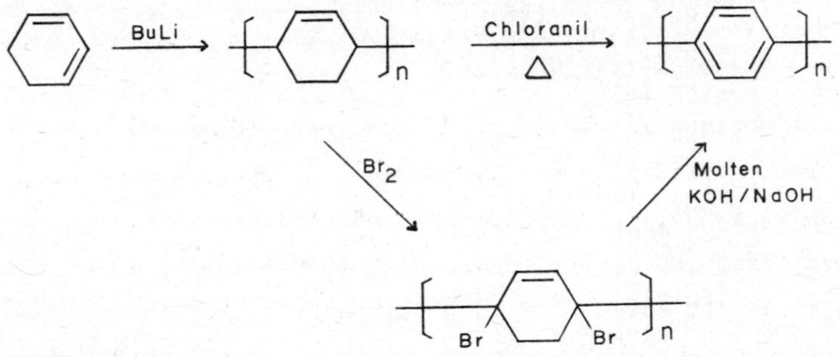

Method [1] *Polymerization* Into a nitrogen-filled, 25-ml syringe bottle is injected 5 ml of dry benzene, 44 g of freshly distilled cyclohexadiene, and 1.0 ml of 1.5 M n-butyllithium solution in hexane. The mixture is shaken periodically for 120 hr to give a viscous yellow solution which is poured slowly into 250 ml of stirred methanol and dried to yield 4.1 g (93%) of poly(cyclohexadiene) with an inherent viscosity of 0.29 dl/g in benzene.

Aromatization In a 250-ml flask equipped with dropping funnel stirrer and condenser is placed 100 ml of 1,1,2-trichloroethane and 10 g of poly(cyclohexadiene) and the solution is heated to reflux. A solution of 40 g of bromine in 50 ml of the solvent is added over 24 hr and reflux is continued for an additional 48 hr. The solution is poured into 1 liter of methanol, and the brown precipitate is collected, washed with methanol, and dried. This solid is then heated under vacuum at 200°C for 15 hr then at 300°C for 27 hr to yield a black residue which is pulverized, extracted with benzene, then methanol, and dried to give a yield of about 85% from poly(cyclohexadiene).

Additional work was done at this point in an effort to improve upon the thermal stability of polyphenylene by crosslinking in the Bakelite manner. It was interesting that in the course of this research a partially soluble intermediate was produced (the sulfonated polymer):

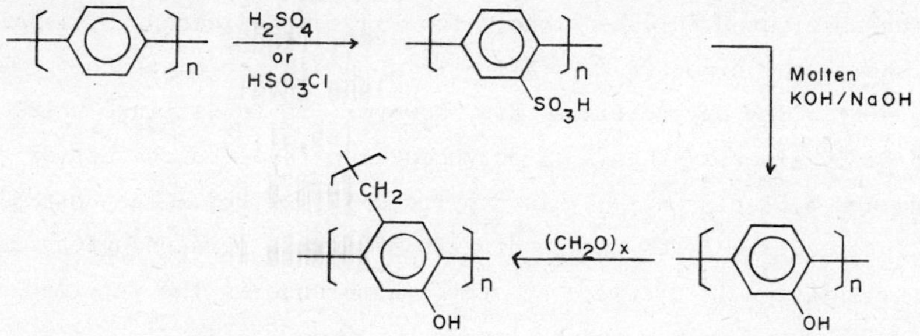

The final product displayed no weight loss at 400°C and a 20-30% weight loss at 900°C by TGA.

Method [2] *Sulfonation* Into a three-necked, round-bottomed flask equipped with a mantle, thermometer, mechanical stirrer, and gas inlet and outlet is placed 50 ml of sulfuric acid and 1.0 g of starting material. This is brought to 100°C under nitrogen atmosphere, and the polyphenylene is then added and the reaction is held at 100°C for 24 hr. The black mixture is cooled to room temperature and poured over ice. The black solid is isolated by suction filtration and washed well with water and dried in a vacuum oven at 100°C. Excessive washing with water results in a loss of product. The degree of sulfonation is 50%; i.e., there is an average of one sulfonic acid group for two phenylene nuclei.

Alkali fusion In an iron crucible are mixed 5.0 g of potassium hydroxide and 5.0 g of sodium hydroxide under nitrogen, and heat is applied to produce a melt (300°C). To this is added with stirring 1.0 g of sulfonated polyphenylene. After 12-24 hr the mixture is cooled, dissolved in water, and acidified with 6 N HCl. The black precipitate is isolated by filtration, washed well with concentrated HCl to remove iron impurities, washed with water, and dried in a vacuum oven at 100°C, to give 0.6 g of product.

Crosslinking Into a 50-ml flask equipped with a magnetic stirrer and a reflux distilling head is placed 10 ml of 37% aq. formaldehyde and 0.10 g of technical barium hydroxide (oxalic acid is used for acidic catalysis), and the solution is heated to reflux and 1.0 g of poly(hydroxyphenylene) is added. After heating for 24 hr, the mixture is cooled and the product is removed by filtration. The black powder is triturated and washed well with water and dried in vacuo. The sample is then heated at 300-400°C for 38 hr under vacuum prior to TGA to yield 1.04 g of product.

2.1-3 By a Diels-Alder Reaction

A totally new approach to polyphenylenes involves a four-center addition by a Diels-Alder mechanism. Bisacetylenes as dienophiles condense with either bispyrones or biscyclopentadienones as dienes with subsequent loss of CO_2 or CO, respectively [13,16]. The polymers are noncrystalline, hence soluble, as well as colorless. Yields are quantitative. The experimental procedure follows closely that for the nonphenylated version.

The unsubstituted polyphenylene which was synthesized by Diels-Alder addition showed the expected crystallinity although the black color was no longer present. The crystallinity did, however, render it insoluble. The method of Diels-Alder synthesis of polyphenylene involved the use of diethynylbenzene and 5,5'-p-phenylene-bis-2-pyrrone [20,21,26]. The unstable bicyclic intermediate loses carbon dioxide to form a para-phenylene unit. The para orientation is dictated by steric hindrance in the intermediate and is predicted by model compound studies.

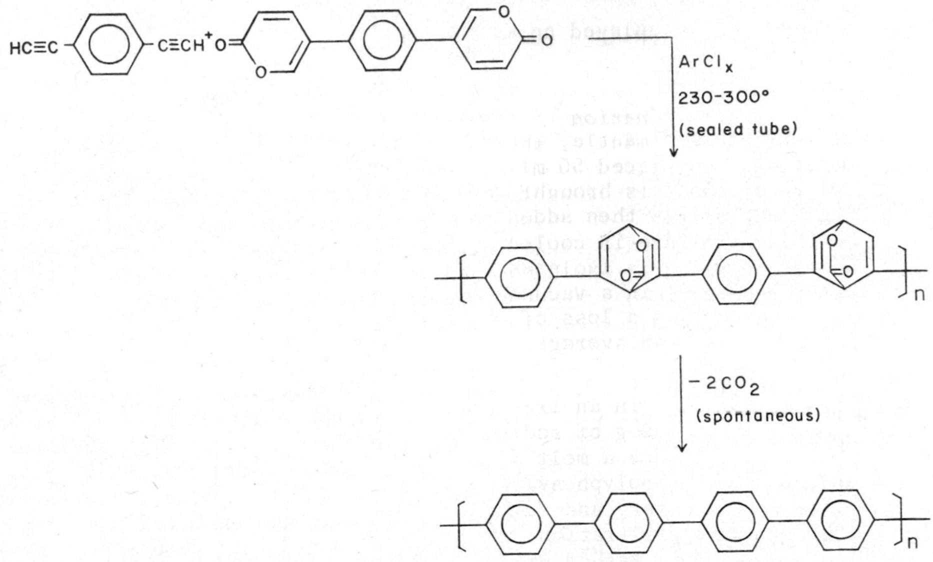

HC≡C–C₆H₄–C≡CH + O=(pyrone)–C₆H₄–(pyrone)=O

$ArCl_x$
230-300°
(sealed tube)

$-2CO_2$
(spontaneous)

The yellow products are obtained in quantitative yield, are insoluble and crystalline, and demonstrate a 10% weight loss by TGA at 500°C in air and 650°C in nitrogen. As illustrated earlier [2], the product can be sulfonated to introduce solubility.

Method [26] Into a glass ampoule is placed 30 ml of 1,2-dichloroethane solvent and 1 mmol of each monomer (0.27 g of 5,5'-p-phenylene-bis-2-pyrrone and 0.13 g of p-diethylnylbenzene), and the solution is degassed by three freeze- (liquid nitrogen) evacuate-thaw cycles. The ampoule is sealed under vacuum and placed in a 500 ml chamber with 100 ml of solvent. The system is heated to 230°C for 140 hr, cooled, and the ampoule contents poured into methanol. The polymeric precipitate was isolated by filtration and dried in vacuo at 80°C for 24 hr.

A phenyl-substituted polyphenylene was prepared from pyrones much like the foregoing example; but using a phenylated bispyrone [22,24]:

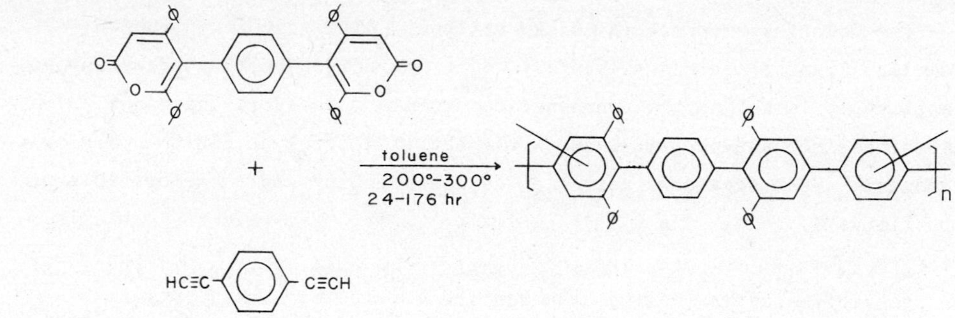

toluene
200°–300°
24–176 hr

The phenyl pendant groups serve to decrease crystallinity and thereby to induce solubility.

These polymers showed stabilities quite similar to the polyphenylenes from Diels-Alder addition of cyclopentadienones (vide infra) [22,24]. A break in the TGA curve in air at 400°C is indicative of a pyrone terminal function, and one at 500°C in air or nitrogen is due to loss of phenyl substitution [23]. Note, however, that the latter type of weight loss does not affect backbone integrity, and thus the 550°C is not necessarily a use-temperature limit.

Another approach to the placement of phenyl substituents on polyphenylene is to use a cyclopentadienone as diene rather than pyrone. Diethynylbenzene will condense with a bis-phenylated cyclopentadienone to give an unstable bicyclic ketone which loses carbon monoxide to yield a phenylated phenylene nucleus quantitatively [21,23]. Extremely pure monomers and proper ratios are required in order to obtain high molecular weights. A monomer ratio imbalance of 2% causes molecular weight to drop to one-half of optimum.

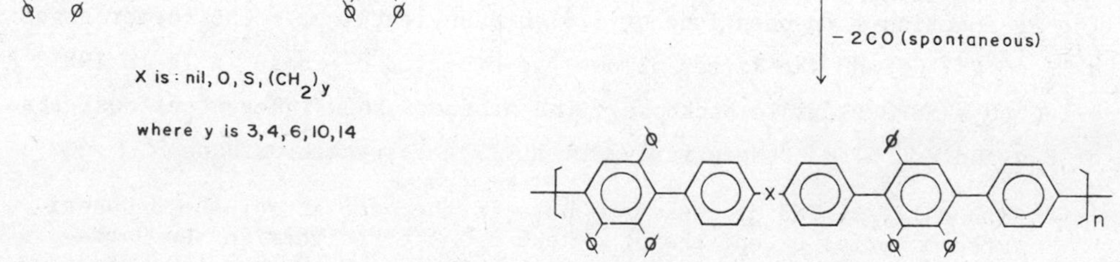

(also meta)

toluene
225°
24 hr

(Unstable)

$-2CO$ (spontaneous)

X is: nil, O, S, $(CH_2)_y$
where y is 3, 4, 6, 10, 14

The colorless product is noncrystalline, quite soluble in common organic solvents, and has molecular weights of 30,000-60,000. The thermal stability is less than the nonphenylated version, however, again due to loss of the pendant phenyl groups with heat (TGA break at 550°C in air or nitrogen). Heat treatment also induces crosslinking and therefore loss of solubility.

Method [23] Into a 25-ml polymerization tube is placed a 10% solution of monomers (equimolar amounts) in toluene. The solution is degassed by three freeze-thaw cycles in liquid nitrogen and the tube is sealed under vacuum. Then, with toluene used as a heat transfer fluid in a Parr pressure reactor, the tube is heated to 225°C for 30 hr. The cooled solution is poured into 200 ml of acetone to precipitate the polymer which is isolated by filtration, purified by reprecipitation, and dried; intrinsic viscosity 0.73 dl/g for para isomer, 0.32 dl/g for meta in toluene.

A third type of substituted polyphenylene from a Diels-Alder reaction synthesized by Harris and his group has both the phenyl and carboethoxy pendant groups with the purpose of improving solubility, mechanical properties, and adhesion to metals and fibers [7]:

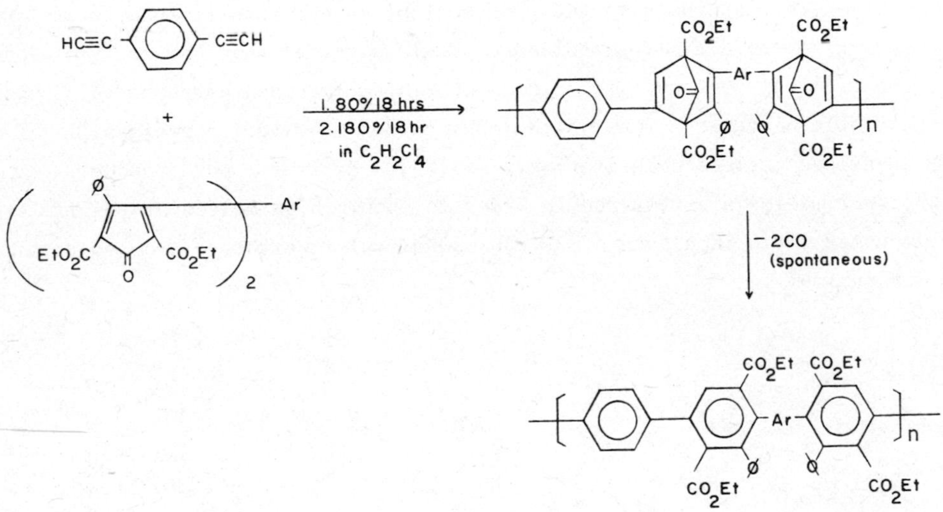

The Ar function is p-phenylene or 4,4'-diphenyleneether. The former gives a Tg of 257°C, and the latter allows the expected decrease in Tg to 198°C owing to a more flexible backbone. The products show inherent viscosities of 3.0 and 2.8 dl/g, respectively, in 1,1,2,2-tetrachloroethane.

Method [7] The process used here is the same as for the unsubstituted material except that the heat transfer solvent in the Parr bomb is xylene and the heat cycle is 80°C for 18 hr and 180°C for 18 hr.

2.2 Polybenzyls

Polybenzyls are the next logical consideration following polyphenylenes. Polybenzyls have been known for some time as products of Friedel-Crafts reactions of unsubstituted aromatic compounds with benzyl halide, alcohol, or ether. High-melting products, however, exhibited low solubility and intractibility. Further, considerable side chain formation was shown to be unavoidable due to ring reactivity and steric hindrance in the step growth mechanism.

An inherent thermal weakness is, of course, the methylene group, in this case benzylic, which is susceptible to oxidation and therefore promotes chain degradation. The advantage of this weak link is that it promotes increased flexibility and solubility, thereby improving processability over polyphenylene.

An anthracene unit incorporated into the backbone contributes both a higher melting point and better solubility. The reaction to incorporate anthracene units is again essentially a Friedel-Crafts condensation using either anthracene with bis(chloromethylbenzene) or bis(chloromethylanthracene) with durene [12]. Numerous variations are reported, by two general reaction schemes:

CH_2Cl / CH_2Cl + H-Ar-H $\xrightarrow[C_2H_5NO_2]{SnCl_4}$ $-[CH_2-$(anthracene)$-CH_2-Ar]_n-$

Type A

(anthracene) + $ClCH_2-Ar-CH_2Cl$ $\xrightarrow[C_2H_5NO_2]{SnCl_4}$ $[$(anthracene)$\ CH_2ArCH_2]_n$

Type B

The Ar functions for type A are durene, anthracene, p-xylene, mesitylene, and bis(2,3,5,6-tetramethylphenyl)methane. For the type B they are α,α'-dichloro-p-xylene; 1,3-bis(chloromethyl)-2,4,6-trimethylbenzene; 1,4-bis(chloromethyl)-2,3,5,6-tetramethylbenzene; p,p'-bis(chloromethyl)diphenylmethane and the octamethyl derivative of the latter. Type B polymers possess 1,4 or a mixture of 1,4 and 9,10 disubstitution.

Softening points were in the region of 400°C, with molecular weights as high as 6250; and the materials were soluble in benzene. The products show intense fluorescence at 440 nm.

> *Method* [12] In a flask is placed 50 ml of nitroethane and an equimolar mixture of the monomers, and the system is maintained under nitrogen. The stannic chloride catalyst is added with stirring to a concentration of 20-25% of the total monomer, and the mixture is heated to 65-70°C for 10-90 min depending on reactants. The deep green reaction mixture is then poured into 1 liter of methanol to precipitate the polymer which is filtered, washed with HCl, and dried. Conversions are commonly 85%.

2.3 Poly(p-xylylene)

The last in the series of carbon-only backbones is the most well developed commercially and the least thermooxidatively stable. This material has been thoroughly reviewed [5,25]. This polymer, poly(p-xylylene) or poly(p-xylelene) or parylene, is formed from a p-xylene dimer, paracyclophane or its quinoid form, which is in turn made by pyrolysis of xylene. The exact mechanisms of dehydrogenation and polymerization are still unknown, and both the dimer and quinoid are reported as intermediates. However, the cyclic dimer has been found as a byproduct in the polymer.

CH_3–C₆H₃(X)–CH_3 → (800-1000° C, vacuum) → [–CH_2–C₆H₃(X)–CH_2–]$_2$ or CH_2=C₆H₃(X)=CH_2

X = H, Cl_{1-3}, Br_{1-2}, or $alkyl_{1-2}$

↓ 550° C, via diradical triplet

$[\dot{C}H_2$–C₆H₄–$\dot{C}H_2]$ → (on surface, <70° C) → $\{CH_2$–C₆H₄–$CH_2\}_n$

The polymerization is carried out by vacuum vapor deposition where the activated species emerges from a heated quartz tube and condenses on a cooler surface. Contributing to the high crystallinity is the lack of chain transfer and termination processes during the surface polymerization; i.e., there is no mobility of the backbone and no liquid state so that neighboring groups cannot suffer hydrogen exchange. As a result free radicals are trapped in the polymer, a fact which has been confirmed by ESR measurements [4].

An intermediate can be trapped by passing the p-xylene pyrolysis product into a solvent at -78°C. When a warmer (room temperature) object is then introduced to the cold solution, polymerization occurs on the surface [3,6].

Paracyclophane can be copolymerized with vinyl-type monomers such as maleic anhydride, acrylonitrile, and styrene in a 1:1 ratio. Further, its copolymerization with chloranil gives a chlorinated polyphenylene ether unit and with sulfur dioxide gives a polysulfone.

The unique preparation method has advantages for its application. It is used to coat electronic parts to provide a thin (0.025 mm, 0.001 in), transparent, void-free surface. There are no residual catalyst or solvent species to cause side reactions or aging problems. The polymer is highly crystalline and solvent-resistant, and melts at 300-400°C. A coating is resistant to hydrolysis and is stress-free, the latter because it is formed in place and is very thin. The unsubstituted polymer does suffer thermo-oxidation above 100°C, a problem which is accelerated by sunlight. However, the chlorinated version of the polymer is more stable and is therefore preferred commercially.

The principal disadvantage is that it cannot be melt-worked (molded, extruded, etc.). Of course, in order to introduce this processability to parylene, one would have to compromise most of its advantages.

Typical properties of poly(p-xylylene) and its chlorinated version are, respectively, tensile strength 480/71 MPa (68,000/10,600 psi), tensile modulus 2.41/3.17 GPa (350,000/460,000 psi), elongation at break 15/220%, Tm 400/290°C, and Tg 80/80°C. They also demonstrate excellent electrical properties as insulators. The brominated and alkylated analogs of parylene have properties similar to the chlorinated version, the only exception being that the elongation of the brominated polymer is 30%, closer to the unsubstituted polymer.

2.4 Recent Developments

2.4-1 Polyphenylenes

Research on polyphenylenes has continued in synthesis, property investigation, and processing. A low-temperature (50°C), longtime (7 days) Friedel-Crafts condensation of benzene in the presence of oxygen and water gave low yields of low molecular weight and irregular structure polymer [27,28]. This product contained dihydro, quinoid units, etc. The synthesis using aluminum chloride-cupric chloride has been shown to be ionic [29].

The condensation of 4,4'-biphenylene diazonium salts with cuprous or ferrous chloride gave a soluble polyphenylene via diradical coupling [36]. This method was also applicable to analogs such as the biphenylene ether, methane, sulfone, sulfide, and amine.

A highly crosslinked polyphenylene was produced by the cyclization (at 220-240°C) of a tetraacetylene-substituted oligomer of polyphenylene [30]. The product was hard, nonporous, and resistant to acids and bases. The use temperature was >10,000 hr at 200-220°C in air or at 400°C in nitrogen. It was tested as a coating for metals although adhesion was only fair to poor (2.4-3.4 MPa or 350-500 psi).

Dilute solution properties of phenyl-substituted polyphenylenes were studied after fractionation of products from three different synthetic methods [41]. Techniques used were light scattering, osmometry, viscometry, fluorescence depolarization, and exclusion chromatography. Of every five phenylene functions, three were found to be para, the other two being meta or para with para predominant. Rigid segments were 7.0-7.5 nm (70-75 Å°) in length.

Processing of polyphenylene was also studied. Fabrication of low- to moderate-crystallinity polymer by powder metallurgical techniques resulted in tensile strengths up to 35 MPa (5000 psi) [31]. The powdered polymer was compacted at 520 MPa (74,000 psi) at room temperature and then free-sintered in nitrogen at 580-615°C for 1 hr. The properties fall between those for polyimide and graphite with excellent high-temperature hydrolysis resistance and thermal aging. It was suggested that a purer starting material would improve the properties even further [32].

A new term has been coined, "ablapaction," as a contraction of ablation-compaction, a process used on polyphenylene [33]. This describes the extreme shrinkage (20-80%) caused by controlled ablation at 590°C for 1 hr in

200 MPa (2000 atm) of hydrogen. Off-gases contain methane and ethane. Other polymers such as poly(p-xylylene), polyimide, and perchloropolyphenylene do not demonstrate this effect.

2.4-2 Polybenzyls

Linear polybenzyls were prepared quantitatively by the condensation of benzyl chloride using one of the following catalyst-solvent pairs at the temperature given [38]:

$TiCl_4 + CH_3Cl$ at -100°C
$SbF_5 + HSO_3F/SO_2$ at -78°C
$AlCl_3 + C_2H_5Cl$ at -78°C
$TiCl_4 + CH_3NO_2$ at +25°C

Number average molecular weights were 3000-6000. There was a tendency toward branching which decreased the thermal stability, which was generally in the 400-480°C range (10% weight loss by TGA).

2.4-3 Allylic Polymers

Several bisallylic-substituted phthalimide compounds have been polymerized with peroxides to give crosslinked resins which give no weight loss up to 360-390°C [34,35] in nitrogen by TGA. Additional allylic sites improved the stability [37].

2.4-4 Biphenylene Crosslinking

Biphenylene has been discovered as a unit which can be placed in a backbone and opened thermally to crosslink with no off-gas:

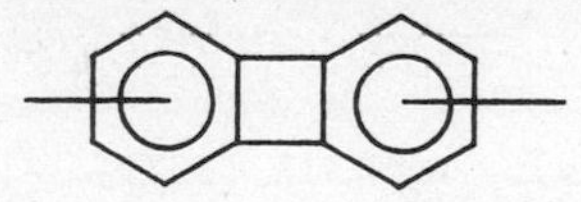

This has been accomplished with poly(aromatic ether ketone) resins [40] and with polyamides, polybenzimidzoles, and polyquinoxalines [39]. The formula requires a 305°C cure for 20 hr and yields a polymer which loses only 0.9% of its weight in air by isothermal gravimetric analysis at 300°C and an additional 1% at 350°C. The latter series is cured for 3 hr at 380°C concommitant with a disappearance of Tg and solubility but with higher moduli being developed.

References

1. P. E. Cassidy and C. S. Marvel, *Macromolecular Synthesis*, Vol. 4 (W. J. Bailey, Ed.), John Wiley and Sons, New York, 1972, pp. 7-8.

2. P. E. Cassidy, C. S. Marvel, and S. Ray, *J. Polym. Sci., A-3*, 1553 (1965).

3. L. A. Errede and B. F. Landrum, *J. Amer. Chem. Soc., 79*, 4952 (1957).

4. W. F. Gorham, *J. Polym. Sci., A-1, 4*, 3027 (1966).

5. W. F. Gorham, *Adv. Chem. Series*, No. 91, 643-659 (1966).

6. W. F. Gorham, R. S. Gregovian, and J. M. Hoyt, *J. Amer. Chem. Soc., 82*, 5218 (1960).

7. F. W. Harris and B. A. Reinhardt, *Polym. Preprints, 15*(1), 691 (1974).

8. S. Hayama and S. Niino, *J. Polym. Sci.: Polym. Chem. Ed., 12*, 357 (1974).

9. P. Kovacic and F. W. Koch, *Encycl. Polym. Sci. Technol., 11*, 380-389 (1969).

10. P. Kovacic, F. W. Koch, and C. E. Stephan, *J. Polym. Sci., A-2*, 1193 (1964).

11. P. Kovacic and A. Kyriakis, *Tetrahedron Lett.*, 467 (1972).

12. G. Montaudo, P. Finocchiaro, and S. Caccamese, *J. Polym. Sci., A-1, 9*, 3627 (1971).

13. H. Mukamol, F. W. Harris, and J. K. Stille, *J. Polym. Sci., 5*, 2721 (1967).

14. S. Niino, K. Tsubaki, and S. Hayama, *J. Polym. Sci., A-1, 11*, 683 (1973).

15. G. K. Noren and J. K. Stille, *J. Polym. Sci. D: Macromol. Rev., 5*, 385-420 (1971).

16. W. Reid and D. Freitag, *Naturwisenschaften, 53*, 306 (1966).

17. A. F. Shepard and B. F. Donnels, *J. Polym. Sci., 4*, 511 (1966).

18. C. Simonsev, N. Asandei, and F. Denes, Romanian Patent 50,976 (1968) [*Chem. Abstr., 70*, 20527 (1969)].

19. J. G. Speight, P. Kovacic, and F. W. Koch, *J. Macromol. Sci.: Rev. Macromol. Chem., 5*(2), 295-386 (1971).

20. J. K. Stille and Y. Gilliams, *Macromolecules, 4*(4), 515 (1971).

21. J. K. Stille, F. W. Harris, H. Mukamol, R. O. Rakutis, C. L. Schilling, G.K. Noren, and J. A. Reeds, *Advan. Chem. Series*, No. 91, 628 (1969).

22. J. K. Stille, F. W. Harris, R. O. Rakutis, and H. Mukamol, *J. Polym. Sci., B, 4*, 791 (1966).

23. J. K. Stille and G. K. Noren, *Polym. Lett., 7*, 525 (1969).

24. J. K. Stille, R. O. Rakutis, H. Mukamol, and F. W. Harris, *Macromolecules, 1*, 431 (1968).

25. M. Szwarc, *Polym. Eng. Sci., 16*(7), 473 (1976).

26. H. F. Vankerckhoven, Y. K. Gilliams, and J. K. Stille, *Macromolecules*, *5*(5), 541 (1972).

27. J. E. Durham and P. Kovacic, *J. Polym. Sci.: Polym. Lett. Ed.*, *14*, 347 (1976).

28. J. E. Durham, K. N. McFarland, and P. Kovacic, *J. Polym. Sci.: Polym. Chem. Ed.*, *16*, 1147 (1978).

29. G. G. Engstrom and P. Kovacic, *J. Polym. Sci.: Polym. Chem. Ed.*, *15*, 2453 (1977).

30. J. E. French, *Org. Coat. Plast. Preprints*, *35*(2), 72 (1975).

31. D. M. Gale, *J. Appl. Polym. Sci.*, *22*, 1955 (1978).

32. D. M. Gale, *J. Appl. Polym. Sci.*, *22*, 1971 (1978).

33. D. M. Gale, *J. Polym. Sci.: Polym. Lett. Ed.*, *15*(7), 439 (1977).

34. S. Hara, K. Iwata, and Y. Yamada, *Makromol. Chem.*, *176*, 127-141 (1975).

35. S. Hara, K. Iwata, and T. Yamada, *Makromol. Chem.*, *177*, 951-955 (1976).

36. S. Hayama, S. Niino, and A. Sakamoto, *J. Polym. Sci.: Polym. Chem. Ed.*, *15*, 1585 (1977).

37. K. Iwata, M. Ogasawara, and S. Hara, *Makromol. Chem.*, *179*, 1361-1364 (1978).

38. J. Kuo and R. W. Lenz, *J. Polym. Sci.: Polym. Chem. Ed.*, *14*(11), 2749 (1976).

39. A. Recca and J. K. Stille, *Macromolecules*, *11*(3), 479 (1978).

40. R. J. Swedo and C. S. Marvel, *J. Polym. Sci.: Polym. Lett. Ed.*, *15*(11), 683 (1977).

41. J. L. Work, G. C. Berry, E. F. Casassa, and J. K. Stille, *J. Polym. Sci.: Polym. Symp.*, *65*, 125 (1978).

Chapter 3

Nonheterocyclic Polymers: Backbones Containing Carbon and Oxygen and their Thio Analogs

The two principal structures covered in this chapter are polyesters and polyethers. Also discussed are their thio analogs. These polymers are outlined in Table 3.1.

3.1 Polyesters

Although polyesters are not usually considered to be in a category of thermally stable materials, some significant advances have been made toward this end. There is even a high-temperature thermoplastic polyester which was developed by the Carborundum Company under the name Ekonol, poly(hydroxybenzoic acid). This polymer is synthesized by the high-temperature self-condensation of either p-acetoxybenzoic acid or phenyl-p-hydroxybenzoate, and is occasionally referred to as a p-oxybenzoyl polymer [8]. The key to this synthesis is evidently proper temperature control through the heat transfer medium, with single crystals being possible. This poly(p-hydroxybenzoic acid) is stable to 455°C (850°F) by TGA or to 325°C (600°F) for continuous use [6,7]. It does display a reversible crystalline transition at 325-360°C. It is processed by compression sintering, but is insoluble. Economy and his co-workers have found a high thermal conductivity for this substance in comparison to other polymers along with a high elastic modulus, 7 GPa (1×10^6 psi) and flexural strength of 75 MPa (11,000 psi). It is, of course, somewhat susceptible to inorganic bases and, to a lesser extent, acids.

$$CH_3\overset{O}{\overset{\|}{C}}\text{-O-}C_6H_4\text{-}CO_2H \xrightarrow{220-280°C} \left[\text{O-}C_6H_4\text{-}\overset{O}{\overset{\|}{C}}\right]_n \text{ "Ekonol"}$$

$$HO\text{-}C_6H_4\text{-}CO_2C_6H_5 \xrightarrow{34-360°C} \left[\text{O-}C_6H_4\text{-}\overset{O}{\overset{\|}{C}}\right]_n$$

Table 3.1 Polyesters and Polyethers

Polymer	Structure	Comments
Polyester	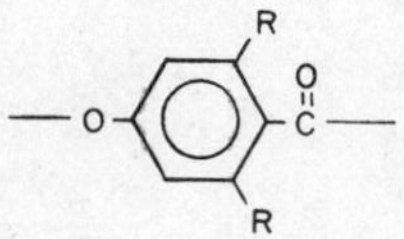	Use temperature 325°C. Fabrication by compression sintering. Crystalline. Insoluble. High flexural modulus.
Poly(phenylene ether), PPO	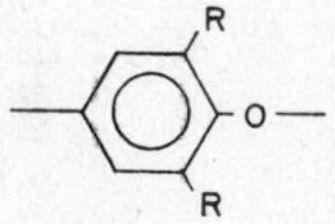	Self-extinguishing. HDT 100-150°C. Tg 90-220°C. Tm 262-484°C.
Poly(perfluoroalkylether) (crosslinked)	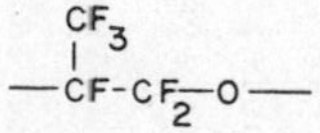 $-CF(CF_3)-CF_2-O-$	Tg -43 to -50°C. Use temperature 230-260°C.
Copoly(arylalkylether)	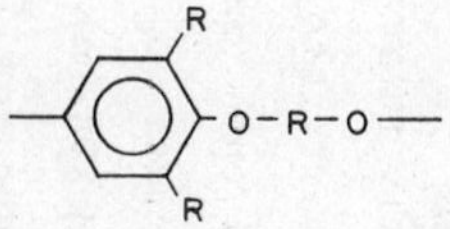	TGA stability 350-450°C. Thermoplastics. Tg 120-320°C.
Poly(ethersulfone)	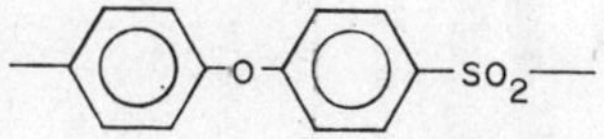	Tg 190-285°C. Use temperature 150-250°C. TGA stability 500°C.
Poly(phenylene sulfide), PPS	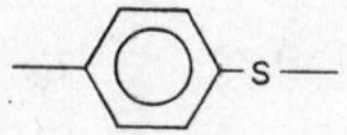	Moldable. Tg 193-204°C. Tm 285°C. Crystalline. Insoluble below 200°C. Crosslinks thermally. TGA stability 500°C. Use temperature 233°C.
Poly(imidothiother)	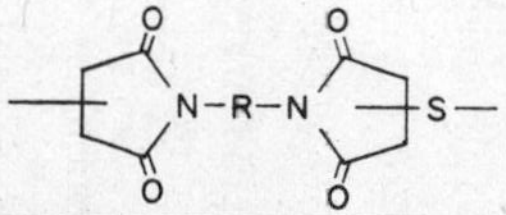	TGA stability 325°C. Soluble. Thermoplastic elastomer.

Method [8] *From p-acetoxybenzoic acid* One mol (180 g) of monomer and 0.018 g of magnesium turnings is mixed in a 1-liter vessel equipped for inert gas flushing and vacuum. With an argon atmosphere at 60 torr, the mixture is stirred and heated according to the following schedule: 220-240°C for 1 hr, 240-250°C for 1/4 hr, and 250-280°C for 0.5 hr. The viscous product weighs 115 g.

From phenyl p-hydroxybenzoate The monomer (1 mol, 214 g) is stirred in 500 g of Therminol 66 at 340-360°C under argon. Heating is discontinued after the theoretical amount of phenol distillate is obtained. Acetone is added and the polymer is isolated by filtration. The product is washed with hot acetone and dried.

Further research by Mulvaney in wholly aromatic polyesters has followed the approach used with aromatic polyamides; i.e., the thermal stability, melting point, and mechanical properties can be varied by mixing ortho- and para-substituted rings and by building order into a copolymer backbone [21]. For example, a substituted poly(p-hydroxybenzoate) can be prepared quantitatively from the acyl chloride with a tertiary amine catalyst:

$$\text{3,5-di-}t\text{-Bu-4-HO-C}_6\text{H}_2\text{COCl} \xrightarrow[\phi\text{-NO}_2,\ 48\text{ hr}]{Et_3N} \left[-O-C_6H_2(t\text{-Bu})_2-\overset{O}{\overset{\|}{C}}- \right]_n$$

The product is white, of low molecular weight (inherent viscosity 0.1), high melting (480°C), fairly thermally stable (10% loss by TGA in N_2 at 420°C), and for the most part insoluble; i.e., it is soluble in trichloroethylene and in exotic solvents such as hot sulfuric, chlorosulfonic, and trifluoromethylsulfonic acids.

A regularly ordered polyaromatic ester backbone can be constructed, much in the same way as done for ordered copolyamides, to achieve some desirable, practical properties. With the ordered backbone shown below, compared to the simple poly(p-hydroxybenzoic acid), solubility improves (chloroform, nitrobenzene, and chlorinated benzenes) along with molecular weight (inherent viscosity 1.05 dl/g). However, the melting point decreases to 349°C and the thermal stability suffers somewhat (TGA: 10% loss at 380°C/N_2). This is yet another example of the sacrifice which seemingly is required in order to achieve processability in a thermally stable polymer.

Tough, flexible films can be cast with this polymer from trichloroethylene. The melting point can be raised in the above material without sacrifice of solubility or thermal stability but with some decrease in molecular weight by replacing half the resorcinol in the last step with hydroquinone. This 50% change of meta substitution to para raises the melting point by 100°C.

> *Method* [21] *Monomer* The disodium salt of bis(2,6-di-tert-butyl-4-carboxyphenyl)terephthalate [25] (2.50 g, 0.0037 mol) is dried at 120°C at 1 torr and suspended in 40 ml of benzene. A cold suspension of 1.5 g (0.012 mol) of oxalyl chloride in 30 ml benzene is added, and the mixture is stirred for 12 hr and filtered. The mixture is purified by lyophilizatior to yield 1.76 g (67.5%) of white bis(2,6-di-tert-butyl-4-chlorocarboxyphenyl)terephthalate.

> *Polymer* The above diacid chloride (1.73 g, 0.0026 mol), dissolved in 15 ml of chloroform, is added quickly to a blendor containing 0.286 g (0.0026 mol) resorcinol, 0.17 g sodium lauryl sulfate detergent, a trace of sodium hydrosulfite antioxidant, and 0.208 g (.0052 mol) of sodium hydroxide in 40 ml of deoxygenated water. After 2 min 0.16 g of tetra-n-butylammonium iodide accelerator in 5 ml of water is added, and the reaction is stirred for 10 min when 10 ml of chloroform is added. In another 5 min the polymer is precipitated by pouring the mixture into 600 ml of acetone. Purification is conducted by dissolving the white fibrous polymer in trichloroethylene and reprecipitating it into methanol to yield 1.69 g (92.5%), softening point 330-360°C, inherent viscosity 1.05 (0.5% solution in trichloroethylene).

3.2 Polyethers

The ether function is an excellent candidate for inclusion in the backbone of high-temperature polymers. The ether bond has not only a high-energy but also a chemical inertness unknown for many heteroatomic systems. It is unreactive except toward only a few reagents: a hydroiodic acid-acetic acid mixture (the Zeisel alkoxyl reagent) [20] and, more recently discovered, butyl lithium [26]. Furthermore, the ether link has the ability to induce flexibility into a backbone by virtue of its rotational freedom. Frequently, a diphenylether moiety will be used in a monomer where one desires to flexibilize a system without sacrificing thermal or chemical stability.

3.2-1 Aromatic Polyethers

The simplest ether in this case is poly(phenyleneoxide) (PPO), and it has been successfully commercialized as Noryl by General Electric. The commercial PPO is commonly compounded with polystyrene to facilitate its processing as a high-temperature engineering plastic.

A large number of papers have been published regarding the synthesis of various poly(phenylene ethers) from phenols by a copper salt complex with amines. A review of the oxidative coupling of substituted phenols also discussed the mechanisms involved [3].

O_2, CuCl / pyridine

R is: H, CH_3, C_6H_5

X is: H, Br, I, Cl

The nature of the 2,6 substituent affects the thermal transitions, as expected. Both methyl and phenyl groups raise the Tg of the polymer, while only phenyl raises the Tm, as shown in the following table [1]:

2,6 Substituent on PPO	Tg(°C)	Tm(°C)
H-	90	262
CH_3-	207	262
C_6H_5-	220	484

The polymerization is usually carried out by using a one-electron oxidation by amine of the phenolic OH with subsequent aromatic substitution at the para position. The unfilled product has an HDT of 100-150°C (212-300°F) and tensile strength up to 32 MPa (12,000 psi) depending on grade. Further, its flexural modulus is 2.86 Pa (400,000 psi). Higher values for these properties result when a glass filler is incorporated into PPO. These materials are self extinguishing and can withstand continuous use at 105°C. The above properties allow PPO to see service in numerous electrical and mechanical situations, many of which require contact with boiling water.

PPO cannot be crystallized thermally; but a 30% crystalline material can be obtained from solvent or solvent vapor.

> *Method* [1,11] To a flask equipped with stirrer, oxygen inlet bubbler, and thermometer is added 135 ml of pyridine and 1 g (0.01 mol) of cuprous chloride. The oxygen is passed into the vigorously stirred mixture which turns dark green; and then 5.0 g (0.041 mol) of 2,6-dimethylphenol is added, whereupon the color turns dark orange in a few minutes. The temperature rises by 20°C; the viscosity then increases, and color reverts to dark green. The mixture is poured slowly into 500 ml of methanol, and the polymer is isolated by filtration and washed with dilute HCl in methanol. The polymer is purified by dissolving it in chloroform and reprecipitating it into methanol. It is then isolated and dried at 110°C and 3 torr to yield 4.2 g (85%); intrinsic viscosity 0.72 dl/g in chloroform, PMT >300°C.

A tetramethyl-substituted PPO has been prepared in low molecular weight (1300) but with a melting point of 290-310°C. Evidently the extra substituents in the 3 and 5 positions serve to restrict considerably the movement about the ether linkage. This same phenomenon was observed for polybenzyls [17].

$$\text{2,3,5,6-}(CH_3)_4C_6H\text{-}OH \xrightarrow[25^\circ]{Cu,\ pyridine} \left[C_6(CH_3)_4\text{-}O \right]_n$$

> *Method* [16] Pyridine (70 ml) is stirred vigorously with 0.5 g of copper(I) chloride to give a light blue solution to which 3 g of hydroxydurene is added to result in a green color. After 45 min the reaction mixture is slowly poured into a solution of 700 ml of water and 10 ml of HCl with stirring. The precipitate is collected, washed with water, dried, and dissolved in $CHCl_3$. This solution is filtered and the polymer reclaimed by precipitation into methanol and dried in vacuo to yield 2.1 g (70%) of poly(tetramethylphenylether).

3.2-2 Aliphatic Polyethers

Aliphatic portions in polyether backbones detract considerably from thermal stability. The only case where this is not so is where a perfluoroaliphatic system exists. This is graphically demonstrated in simple vinyl systems where one can compare EPR (ethylene-propylene rubber) to its partly fluorinated analog, Viton (duPont). EPR can tolerate 200-250°C maximum, while Viton can sustain itself through thermal excursions to 500°C. The effects of pendant groups and ether linkages in a fluoroalkylene backbone are discernable in Table 3.2 [19]. The objectives in the preparation of these types of materials were to incorporate into elastomers chemical and thermal resistance, low-temperature flexibility, and good mechanical properties over a wide temperature range. While pendant CF_3 or CF_3O groups served to lower Tg, the most pronounced effect of this type was realized by injection of the ether function into the backbone. Unfortunately, the service temperature is also lowered in this process but usually not as much as Tg. The mechanical and thermal properties limit the above materials to narrow application.

Hexafluoropropylene oxide can be prepared and polymerized anionically to liquids of molecular weights above 4000. End capping of these polymers

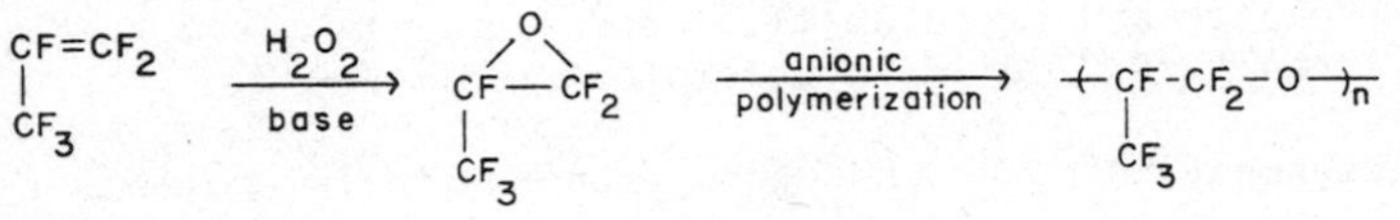

$$CF(CF_3){=}CF_2 \xrightarrow[\text{base}]{H_2O_2} \underset{CF_3}{CF}\overset{O}{—}CF_2 \xrightarrow{\text{anionic polymerization}} {-}\!\!\left(\underset{CF_3}{CF}{-}CF_2{-}O\right)\!\!{-}_n$$

provides materials which are commercial hydraulic fluids and polymerization solvents. Similar oligomers have undergone chain extension to give copolymers

Table 3.2 Effect of Pendant Groups and Ether Linkages in a Fluoroalkylene Backbone

Polymer	Service Temp. (°C)	Tg (°C)
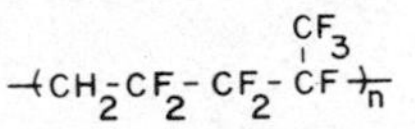 $-(CH_2-CF_2-CF_2-CF(CF_3))_n-$ FEP, Teflon FEP, Viton	200	-10
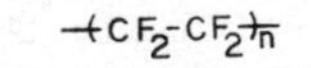 $-(CF_2-CF_2)_n-$ PTFE, Teflon	270	127
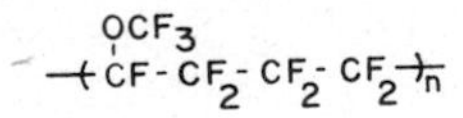 $-(CF(OCF_3)-CF_2-CF_2-CF_2)_n-$	250	-12
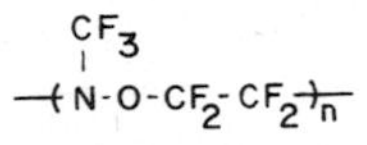 $-(N(CF_3)-O-CF_2-CF_2)_n-$	180	-50
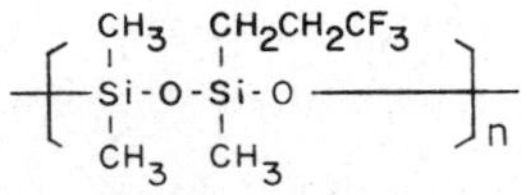 $-[Si(CH_3)_2-O-Si(CH_3)(CH_2CH_2CF_3)-O]_n-$	220	-65
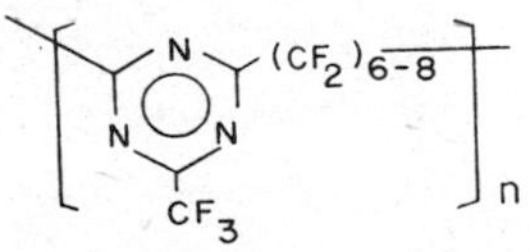 $-[C_3N_3(CF_3)-(CF_2)_{6-8}]_n-$ (triazine ring bearing CF_3)	260	-10

which have been proposed as high-temperature sealants for aerospace applications [19]. This chain extension takes place through the cyclization of nitrile groups to give triazines or the reaction of nitriles with nitrile oxides to yield oxadiazoles. The triazine system possesses a Tg -50°C and a use temperature of 260°C. However, the oxadiazole polymer has better mechanical properties although its thermal data are not as impressive (Tg -43°C, use temperature 230°C).

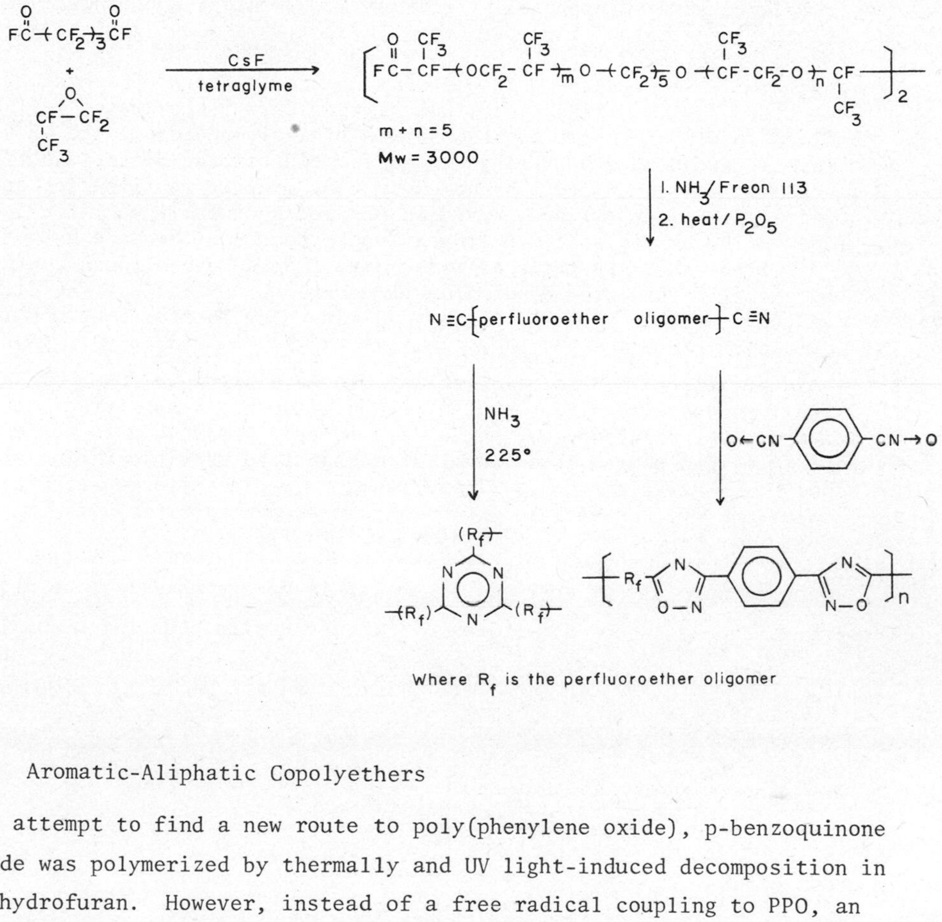

$$FC(=O){-}(CF_2)_3{-}C(=O)F + CF(CF_3){-}CF_2 \text{(epoxide)} \xrightarrow[\text{tetraglyme}]{CsF} \left[FC(=O){-}CF(CF_3){-}(OCF_2{-}CF(CF_3))_m{-}O{-}(CF_2)_5{-}O{-}(CF(CF_3){-}CF_2{-}O)_n{-}CF(CF_3){-} \right]_2$$

m + n = 5

Mw = 3000

1. NH_3/Freon 113
2. heat/P_2O_5

$N{\equiv}C{-}[\text{perfluoroether oligomer}]{-}C{\equiv}N$

NH_3, 225° → triazine network: (R_f)-substituted 1,3,5-triazine

$O{\leftarrow}CN{-}C_6H_4{-}CN{\rightarrow}O$ → $\left[R_f{-}\text{(1,2,4-oxadiazole)}{-}C_6H_4{-}\text{(1,2,4-oxadiazole)} \right]_n$

Where R_f is the perfluoroether oligomer

3.2-3 Aromatic-Aliphatic Copolyethers

In an attempt to find a new route to poly(phenylene oxide), p-benzoquinone diazide was polymerized by thermally and UV light-induced decomposition in tetrahydrofuran. However, instead of a free radical coupling to PPO, an ionic 1:1 copolymerization occurred with solvent [22,23].

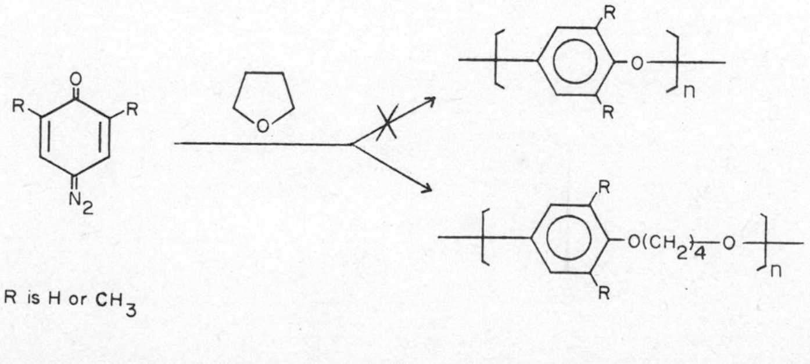

p-Benzoquinone diazide (2,6-R, 4-N_2) + THF ↛ $\left[C_6H_2R_2{-}O \right]_n$

→ $\left[C_6H_2R_2{-}O(CH_2)_4{-}O \right]_n$

R is H or CH_3

This reaction also occurred with other cyclic ethers such as 2,2-bis(chloromethyl)oxetane (Penton monomer, Hercules) and bicyclo-[2.2.1]-7-oxoheptane. The products are low molecular weight (6000), white, very soluble powders; but of course the thermal and oxidative stability is lessened by inclusion of aliphatic units. The TGA analysis of the THF copolymer shows that it is stable to 350°C.

Method [22,23] *Monomer preparation* A solution of 20 g (0.12 mol) of 2,6-dimethyl-p-aminophenol hydrochloride, 20 ml of conc. HCl, and 350 ml of absolute ethanol is cooled to -5°C and 18 ml (15.6 g, 0.13 mol) of freshly distilled isoamylnitrite is added dropwise over 20 min. The mixture is stirred for 2 hr and poured into 2 liters of anhydrous ether at 0°C. The resulting diazonium chloride salt is collected by filtration, suspended in 110 ml of absolute ethanol, and cooled to -5°C; to this is added 55 g of freshly prepared moist silver oxide, and the suspension is stirred for 3 hr at -5°C. The mixture is filtered, added to 1 liter of ether, and stored in Dry Ice/acetone (-78°C) for 1 day. The yellow crystals of 2,6-dimethyl-p-benzoquinone diazide hydrate are collected and recrystallized twice more from an ethanol/ether mixture to yield 35% of the red anhydrous monomer, mp 109-111°C. The diazide is very unstable to heat, light, and water and is stored at 0°C in the dark over P_2O_5.

Polymerization A solution of 1 g of the diazide in 20 ml of THF is deoxygenated with nitrogen at 0°C in a Vycor flask and irradiated with a mercury vapor lamp (G.E. AH-4) for 36 hr. The volume is concentrated to 10 ml and poured into 150 ml of methanol to precipitate the white polymer. Purification is accomplished by reprecipitation of a benzene solution into methanol to yield 45%; inherent viscosity 0.13 in chloroform.

Another more direct approach is available to arylene-alkylene ether copolymers and analogous structures by Friedel-Crafts coupling of bis(chloropropyl)arylenes with active aromatic compounds [10].

$$Cl\text{-}C(CH_3)_2\text{-}Ar\text{-}C(CH_3)_2\text{-}Cl + H\text{-}Ar'\text{-}H \xrightarrow[RNO_2,\ \phi_3\overset{+}{C}\overset{-}{MF_x}]{AlCl_3,\ RCl_x\ \text{solvent}} \left[C(CH_3)_2\text{-}Ar\text{-}C(CH_3)_2\text{-}Ar' \right]_n$$

Ar is: m-or p-C_6H_4

Ar' is: Naphthalene

$-C_6H_4-O-(CH_2)_2-O-C_6H_4-$

$-C_6H_4-X-C_6H_4-$

Where X is: O, S, NH, CH_2 or $C(CH_3)_2$

$-C_6H_4-C(CH_3)_2-C_6H_4-$

MF_x is: AsF_6, SbF_6 or BF_4

The nitro compound (nitrobenzene or nitromethane) is necessary to allow formation of high molecular weight polymers, and the trityl metal halide prevents reaction reversal. Of course the Friedel-Crafts reaction must have available an activated nucleus for the electrophilic aromatic substitution mechanism to be operable; hence electron withdrawers in the Ar' system, ring deactivators such as C=O and SO_2, obviate any substitution.

The clear, polymeric products have Tg values from 120 to 320°C depending on rigidity of structure, can be fabricated as thermoplastics, and show initial weight loss at 450°C in air and 500°C in nitrogen. Other properties are tensile strength 79.3 MPa (11,500 psi), tensile modulus 2.5 GPa (360,000 psi), and ultimate elongation 20-30%.

Method [10] A mixture of 16.8 ml of nitrobenzene, 51.0 ml of phenylether, 76.5 g of p-bis(2-chloropropyl)benzene, 0.350 g of triphenylmethylhexafluoroarsenate, and 400 ml of methylene chloride is cooled to -42°C, and 3.75 g of $AlCl_3$ is added. The red-brown mixture is stirred for 2.5 hr, during which time 300 ml more of solvent is added. To terminate the chain, 5 ml of phenylether is added as a capping agent, and then 5 ml of methanol and 100 ml of solvent is added to stop the reaction. An additional 300 ml of methylene chloride is added to the mixture, which is poured into 5 liters of acetone to precipitate the polymer. The solid is collected, washed with methanol, and dried at 100°C to yield 97% polymer; inherent viscosity 1.3 in o-dichlorobenzene.

3.2-4 Aromatic Poly(ether sulfones)

The sulfone function is another type of backbone inclusion which can maintain stability of a backbone while contributing flexibility. The diphenylsulfone group is used as a linkage in otherwise stiff polymer backbones.

Poly(phenylenesulfones) are marketed by a number of companies as high-temperature, injection-moldable thermoplastics. The comparison of structures and thermal properties is given in Table 3.3 [18]. From these data it again is apparent that both ether and isopropylyl functions in the backbone induce a free rotation witnessed by lower Tg's. Yet they can be used at temperatures of 150-250°C even under stress. A static thermal test shows no embrittlement and only slight discoloration of these materials when heated in air at 225°C for 170 days. TGA shows a 10% weight loss at 500°C in air and 550°C in argon. These values are truly indicative of its stability since the 10% loss value occurs just following the downward break in the curve. Although the polysulfones can be made to crystallize from solution, they are amorphous when cooled from the molten state; therefore, the Tg is the most important parameter for this class of polymers.

Table 3.3 Comparison of Structures and Thermal Properties of Poly(phenylenesulfones)

Structure	Name	Company	Tg (°C)
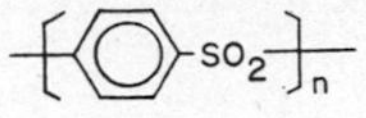 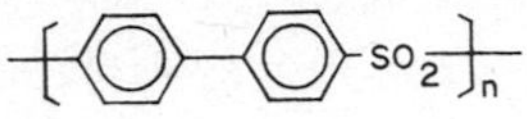 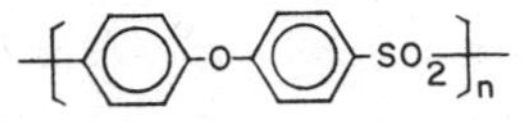 $\left[C_6H_4-SO_2 \right]_n$ (mostly)	Astrel 360	3M	(mp > 500)
and $\left[C_6H_4-C_6H_4-SO_2 \right]_n$			
and $\left[C_6H_4-O-C_6H_4-SO_2 \right]_n$			
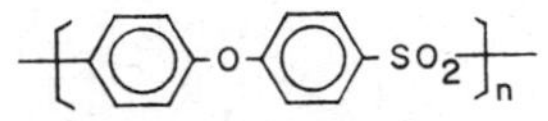 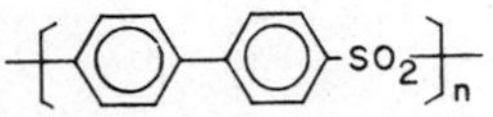 $\left[C_6H_4-O-C_6H_4-SO_2 \right]_n$ (mostly)	720P	ICI	285
and $\left[C_6H_4-C_6H_4-SO_2 \right]_n$			
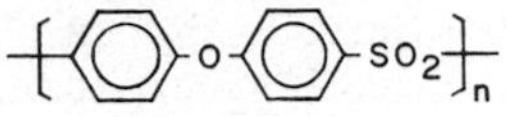 $\left[C_6H_4-O-C_6H_4-SO_2 \right]_n$	200P	ICI	230
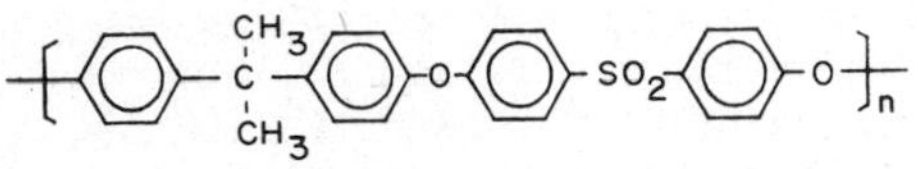 $\left[C_6H_4-C(CH_3)_2-C_6H_4-O-C_6H_4-SO_2-C_6H_4-O \right]_n$	UDEL	Union Carbide	190

Poly(ether sulfones) are resistant to creep at 20°C, as would be expected of a glassy substance. Further, they are also generally impact-resistant even below 0°C. These properties are, of course, structure-dependent, i.e., para catenation and lack of pendant groups being necessary inclusions.

Poly(ether sulfones) can be prepared in two ways, by a nucleophilic ether synthesis (type 1) or an electrophilic (Friedel-Crafts) sulfone coupling (type 2):

Type 1

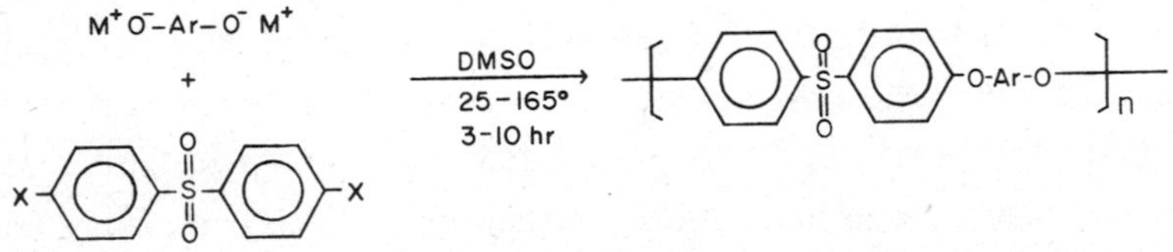

$$M^+\,{}^-O-Ar-O^-\,M^+ \;+\; X-C_6H_4-S(=O)_2-C_6H_4-X \xrightarrow[\text{25-165°, 3-10 hr}]{\text{DMSO}} \left[C_6H_4-S(=O)_2-C_6H_4-O-Ar-O \right]_n$$

X is F or Cl

M is Na or K

Type 2

$$ClSO_2-Ar-SO_2Cl + \text{C}_6\text{H}_5\text{-O-C}_6\text{H}_5 \xrightarrow[\substack{120-230^\circ \\ 4-20\text{ hr}}]{FeCl_3} \left[\text{C}_6\text{H}_4\text{-O-C}_6\text{H}_4\text{-}SO_2\text{-Ar-}SO_2 \right]_n$$

The aromatic nuclei (Ar) which can be used here are biphenyl or naphthalene, both of which induce higher Tg's due to chain rigidity. The use of bisphenol A as a phenolic salt in the former reaction gives the lowest Tg of any (190°C). The use of α,α'-bis(4-hydroxyphenyl)-p-diisopropylbenzene as a comonomer also gives a lower Tg product, but, in addition, gives a lower melt viscosity to aid processing.

In the latter reaction the diphenylether can be replaced with other aromatic systems, provided that ring deactivators are not present; i.e., compounds such as benzophenone and diphenylsulfone would prohibit reaction.

Method Nucleophilic substitution [13,18] In a 250-ml flask equipped with stirrer, condenser, and azeotrope water trap is placed 2,2-bis(4-hydroxyphenyl)propane (0.05 mol), 13.1 g of a 42.8% potassium hydroxide solution (0.1 mol KOH), 50 ml of dimethylsulfoxide, and 6 ml benzene, and the apparatus is purged with nitrogen. The mixture is heated at reflux for 3-4 hr, while removing the water from the reaction mixture as an azeotrope with benzene and distilling off enough of the latter to give a refluxing mixture at 130-135°C, consisting of the dipotassium salt of the 2,2-bis(4-hydroxyphenyl)propane and dimethylsulfoxide. The mixture is cooled and 14.35 g (0.05 mol of 4,4'dichlorodiphenylsulfone) is added followed by 40 ml of anhydrous dimethylsulfoxide, all under nitrogen. The mixture is heated to 130°C and held at 130-140°C with stirring for 4-5 hr. The viscous, orange solution is poured into 300 ml water with vigorous stirring, and the finely divided white polymer is filtered and dried in a vacuum oven at 110°C for 16 hr. The yield is 22.2 g (100% yield). The reduced viscosity in chloroform is 0.59. Films are prepared by compression molding the powder.

Method Electrophilic substitution [2,27] Equimolar amounts of diphenylether and diphenylether-4,4'-disulfonyl chloride are added together carefully in nitrobenzene while using a slight theoretical excess of aluminum chloride. The temperature is raised gradually to 120°C until HCl gas is no longer evolved. The solution is cooled to room temperature and poured into an excess of methanol to precipitate the polymer. The product is purified by twice more dissolving it in dimethylacetamide (DMAc) and precipitating it into a methanol-water mixture. Yields are nearly quantitative, and inherent viscosities range up to 0.35 dl/g in DMAc. The product is also soluble in NMP (N-methyl-2-pyrrolidine), DMSO, pyridine, and phenolic solvents.

A similar product can be obtained by the condensation of the disulfonic acid with diphenylether with polyphosphoric acid; however, molecular weights are quite low (1000); inherent viscosity 0.1 dl/g.

A process has been patented [24] whereby one can obtain a phenylene sulfone linkage by quantitative oxidation of a phenylene sulfide system with H_2O_2 in acetic acid and chloroform or with aqueous potassium permanganate.

$$\left[\!-C_6H_4-O-C_6H_4-S-C_6H_4-O-C_6H_4-SO_2-\right]_n \xrightarrow{\text{oxidation}} \left[\!-C_6H_4-O-C_6H_4-SO_2-\right]_{2n}$$

The starting material was prepared by the nucleophilic substitution method using bis(p-chlorophenyl)sulfone and the sodium salt of bis(p-hydroxyphenyl)sulfide.

3.2-5 Poly(phenylene sulfides)

Another of the polyethers to be commercialized as an engineering plastic is the thio analog of PPO or PPS. This product is marketed by Phillips under the name Ryton [12].

The first synthesis of poly(phenylene sulfide) was from p-dichlorobenzene, but this method allowed more than a monosulfide bridge [15]:

$$S + Na_2CO_3 + Cl-C_6H_4-Cl \xrightarrow[\text{sealed tube}]{\Delta} \left[\!-C_6H_4-S_x-\right]_n$$

The commercial process uses sodium sulfide in a polar solvent [15]:

$$Na_2S + Cl-C_6H_4-Cl \longrightarrow \left[\!-C_6H_4-S-\right]_n$$

PPS and other substituted versions can be prepared on a laboratory scale from thiophenol by a third process which gives a 95-99% yield overall.

$$C_6H_5SH \xrightarrow{Br_2} Br-C_6H_4-S-S-C_6H_4-Br \xrightarrow[C_6H_{13}OH]{Cu} Br-C_6H_4-\bar{S}\overset{+}{Cu}$$

$$Br-C_6H_4-\bar{S}\overset{+}{Cu} \xrightarrow{\text{200° for 5 days or 200° in pyridine for 24 hr}} \left[\!-C_6H_4-S-\right]_n$$

Ryton is used as a high-temperature molding resin (injection or impression) and corrosion preventive coating and displays a Tg of 193-204°C and a melting point (Tm) of 285°C. It is a crystalline material which is soluble only above 200°C in aromatic and chlorinated aromatic solvents. It has the ability to crosslink at elevated temperatures, thereby providing an irreversible cure. The TGA shows no weight loss below 500°C in air but demonstrates complete decomposition by 700°C. In an inert atmosphere it suffers 60% loss at 1000°C. By static test its properties are maintained after 4 months at 233°C (450°F) in air. Ryton's tensile properties in the unfilled form are strength 71.6 MPa (10,800 psi) and modulus 3.36 Pa (480,000 psi).

Method [9] *Bis(p-bromophenyl)disulfide* Bromine (65.6 g, 0.410 mol) is placed in a 250-ml three-necked flask fitted with a stirrer, a condenser, and a dropping funnel. The flask and contents are cooled to 10°C and 15.1 g (0.137 mol) of the thiophenol is added dropwise over a 20- to 30-min period to keep the temperature at 10-15°C. After the addition, the reaction mixture is allowed to warm to room temperature; after 2-2.5 hr, the contents are heated to 50-55°C, and the excess bromine is removed by distillation. The residual solid is washed with a 10% aq. solution of sodium bisulfite, then dissolved in boiling ethanol. The solution is cooled to yield 25.6 g (99.4%) of bis(p-bromophenyl)disulfide, light cream to white plates, mp 91-92°C.

Copper p-bromothiophenoxide A solution of 20 g (0.053 mol) of bis(p-bromophenyl)disulfide in 250 ml of n-hexanol is placed in a 500-ml, three-necked flask equipped with a stirrer, a gas inlet, and a condenser to which is attached a gas outlet. A flow of inert gas through the apparatus is maintained throughout the reaction. Copper metal, electrolytic dust purified (5.4 g, 0.084 mol), and 5.25 ml (0.1 mol) of pyridine are added to the reaction vessel. The reactants are heated to a gentle reflux with stirring for 4-6 hr until the copper metal is completely consumed. The copper salt precipitates out as a bright yellow solid, which is filtered, washed with methylene chloride and then acetone, and dried to give a nearly quantitative yield of copper p-bromothiophenoxide, mp 260-275°C.

Polymerization of copper p-bromothiophenoxide For solid state polymerization, 10 g (0.0397 mol) of copper p-bromothiophenoxide is placed in a 75-ml, heavy-walled Pyrex ampoule. The ampoule and monomer are evacuated and flushed at least three times with an inert gas (argon or purified nitrogen) and sealed with a torch at atmospheric pressure, the atmosphere in the ampoule being inert. The ampoule is heated at 200°C for 5 days. (If 20 ml of pyridine is used as a solvent, a 125-ml ampoule is used and heating is maintained for 24 hr.)

After cooling to room temperature, the ampoule is cooled to Dry Ice temperature, opened, and the solids are ground if necessary. The solids are washed with 200 ml of conc. HCl for 1 hr, then with water, and finally twice more with HCl. The light grey to dark grey solid phenylene sulfide polymer is oven-dried to yield 95-100%.

The product softens at 250-280°C and has a Tm of 280-287°C. It can be fractionated into three portions in the following order: toluene-soluble (0-10%), diphenylether-soluble (60-100%), insoluble residue (0-40%). All extracts are recovered by precipitation into methanol. TGA in air or nitrogen shows weight loss beginning at 425°C, but in nitrogen this loss levels off at 50% [14].

Yet a fourth process has been discovered which uses a simple aromatic compound with sulfur chloride and an iron catalyst under quite mild conditions (25°C for 1-3 hr):

$$H{-}Ar{-}H + SCl_2 \xrightarrow[CHCl_3]{Fe} {-}\!(Ar{-}S)_n\!{-}$$

-Ar-	Tm (°C)
$-C_6H_4-S-C_6H_4-$	125
$-C_6H_2(CH_3)_2-$	270-285
$-C_6(CH_3)_4-$	291
$-C_6H_2(CH_3)_2-CH_2-C_6H_2(CH_3)_2-$	240-250

Because of the low temperatures used in the syntheses, the products precipitated at low molecular weights (950-1500), and the yields were only moderate (50-70%). An increased reaction temperature resulted in aromatic halogenation. There existed about 10% disulfide bridges by this process. The methylene bridge from the substituted diphenylmethane monomer caused a lower Tm by about 30°C when incorporated with sulfide bridges. Interestingly, however, homopolymers with methylene bridges have nearly the same Tm as those with sulfide bridges.

Poly(phenylene disulfide) can also be prepared but in a simpler process from thiohydroquinone [16]. The yield of the reaction was 75%, and

$$HS{-}C_6H_4{-}SH \xrightarrow[40^\circ]{I_2,\,Ethanol} {-}\!\!\left[S{-}C_6H_4{-}S\right]_n\!\!{-}$$

the product was only partially soluble and had a melting point of 198-202°C. The disulfide linkage is responsible for a considerably lower melting point than the monosulfide, as is evident when this product is compared to PPS (280-287°C). One might suspect then that the thermal stability of the poly(phenylene disulfide) is also below that of PPS.

Unlike the monosulfide, the disulfide polymers are amorphous. When both types of links are incorporated into a backbone, the disulfide groups are segregated into the amorphous region.

In this system extra pendant methyl groups are seen again to raise the Tm. Also oxyben and sulfur links are similar in Tm, but a -CH_2- link gives a higher value.

> *Method* [16] A solution is prepared by dissolving 0.69 g of dithiohydroquinone in 20 ml of ethanol, and to this is added slowly, with stirring, a solution of iodine in ethanol until the reaction mixture becomes violet. The yellow precipitate is isolated, washed with aq. sodium bisulfite, water, ethanol, and ether, and dried to yield 75% product (0.51 g).

3.2-6 Poly(imidothioethers)

A copolymer containing the thioether function can be obtained by a very fast hydrogen-addition process with H_2S [4]. The electron-poor maleimide double

Cresol, 25°

R is: o-, m-, or p-C_6H_4

or

or

where X is: nil, O, CH_2, SO_2

bonds are extremely susceptible to reduction and attack by the strong thiol nucleophile. The use of phenolic solvents prevents the formation of an insoluble crosslinked gel by providing protons which determine the competing anionic polymerization.

The white or yellow powdery products are soluble only in strongly polar solvents (DMF, DMAc, DMSO, or hexafluoroacetone) and show an initial weight loss in TGA at 325°C with catastrophic decomposition at 450-500°C.

The advantage of such a copolymer lies in combining the thermal stability of the imide with the flexibility induced by the thioether. However, thermal stability in air is limited by oxidation of the sulfide group. This reaction is shown by strongly exothermic peaks in the DSC with subsequent catastrophic weight loss in many sulfur-containing systems.

> *Method* [4] In a 4 x 20 cm tube with stirrer and gas bubble system is placed 5 g of the bismaleimide, 50 ml of cresol, and 2 drops of N,N,N',N'-tetramethylethylenediamine. A 1:5 mixture of H_2S and N_2 is passed slowly into the solution for 2 hr. The viscous mixture is poured into methanol which contains a small amount of HCl to precipitate the fibrous polymer which is collected, extracted with ethanol, and dried. Inherent viscosities range up to 0.53 in DMF.

A thermoplastic elastomer with 800% elongation and 10.3 MPA (1500 psi) tensile strength can be obtained in a similar manner to that described above [4,5]. This is derived from reaction first of an excess (10 times) of bismaleimide with oligomeric (Mn 850-3500) polythioalkylether dithiols, and these maleimide-terminated thio blocks are joined with imidothioether blocks by the addition of H_2S. Polythioethers are known commercial elastomers (Thiokol) and contribute a "soft" block to the imide backbone. Further, the polysulfide portion imparts solvent resistance and low-temperature flexibility. The contribution of the imidothioether block is to minimize thermal degradation of the thioether block during processing at elevated

N–R–N (bismaleimide) + HS–R'–SH → (1. Base, cresol; 2. H_2S) → [–succinimide–N–R–N–succinimide–S–R'–S–]$_n$ + [–succinimide–N–R–N–succinimide–S–]$_n$

where R is m-or p-C_6H_4– or Ø-CH_2-Ø

and R' is:

$-(CH_2)_x-$ and x = 2, 3, 6, or 10

$-(CH_2CH_2-S)_y-CH_2CH_2$ and y = 15 to 55

$-CH_2CH_2OCH_2CH_2-$

$-CH(CH_3)-CH_2-$

$-CH_2-C(=O)-O-CH_2-CH_2-O-C(=O)-CH_2-$

temperatures. Since the blocks are short, however, only one Tg is observed. TGA data show that these blocks are less stable than the simple thioimido-ethers [4] in that a 10% weight loss is observed from 248 to 290°C.

> *Method* [5] Into a 20 x 2.5 cm polymerization tube equipped with a stirrer and a thermometer is placed 5 g (0.0012 mol) of Thiokol LP-12 (a polysulfide, Mn = 4000), 35 ml maleimido-4,4'-diphenylmethane, and 2 drops of tri-n-butylamine, and the mixture is stirred for 30 min. Then H_2S is slowly bubbled into the mixture for 30 min to produce a viscosity and temperature increase. The mixture is then stirred for 1 hr and the light yellow, viscous solution is poured into methanol which contains acetic acid. The pale cream-colored powder is isolated, washed, and dried for a 95% yield. (Polysulfide oligomers of lower molecular weight can be synthesized from 2-chloroethylformal and sodium sulfide.)

3.3 Recent Developments

3.3-1 Polyesters

An aromatic polyester, poly-p-hydroxybenzoic acid or EKONOL (Carborundum), has been thoroughly described in the literature [40]. It is synthesized from either the phenylester or the p-acetoxy acid of the parent compound. A double-helix crystalline structure was found with a reversible crystalline transition temperature at 325-360°C. The polymer is fabricated by high-temperature compression sintering and is self-lubricating. It has a high flexural modulus, thermal conductivity, solvent resistance, and electrical insulating characteristics. The product can also be applied by plasma spraying. A plasticizer used with the polyester is PTFE.

3.3-2 Polyethers

Aromatic polyethers have been synthesized recently by a phase transfer catalyzed condensation of dichloroaromatic compounds with bisphenols [46]. This was accomplished by the use of quaternary ammonium salts, crown ethers, and poly(ethylene glycols) in methylene chloride and aqueous alkali.

The mechanism of thermal degradation of poly(2,6-dimethyl-1,4-phenylene oxide) was found to occur by a Fries-like rearrangement to a polybenzyl structure as a first step. Then scission of the benzyl bonds occurred to yield o-cresol and other methyl-substituted phenols [47].

A phosphorus-containing polyether has been produced in 85% yield from the sodium salts of bisphenols and bis(p-chlorophenyl)phenylphosphine oxide [42]. The product shows no decomposition in nitrogen up to 300°C. A 10% weight loss occurs by TGA at 460°C in either air or nitrogen.

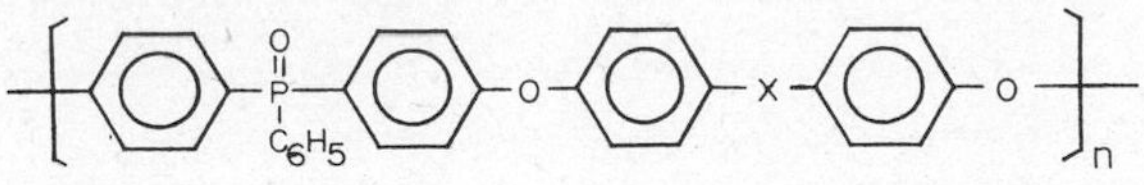

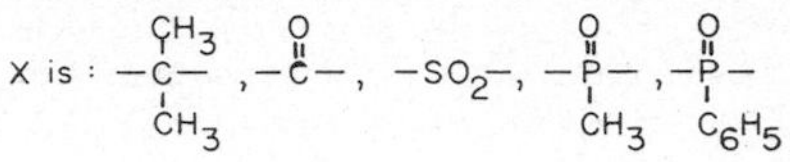

$$\left[C_6H_4-P(O)(C_6H_5)-C_6H_4-O-C_6H_4-X-C_6H_4-O \right]_n$$

X is : $-C(CH_3)_2-$, $-C(O)-$, $-SO_2-$, $-P(O)(CH_3)-$, $-P(O)(C_6H_5)-$

A fluorinated polyether which is elastomeric and moderately thermally stable has the following structure [48]:

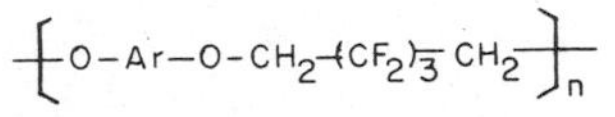

$$\left[O-Ar-O-CH_2-(CF_2)_3-CH_2 \right]_n$$

Ar is : $-C_6F_4-$, $-C_6F_4-C_6F_4-$, -perfluoropyridinylene,

perfluoro-1,3-pyrimidinylene,

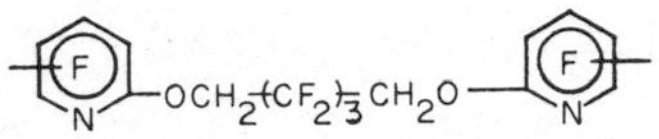

$$\left[C_5F_2N-OCH_2(CF_2)_3CH_2O-C_5F_2N \right]$$

The condensation of monomers (aliphatic diols and perfluoroaromatics) was effected by sodium hydride in THF and tetramethylene sulfone. Glass transition temperatures range from -10 to +2°C and thermal stabilities are 315-375°C in oxygen and 355-445°C in nitrogen (10% weight loss by TGA). The best stability in oxygen is demonstrated by the biphenylene unit, and the best elastomer is produced with substituted pyridine in the backbone.

Poly(perfluoroether)oxadiazole elastomers which are soluble in Freon-TF and have pendant nitrile groups as curing sites are known [39]. These are crosslinked with N,N'-terephthalonitrile oxide (TPNO) at 95°C to give Shore A hardness values of 40-70, tensile strengths up to 166 MPa (2425 psi), and elongation-to-break of 350-835%.

A ladder-like polyether, a polyspiroketal, was reported to result from the reaction of quinone with pentaerythritol [32]:

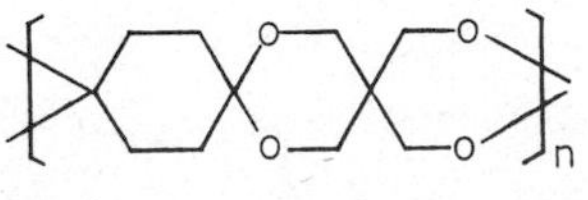

This polymer was 95% crystalline and decomposed at 350°C with no melting.

3.3-3 Poly(ether sulfones) and Analogs

Several papers have appeared which deal with poly(arylene ether sulfones) [28-31]. The alkali metal salts of several halogenophenylsulfonyl phenoxides were prepared [28] and condensed [29] in the melt and solution (sulfolane). The polymers were amorphous with Tgs of 200-240°C, generally, with some down to 165°C or up to 298°C, depending on structure. Also measured were impact strength, tensile strength, creep, modulus, and density. Considerable toughness was observed down to temperatures of -180°C. However, significant degradation of properties was noted after storage of the material at 150°C.

Marvel and co-workers have published numerous papers on poly(ether sulfone ketones) as laminating resins [33,34,36,41,45,50,54]. Numerous soluble polymers were produced with various crosslinking moieties incorporated (acetylene, disulfide, and cyano). These resins have shown promise as glass-laminating materials with good isothermal stabilities to 300°C in air.

A sulfonated poly(sulfone ether) was produced by sulfonating the polymer either in solution or on the surface of the polymer after fabrication [51]. The pendant sulfonic acid group can then be converted to the salt by treatment with sodium methoxide. This process decreases the gas permeability and increases water absorption of the polymer to provide a candidate for reverse osmosis membranes. A film for this purpose can be cast from DMF solution. At 0.5 sulfonate groups per repeating unit, a 95% rejection ratio can be obtained with good flux. The evaluation of polysulfones as reverse osmosis hollow fibers has been accomplished [37,38].

The mechanical behavior of commercial polysulfone (3 M) [52] and the effect of thermosetting plasticizers on the processing and mechanical properties of polysulfone [53] have been reported.

3.3-4 Poly(phenylene sulfides)

PPS has been characterized, including the effect of crystallinity on properties [35,44]. Virgin PPS is about 65% crystalline, and amorphous material crystallizes readily below 121°C. PPS demonstrates a Tg of 85°C and a Tm of 285°C; therefore, annealing of molded pieces is accomplished at 204°C. Quenching results in 5% crystallinity. With annealing, the flexual strength goes from 103 to 90 MPa (15,000-13,000 psi), and the HDT increases from 100 to 128°C; but the tensile strength decreases.

PPS can be cured, a process which involves (1) chain extension with loss of low molecular weight species, (2) oxidative crosslinking, (3) thermal crosslinking, and (4) oxygen uptake with loss of SO_2 [43].

3.3-5 Polycarbonates

Interfacial condensation of phosgene with bisphenolfluorenone (BPF) and BPF-endcapped silicone oligomers gave silicon block-modified bisphenolfluorenone polycarbonate [49]. These polymers showed potential as tough, transparent, heat-resistant plastics if they contained 15-20% siloxane in the backbone. This material is one which gives two Tg values, one for the siloxane block at -100°C and one for the polycarbonate region at 275°C. Other properties are as expected for such a system; i.e., modulus and yield stress decrease with increasing silicon portion, while elongation and impact toughness increase. HDT values are up to 216°C. The limiting oxygen index of 50 illustrates the low flammability of these polymers.

References

1. H. S. Blanchard, H. L. Zinkheiner, and G. A. Russell, *J. Polym. Sci.*, *58*, 469 (1962).
2. S. M. Cohen and R. H. Young, *J. Polym. Sci.*, *A-1*, *4*, 722 (1966).
3. G. D. Cooper and A. Katchman, *Advan. Chem. Series*, No. 91, 660 (1969).
4. J. V. Crivello, *Polym. Preprints*, *13*(2), 924 (1972).
5. J. V. Crivello and P. C. Juliano, *J. Polym. Sci.: Polym. Chem. Ed.*, *13*(8), 1819 (1975).
6. J. Economy, B. E. Nowak, and S. G. Cottis, *Polym. Preprints*, *11*(1), 332 (1970).
7. J. Economy, B. E. Nowak, and S. G. Cottis, *SAMPE J.*, *6*, 6 (1970).
8. J. Economy, R. S. Storm, V. I. Matkovich, S. G. Cottis, and B. E. Nowak, *J. Polym. Sci.: Polym. Chem. Ed.*, *14*, 2207 (1976).
9. C. E. Handlovits, *Macromolecular Synthesis*, Vol. 3 (W. J. Bailey, Ed.), John Wiley and Sons, New York, 1969, pp. 131-134.
10. A. Fritz, *Polym. Preprints*, *12*(1), 232 (1971).
11. A. S. Hay, *J. Polym. Sci.*, *58*, 581 (1962).
12. H. W. Hill and J. T. Edmonds, *Polym. Preprints*, *13*(1), 603 (1972); and J. T. Edmonds and H. W. Hill, U.S. Patent 3,354,129 (November 21, 1967).
13. R. N. Johnson and A. G. Garnham (Union Carbide Corp.), British Patent 1,078,234 (1967).
14. R. W. Lenz, C. E. Handlovits, and H. A. Smith, *J. Polym. Sci.*, *58*, 351 (1962).

15. A. D. Macallum, *J. Org. Chem., 13,* 154 (1948).

16. G. Montaudo, G. Bruni, P. Maravigna, and F. Bottino, *J. Polym. Sci.: Polym. Chem. Ed., 12*(2), 2881 (1974).

17. G. Montaudo, P. Finocchiaro, and S. Caccamese, *J. Polym. Sci., A-1, 9,* 3627 (1971).

18. J. B. Rose, *Polymer, 15*(7), 456 (1974).

19. R. W. Rosser, J. A. Parker, R. J. Depasquale, and E. C. Stump, Jr., *Polym. Preprints, 15*(1), 216 (1974).

20. R. L. Shriner, R. C. Fuson, and D. Y. Curtin, *The Systematic Identification of Organic Compounds,* 4th Ed., John Wiley and Sons, New York, 1956, p. 116.

21. D. R. Stevenson and J. E. Mulvaney, *J. Polym. Sci., A-1, 10*(9), 2713 (1972).

22. J. K. Stille and P. E. Cassidy, *J. Polym. Sci., B: Polym. Lett., 1,* 563 (1963).

23. J. K. Stille, P. E. Cassidy, and L. Plummer, *J. Amer. Chem. Soc., 85,* 1318 (1963).

24. J. Studinka and R. Gabler, Swiss Patent 491,981 (July 31, 1970).

25. D. S. Tarbell, J. W. Wilson, and P. E. Fanta, *Organic Synthesis Collective,* Volume 3 (E. C. Horning, Ed.), John Wiley and Sons, New York, 1955, p. 267.

26. E. J. Vandenberg, *J. Polym. Sci., A-1, 10,* 2887 (1972).

27. H. A. Vogel, British Patent 1,060,546 (April 1963).

28. T. E. Attwood, D. A. Barr, G. G. Feasey, V. J. Leslie, A. B. Newton, and J. B. Rose, *Polymer, 18,* 354 (1977).

29. T. E. Attwood, D. A. Barr, T. King, A. B. Newton, and J. B. Rose, *Polymer, 18,* 359 (1977).

30. T. E. Attwood, T. King, V. J. Leslie, and J. B. Rose, *Polymer, 18,* 369 (1977).

31. T. E. Attwood, T. King, I. D. McKenzie, and J. B. Rose, *Polymer, 18,* 365 (1977).

32. W. J. Bailey and A. A. Volpe, *Polym. Preprints, 8,* 292 (1967).

33. A. Banihashemi and C. S. Marvel, *J. Polym. Sci.: Polym. Chem. Ed., 15,* 2667 (1977).

34. A. Banihashemi and C. S. Marvel, *J. Polym. Sci.: Polym. Chem. Ed., 15,* 2653 (1977).

35. D. G. Brady, *J. Appl. Polym. Sci., 20,* 2541 (1976).

36. M. Bruma and C. S. Marvel, *J. Polym. Sci.: Polym. Chem. Ed., 14,* 1 (1976).

37. I. Cabasso, E. Klein, and J. K. Smith, *J. Appl. Polym. Sci., 20,* 2377 (1976).

38. I. Cabasso, J. K. Smith, and E. Klein, *Org. Coat. Plast. Preprints, 35*(1), 492 (1975).

39. R. E. Cochoy, *J. Appl. Polym. Sci.*, *20*(4), 1035 (1976).
40. J. Economy, R. S. Storm, V. I. Matkovich, S. G. Cottis, and B. E. Nowak, *J. Polym. Sci.: Polym. Chem. Ed.*, *14*, 2207 (1976).
41. R. L. Frentzel and C. S. Marvel, *J. Polym. Sci.: Polym. Chem. Ed.*, *17*, 1073 (1979).
42. S. Hashimoto, I. Furukawa, and K. Ueyama, *J. Macromol. Sci. Chem.*, *A11*(12), 2167 (1977).
43. R. T. Hawkens, *Macromolecules*, *9*(2), 189 (1976).
44. H. W. Hill and D. G. Brady, *Org. Coat. Plast. Preprints*, *36*(2), 363 (1976).
45. F. Huang and C. S. Marvel, *J. Polym. Sci.: Polym. Chem. Ed.*, *14*(11), 2785 (1976).
46. Y. Imai, M. Ueda, and M. Si, *J. Polym. Sci.: Polym. Lett. Ed.*, *17*, 85 (1979).
47. J. Jachowicz, M. Kryszewski, and P. Kowalski, *J. Appl. Polym. Sci.*, *22*, 2891 (1978).
48. P. Johncock, M. A. H. Hewins, and A. V. Cunliffe, *J. Polym. Sci.: Polym. Chem. Ed.*, *14*, 365 (1976).
49. R. P. Kambour, J. E. Corn, S. Miller, and G. E. Niznik, *J. Appl. Polym. Sci.*, *20*, 3275 (1976).
50. R. Kellman and C. S. Marvel, *J. Polym. Sci.: Polym. Chem. Ed.*, *14*, 2033 (1976).
51. A. Noshay and L. M. Robeson, *J. Appl. Polym. Sci.*, *20*, 1885 (1976).
52. R. J. Wann, G. C. Martin, and W. W. Gerberich, *Polym. Eng. Sci.*, *16*(9), 645 (1976).
53. A. Wereta, G. A. Loughran, and F. E. Arnold, *Org. Coat. Plast. Preprints*, *36*(2), 387 (1976).
54. D. Tm. Worn and C. S. Marvel, *J. Polym. Sci.: Polym. Chem. Ed.*, *14*, 1637 (1976).

Chapter 4

Nonheterocyclic Polymers: Backbones Containing Carbon and Acyclic Nitrogen, Polyamides, and Others

The largest and most important group of linear-acyclic nitrogen polymers is polyamides. The extensive research in this area is due to the success of the aliphatic and, later, the aromatic systems which have been developed into commercial products. Another contributing factor is undoubtedly the simplicity of the system and the plethora of routes to amide functions; i.e., the only monomers necessary are diacids and diamines (or an amino acid), or derivatives.

Polyamides and a few miscellaneous types of linear nitrogen backbones are summarized in Table 4.1.

4.1 Polyamides

Because of the commercial success of aliphatic polyamides, aromatic systems were a predictable extension of research by fiber producers (duPont, Monsanto, etc.). Given this impetus and the desirable properties of the all-aromatic polymers, this class of materials was one of the first exploited on a large scale as thermally stable polymers. In 1974 the aliphatic and aromatic polyamides were given separate generic nomenclature by the Federal Trade Commission. As a result, "nylon" refers to the material in which less than 85% of the amide groups are attracted to aromatic rings. The term "aramid" then refers to the polyamide with at least 85% of the amide function directly adjacent to aromatic rings.

For polyamides other terminology has been generated which designates whether the acid (A) and amine (B) functions originate on the same monomer molecule (i.e., an amino acid), or whether two different monomers were used (a diamine and a diacid). An A-B type of polyamide derives from the polymerization of an amino acid or its derivative, e.g., poly(p-benzamide). An AA-BB polyamide is illustrated by poly(p-phenylene terephthalamide),

Table 4.1 Polyamides and Other Linear, Nitrogen-Containing Polymers

Polymer	Structure	Comments
Poly(m-benzamide)	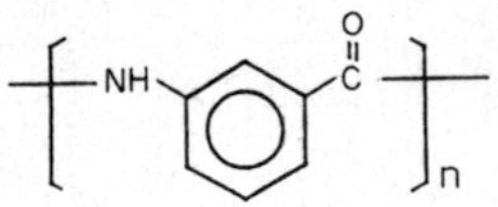	Soluble.
Poly(m-phenylene) (isophthalamide), duPont Nomex	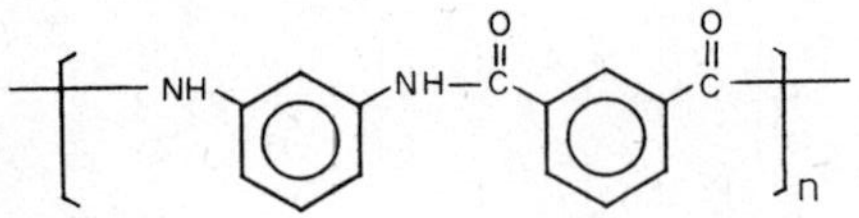	Strong fibers.
Poly(p-benzamide), duPont early Fiber B	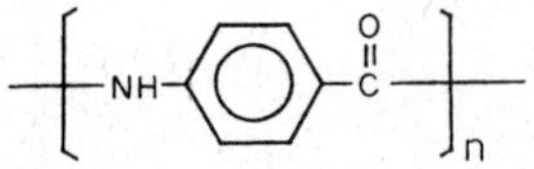	Rigid rod backbone. High solution viscosity.
Poly(p-phenylene terephthalamide), duPont later Fiber B	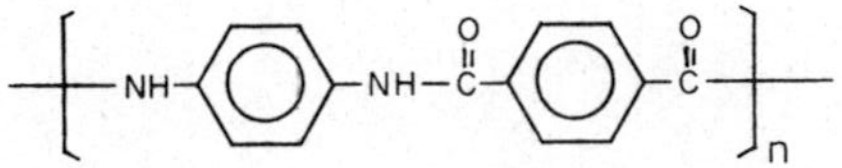	High modulus fiber.
Poly(squaryl amide)	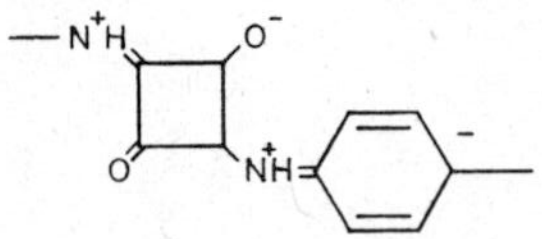	Acid-soluble. Stable at 375-425°C.
Poly(hexahydrobenzodipyrrole amide)	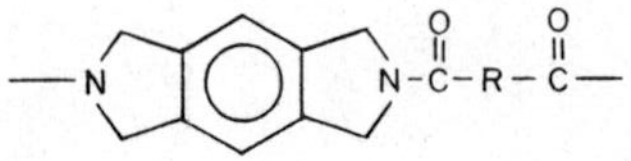	Tm 350-435°C. Stable at 305-405°C. Decomposes in UV light. Soluble.
Poly(diaroylphthalamides)	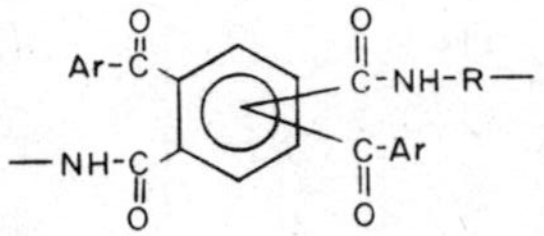	Stable at 300-375°C/air 360-415°C/N_2. Soluble.

Poly(benzamide urea)	—NH-C(=O)-NH-[benzene ring]-C(=O)-NH-R—	Softens at 140-260°C. Stable to 340°C. Soluble.
Poly(amide sulfones)	—NH-C(=O)-[benzene ring, $SO_2\phi$]-C(=O)-NH-R—	Thermal properties are lowered by sulfone group.
Poly(hexahydrobenzodipyrrole sulfonamides)	—N[ring system]N-SO_2-R-SO_2—	Stable to 300°C. Acid-soluble.
Copoly(amide sulfonamides)	—C(=O)-Ar-SO_2-NH-R-NH—	Soluble. Films: amorphous, transparent, tough. PMT 155-275°C. Stable to 340°C.
Poly(oxamides)	—C(=O)-C(=O)-NH$(CH_2)_x$-NH—	Crystalline. Soluble. Tm 195-320°C. Stable to 400°C/N_2.
Poly(dipiperidylamides)	—N[piperidine ring]-$(CH_2)_n$-[piperidine ring]N-C(=O)-R-C(=O)—	Soluble. Film and fiber formers. Stable at 400-465°C. PMT 160-455°C.
Poly(azoamides)	—[benzene ring]-C(=O)-NH-[benzene ring]-N=N—	Yellow, orange. Tough films. Stable to 360°C for 479 hr. Soluble.
Poly(diphenylamine)	(Unknown structure)	Soluble. Softens at 80-100°C. Stable to 465°C. Crosslinkable.
Polyimines or Polyazomethines	=N-[benzene ring]-N=CH-[benzene ring]-CH=	Stable at 400°C. Black. High modulus fibers.
Poly(propiolamide)	—CH=CH-C(=O)-NH— ⇌ —CH=CH-C(OH)=N—	Brown. Water-soluble. Stable to 310°C.

which is the result of the polymerization of a diacid (AA), terephthalic acid, with a diamine (BB), p-phenylene diamine.

Again, owing to the scientific and commercial success of polyamides, both nylons and aramids, research has continued on numerous variations. The ordering and varying of backbone functions have been shown to be extremely important to the final properties. Aromatic silane functions have been introduced to aramids in an effort to seek improved properties [27]. The ease of formation of the amide group makes it widely used as a method to link other kinds of backbone functions, as can be seen in other chapters. Recently, the interest in liquid crystalline solutions has extended to polyamides as is nicely delineated by P. W. Morgan [23], well known because of his work in aramids at duPont.

Because of the excellent review articles available on aramids [7,11, 18,23,25,26], this chapter devotes only one section to their discussion, while 10 sections are used for examining little-known polyamide curiosities.

4.1-1 Aromatic Polyamides

Polyaromatic amides (aramids) are among the oldest members of the class of thermally stable polymers and among the first given to practical application, developed by duPont for its Nomex fiber. This system is essentially a copolyamide of m-phenylene diamine and isophthalic acid, the meta catenation providing flexibility:

$$\left[-NH-C_6H_4-NH-\overset{O}{\overset{\|}{C}}-C_6H_4-\overset{O}{\overset{\|}{C}}- \right]_n$$

The order of occurrence of these units along the backbone is extremely important since the melting point can vary by 100°C from an ordered to a random arrangement.

The use temperature of Nomex is 370°C, and the TGA shows a 10% weight loss at 450°C. The flame-resistant character of this material is attributed to the high melting point and slow thermal decomposition. If a para orientation is used for aramids, one witnesses an increase in modulus by three to six times, lower elongation to break, increased intractability, more crystallinity, improved thermal stability, higher tensile strength, and lower solubility. A product with these characteristics is duPont's Kevlar (formerly Fiber B), which is postulated to contain the following structural units:

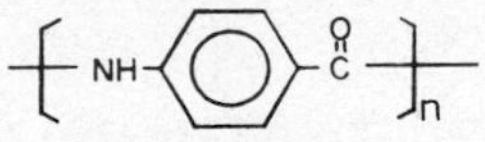

$$\left[NH-C_6H_4-\overset{O}{\overset{\|}{C}} \right]_n \quad \text{and} \quad \left[C_6H_4-NH-\overset{O}{\overset{\|}{C}}-C_6H_4-\overset{O}{\overset{\|}{C}}-NH \right]_n$$

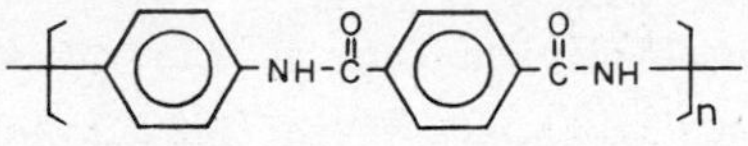

Early Fiber B Later Fiber B (Kevlar)

Monsanto's product (X-500) is similar in that it is a poly(amide hydrazide):

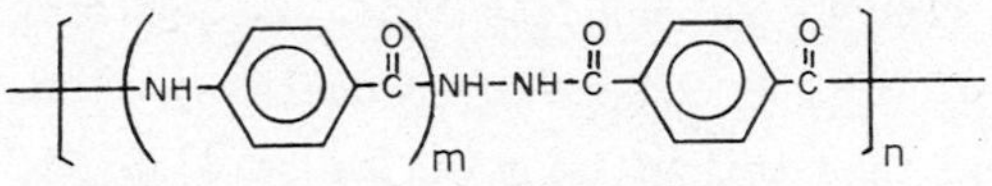

$$\left[\left(NH-C_6H_4-\overset{O}{\overset{\|}{C}} \right)_m NH-NH-\overset{O}{\overset{\|}{C}}-C_6H_4-\overset{O}{\overset{\|}{C}} \right]_n$$

The above properties lead to the use of the para-oriented aramids as tire belt reinforcement (Goodyear's Exten), aerospace composites, electrical insulation, high-speed bearings, and corrosive gas filtration cloth. The advantages over inorganics or metals are lower density, conductivity, and corrosivity, and higher extensibility, abrasion resistance, and strength-to-weight ratio. These commercial materials have been extensively reviewed by Preston and Black [7]. Polyamides and other types of backbones which have been used in high-performance fibers have also been reviewed [19,29,30]. Because of the commercial status and these recent comprehensive reviews available, extensive coverage of these more common aramids will not be made here.

A study has shown the effect on thermal stability of various substituents in the backbones of aramids [17]. First, of course, all products from isophthalic acid were lower in melting point than those from terephthalic acid. However, isophthalic acid polymers demonstrated better oxidative stability and flame resistance than did those from terephthalic acid. Second, for the diamine:

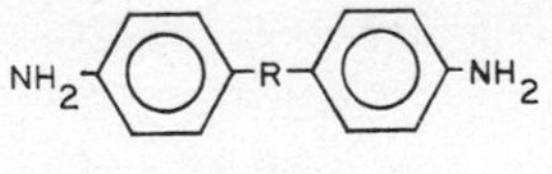

$$NH_2-C_6H_4-R-C_6H_4-NH_2$$

the greatest stability was found when R was simply a single bond, i.e., benzidine. A methylene or sulfone function lowered this property considerably, and phenylmethylene or cyclohexylmethylene caused further loss of thermal stability.

Some work has been done in using other aromatic connecting groups in aramids, this work having been carefully reviewed by Preston [23]. He has tabulated considerable data: viscosity, spinning solvents and conditions, synthetic method, and tensile properties. These backbones include the following, in addition to those given above: 4',4'-biphenylene, 2,6-naphthalylene,

3,3'-dimethyl-4,4'-biphenylene, 4,4'-diphenylene, ether, 4,4'-diphenylene sulfide, 4,4'-diphenylene sulfone, 4,4'-diphenylene ketone, and 4,4'-diphenylene alkylene, where the alkylene is methylene, 1,2-ethylene, or isopropylidene.

Of the methods of condensation polymerization, only two lend themselves to the synthesis of aramids of high molecular weight, solution or interfacial. Low temperature (i.e., <100°C) polycondensation has been well reviewed by Morgan [22]. Because of the much easier processing of the product from solution techniques, this method is given as an example. (The problem in fabrication arises from the difficulty in dissolving the solid aramid once it is precipitated.) The reaction solvents are commonly amides (DMAc, HMPA, and HMPA or NMP) and, of course, an acid acceptor (tertiary amine) is employed as in the interfacial method (similar to the nylon rope trick).

> *Method* *Solution polymerization of m-aminobenzoyl chloride hydrochloride* [32] Under a dry nitrogen atmosphere in a three-necked flask in a cooling bath (-30°C) is placed 471.5 g DMAc and 140 g of m-aminobenzoyl chloride. The mixture is allowed to warm to 25°C and stirred for 5 hr. The solution is poured slowly in portions into water in a high-speed blendor to precipitate the poly(m-aminobenzamide). The solid is collected by filtration, washed in the blendor three times with water and once with acetone, and dried; inherent viscosity 1.25 dl/g.
>
> A spinning solution (19.8%) is prepared by dissolving the polyamide in DMAc which contains 6.4% lithium chloride.

Another, and quite recent, solution (in NMP) method employs aryl phosphites [$(C_6H_5O)_3P$, triphenyl phosphite], a tertiary amine (pyridine), and a metal salt (lithium or calcium chloride) to produce fairly high molecular weight polyaromatic amides [35].

NH_2–C6H4–CO_2H —($P(OAr)_3$, pyridine)→ pyridinium OAr^- / HP(OAr)(OAr)–O–C(=O)–C6H4–NH_2 —(NH_2–C6H4–CO_2H)→ HO_2C–C6H4–NH–C(=O)–C6H4–NH_2 + ArOH + HO-$P(OAr)_2$ —etc.→ [–NH–C6H4–C(=O)–]$_n$

Ar is: C_6H_5 p-C_6H_4Cl,
o-, m- or p-$C_6H_4CH_3$

The reaction is sensitive primarily to temperature (80-120°C), type (lithium or calcium chloride), and concentration (2-12%) of salt, and to type of tertiary amine, pyridine being the only good amine. Most di- or triarylphosphite or phosphinites are suitable for high yields (95-100%) and viscosities (1.05-1.48 dl/g), whereas trialkylphosphites give no polymer. This method is also satisfactory for the AA-BB type polymerization of isophthalic acid and aromatic diamines but not for terephthalic acid systems.

Method *Poly(p-benzamide) by the phosphorylation process* [35] In 40 ml of NMP and 10 ml of pyridine was dissolved 2 g of lithium chloride, 2.74 g (0.02 mol) of p-aminobenzoic acid, and 6.21 g (0.02 mol) of triphenylphosphite, and the solution was heated under nitrogen for 6 hr at 100°C with stirring. The mixture was then cooled and poured into 200 ml of methanol to precipitate the polymer. The solid was isolated by filtration, ground to a powder, washed with methanol, and finally dried to give a 100% yield of poly(p-benzamide) with an inherent viscosity of 1.27 dl/g (in H_2SO_4 at 30°C).

4.1-2 Poly(squaryl amides)

Squaric acid or its ester or acyl chloride will form a polyamide with diamines by being heated in various protic, aprotic, or strongly acidic solvents [glycerol, polyphosphoric acid (PPA), or N-methyl-2-pyrrolidone (NMP) with 5% lithium chloride]. Both 1,2 and 1,3 orientations exist in the product, with the latter favored in more basic solvents [20,24,25]. Maximum 1,2 substitution (and improved solubility) is obtained from squaryl chloride and p-phenylene diamine at low temperatures. The ionic structure of the product has been proposed on the basis of x-ray photoelectric and electronic absorption spectral studies [26].

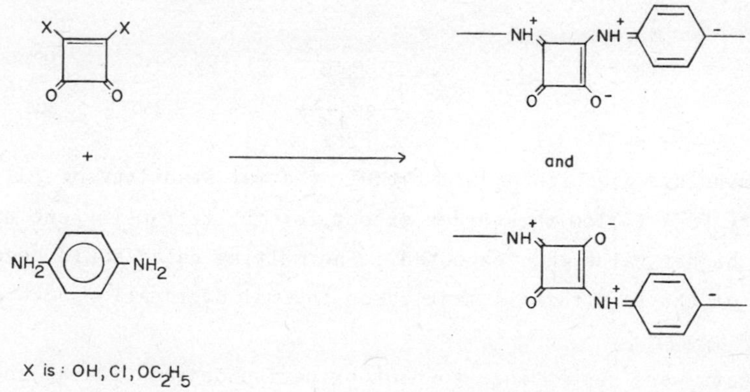

X is: OH, Cl, OC_2H_5

Inherent viscosities of 0.1 to 0.3 dl/g were obtained in 98% sulfuric acid (0.5 g/100 ml). The 1,2 structure, although more soluble, exhibited a lower thermal stability (325-375°C) than the 1,3 product (375-425°C), with residual weights of 30% at 800°C [26].

> *Method* [25] A solution of 0.96 g (6.36 mmol) of squaryl dichloride [10] in 10 ml of NMP containing 0.25 g of LiCl is added dropwise for 15 min to a stirred solution of 0.685 g (6.36 mmol) of p-phenylene diamine in 10 ml of NMP at 0°C under nitrogen. Then 1.12 ml of pyridine is added, and the mixture is stirred for 1 hr at 0°C, then at 23° for 1 hr, and finally at 40°C for 15 min in the dark. The polymer is precipitated by pouring the solution into 100 ml of 1:1 water-methanol. The red-black product is isolated, washed with water and then methanol, and dried at 100°C under vacuum to yield 1.14 g (98%). The polyamide is soluble in 98% sulfuric acid, fluorosulfonic acid, and in NMP which contains 5% LiCl.

4.1-3 Poly(hexahydrobenzodipyrrole amides)

An interfacial condensation process allowed preparation of the following series of polymers generally in yields above 90% [33]:

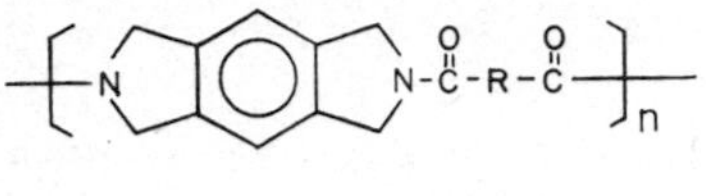

$$\left[N \quad N-\overset{O}{\overset{\|}{C}}-R-\overset{O}{\overset{\|}{C}}- \right]_n$$

where R is:

R	Tm(°C)	TGA(°C)/N$_2$
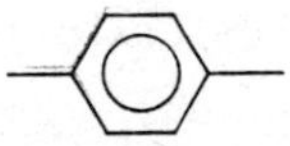 (p-phenylene)	435	405
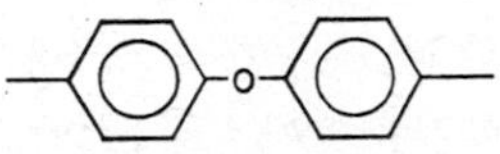 (phenylene)-O-(phenylene)	380	390
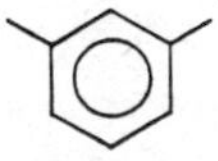 (m-phenylene)	380	360
$-(CH_2)_4-$	420	350
$-(CH_2)_8-$	350	305

The polymers are given in order of their thermal stability by TGA. The Tm data (by DSC) follow this order except for the tetramethylene system, which shows a higher value than expected. The melting data should be evaluated in view of the fact that in most cases thermal degradation occurs below the Tm.

All products were susceptible to photooxidative degradation as shown by a drastic decrease in viscosity (in m-cresol) after exposure to UV light.

Method [12] Into 35 ml of water is placed 0.966 g of diisoindoline dihydrobromide (hexahydrobenzodipyrrole); and with stirring, two other solutions are added simultaneously within 30 sec: (1) 0.480 g of sodium hydroxide in 10 ml of water and (2) 0.609 g of isophthaloyl chloride in 15 ml of methylene chloride. The mixture is stirred for 15 min, and the methylene chloride is evaporated. The white solid is isolated; washed with hot water, cold water, alcohol, and acetone; and dried in vacuo at 60°C. The yield is 90% with an inherent viscosity of 1.52 dl/g in m-cresol at a concentration of 0.5 g/100 ml.

4.1-4 Polyamides from Pseudo Diaroylphthaloyl Chlorides

An interesting pseudo phthaloyl chloride lends itself to ring opening to yield polyamides with aroyl (benzoyl and toluoyl) pendant functions which promote solubility [34]. The orientation of the aroyl groups may be meta or para on the central benzene nucleus.

Ar, Cl, O, O, C, O, O, Cl, Ar

+ NH_2-R-NH_2

25°, 24 hr, Et_3N

Ar-C(=O), C(=O)-NH-R, NH-C(=O), C(=O)-Ar, n

Ar is: C_6H_5, or $CH_3C_6H_4$

R is: $-(CH_2)_2-$, $-(CH_2)_6-$ or $-CH_2-C_6H_4-CH_2-$

The thermal stabilities are generally comparable to aliphatic polyphthalamides in that the para catenation was more stable and the phenyl substituent better than the tolyl. Stabilities by TGA were, with one exception, 300-375°C in air and 360-415°C in nitrogen. By far the most significant difference between stabilities in air and nitrogen occurred with the ethylene diamine-toluoylisophthaloyl system. An improvement of 145°C in thermal stability was demonstrated by using a nitrogen atmosphere. This material was the worst one of the series (by 35°C) in air but was second best (by 5°C) in nitrogen.

Method [34] Solid 2,5-dibenzoylterephthaloyl chloride (2.05 g) is added with stirring to a mixture of 0.58 g of hexamethylene diamine and 1.4 ml of triethylamine in 10 ml of NMP. The mixture is stirred for 24 hr then poured into 1 liter of water. A quantitative yield (2.27 g) of powder is received after the precipitate is dried; inherent viscosity in NMP 0.53.

4.1-5 Poly(benzamide ureas)

Polymers with both amide and urea linkages show good solubilities in protic and polar aprotic solvents (DMF, DMSO, pyridine, m-cresol, and NMP); but they have thermal stabilities of only 200-300°C [21]. The piperazine unit allows the highest thermal stability of 340°C. The softening temperatures are in the range 140-260°C.

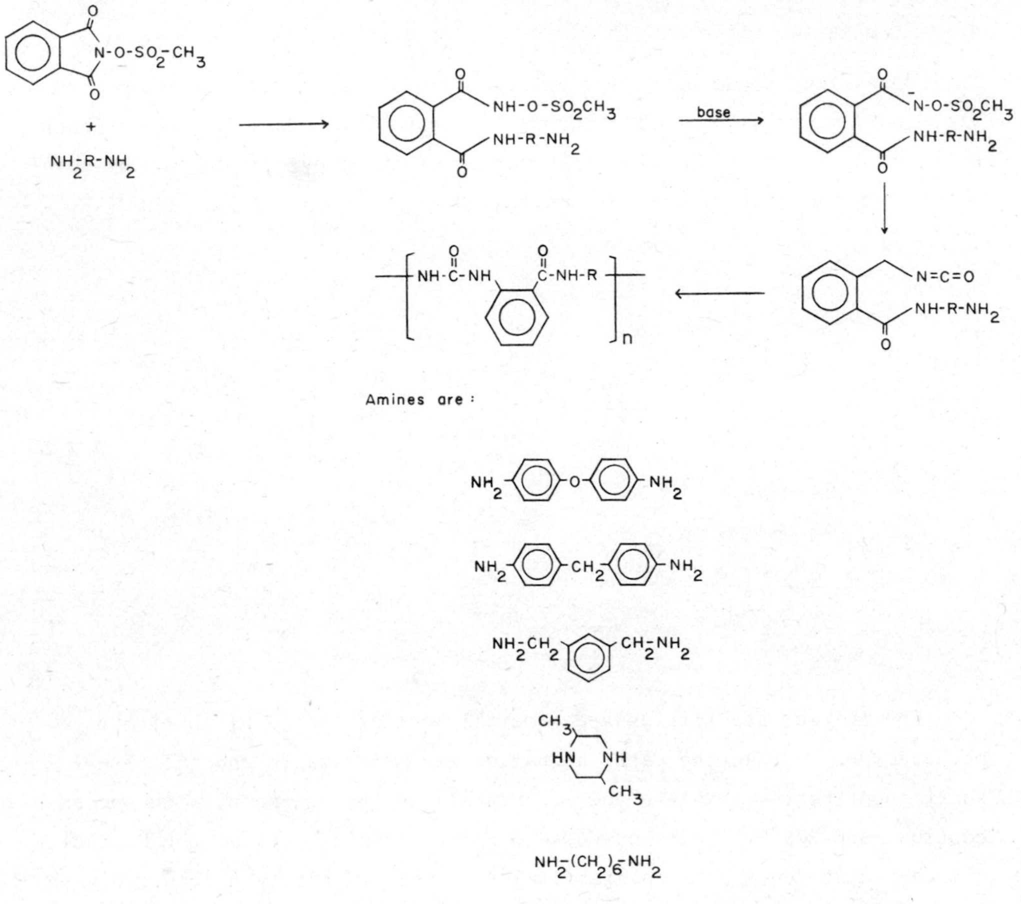

$N-O-SO_2-CH_3$ + NH_2-R-NH_2 → $NH-O-SO_2CH_3$ / $NH-R-NH_2$ —base→ $\bar{N}-O-SO_2CH_3$ / $NH-R-NH_2$ → $N=C=O$ / $NH-R-NH_2$ → $[NH-C(=O)-NH-C_6H_4-C(=O)-NH-R]_n$

Amines are:

$NH_2-C_6H_4-O-C_6H_4-NH_2$

$NH_2-C_6H_4-CH_2-C_6H_4-NH_2$

$NH_2-CH_2-C_6H_4-CH_2-NH_2$

CH_3, HN, NH, CH_3 (2,5-dimethylpiperazine)

$NH_2-(CH_2)_6-NH_2$

IR spectral data suggest that on thermal decomposition, deamination occurs to form a quinazoline-2,4-dione ring as a terminal group. This mechanism would serve to explain the high thermal stability demonstrated by the only bis-tertiary amine used. A piperazine unit would be unable to form the substituted quinazolinedione and would, therefore, force the decomposition to occur by a more energetically demanding route.

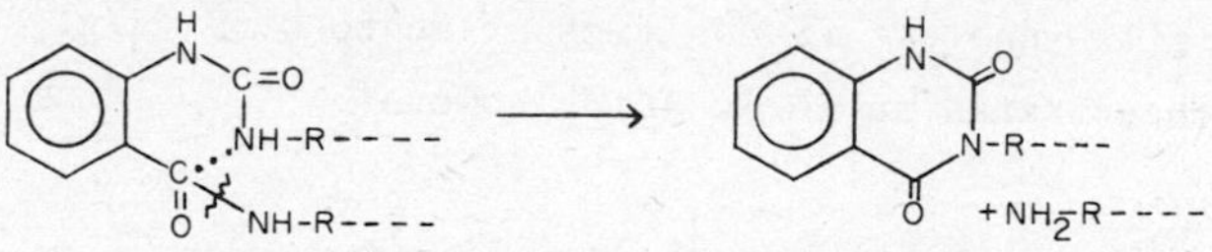

Method [21] To a solution of 5 mmol of the diamine and 5 mmol of pyridine or triethylamine in 5 ml of NMP is added 5 mmol (1.206 g) of N-mesyloxyphthalimide [14] and the mixture is stirred at 25°C for 24 hr whereupon it is poured into 500 ml of water. The polymer is isolated by filtration, washed with water, and dried at 110°C for a 95% yield.

4.1-6 Sulfone-Containing Polyamides

Pendant phenylsulfone groups were studied on poly(isophthalamides) and poly(terephthalamides) with both aromatic and aliphatic diamine connecting functions [1]. The sulfone group prevented formation of a high molecular weight polymer and lowered Tg values by 50-150°C compared to the nonsubstituted analog. Further, thermal stabilities are lessened by this pendant function.

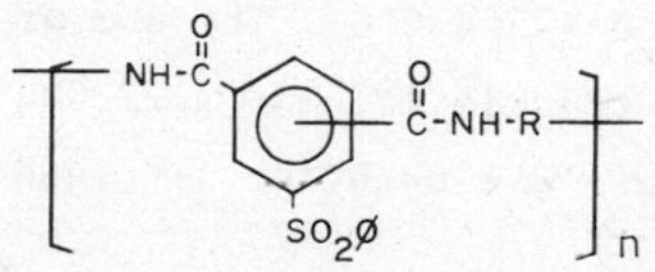

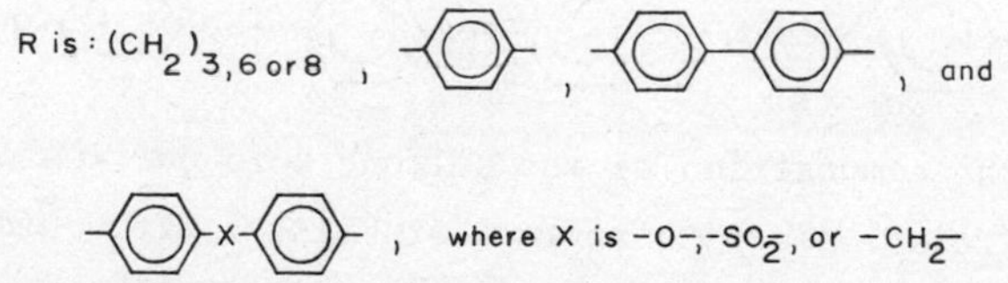

R is: $(CH_2)_{3,6\ or\ 8}$, , , and

, where X is $-O-$, $-SO_2-$, or $-CH_2-$

Method [1] The diamine (0.1 mol) is dissolved in 100 ml of water which contains 1.0 g of sodium carbonate and 0.1 g of sodium lauryl sulfate. The amine solution is added rapidly to a stirred solution of 0.01 mol of the phthaloyl chloride in benzene. The product is recovered by filtration, washed with hot water and acetone, and vacuum-dried at 100°C.

4.1-7 Poly(hexahydrobenzodipyrrole sulfonamides)

Sulfonamide polymers have been prepared with the hexahydrobenzodipyrrole (i.e., diisoindoline) unit [13], as have simple amide polymers [12]. A solution or interfacial process affords products which display a stability of about 300°C maximum. Yields (>95%) and molecular weights obtained are fair by this approach, with inherent viscosities of up to 0.69 dl/g. Solubility is limited to concentrated sulfuric acid, however.

$Cl-SO_2-R-SO_2-Cl$ + diisoindoline (HN, NH) → $-[N(\text{diisoindoline})N-SO_2-R-SO_2]_n-$

R is: *m*-phenylene, 4,4'-biphenylene, and 4,4'-oxydiphenylene

TGA stability: (°C)	300	295	285/308

Polyamides with sulfone ether blocks in the backbone impart improved impact strength, but the effect is more pronounced with more rigid backbones [8]. [See the discussion of poly(sulfone ether imides).] Terephthaloyl chloride condensed with the oligomeric diamine below gave a crystalline polymer with a Tg of 260°C and Tm of 425-430°C. The use of isophthaloyl chloride results in a Tg of 230°C, tensile modulus 2.22 GPa (323,000 psi), tensile strength 76.8 MPa (11,150 psi), elongation 11%, and impact strength 11.2 J (135 ft-lb./in^3).

$NH_2-C_6H_4-O-C_6H_4-SO_2-C_6H_4-[O-Ar-O-C_6H_4-SO_2-C_6H_4]_n-O-C_6H_4-NH_2$

It is apparent that the composition of the sulfone acid portion of the backbone has little or no relationship to thermal stability. It would appear then that decomposition occurs in the diisoindoline nucleus or the sulfonamide linkage. The rapid weight loss observed at 300°C is attributed to loss of SO_2 by IR spectral analysis of the off-gas.

Method *Interfacial* [13] In a 200-ml resin kettle equipped with stirrer, two additional funnels, and a gas (N_2) blanket system are placed 0.966 g (0.003 mol) of diisoindoline dihydrobromide and 50 ml of water. To the stirred solution are added simultaneously a solution of 0.480 g (0.012 mol) of NaOH in 25 ml of water and a solution of 0.825 g (0.003 mol) of 4,4'-oxy-bisbenzenedisulfonyl chloride in 25 ml of benzene. The mixture is stirred for 45 min; the benzene is evaporated; and the white solid is isolated, washed with hot water, cold water, alcohol, acetone, and dried at 60°C to yield 97% product.

Solution [13] In a system equipped like the one above is placed a mixture of 0.530 g of powdered $Ca(OH)_2$, 0.966 g of diisoindoline, and 35 ml of methylene chloride. To this stirred mixture is added slowly a solution of 0.003 mol of the disulfonyl chloride, and stirring is continued for 45 min. The mixture is then poured into excess distilled water, and the solid polymer is added, washed, and dried as described above.

4.1-8 Copoly(amide sulfonamides)

An interfacial polymerization procedure between a diamine and chlorosulfonylbenzoyl chlorides yields a backbone with both amide and sulfonamide links [15]. These are high enough in molecular weight to form transparent, amorphous, tough films, and are soluble in a wide range of solvents.

$$Cl-\overset{O}{\overset{\|}{C}}-Ar-SO_2-Cl + NH_2-R-NH_2 \longrightarrow -\!\!\left(\overset{O}{\overset{\|}{C}}-Ar-SO_2-NH-R-NH\right)_n\!\!-$$

Ar is: m- or p- C_6H_4 R is: $-(CH_2)_6-$

$-CH_2-C_6H_4-CH_2-C_6H_4-$

$-C_6H_4-CH_2-C_6H_4-$

$-C_6H_4-O-C_6H_4-$

The yields of products were above 82% and frequently above 96% with one exception: m-phenylene diamine allows only 23-27%. Similarly, the products showed inherent viscosities of 0.45-0.80 dl/g with the same exception. In view of the low reactivity of benzene sulfonyl chloride compared to benzoyl chloride and the low basicity of aromatic amines, it is interesting that these yields and molecular weights can be obtained with mild conditions (room temperature for 30 min).

The products are generally soluble in 95% H_2SO_4, DMSO, DMAc, pyridine, 10% aq. NaOH and m-cresol, a group covering both acids and bases. The polymer melt temperatures (PMT) range from 155 to 275°C, which are from 43 to 97° below analogous amide-amide systems. These new materials are nearly equal to or higher (by 84°C) than comparable sulfonamide-urea polymers in PMT. When compared to the totally sulfonic acid analog (i.e., sulfonamide-sulfonamide), the PMT of the amide-sulfonamide can be higher or lower, depending on the aromatic character and content.

Generally the wholly aromatic amide-sulfonamides are less thermally stable than pure polyamides due to the inherent weakness of the sulfonamide and its ability to lose SO_2 upon heating. TGA data show that most of these polymers are stable to only 340°C.

> *Method* [15] In a blendor a solution is made consisting of 2.32 g (0.020 mol) of hexamethylene diamine, 4.33 g (0.0408 mol) of Na_2CO_3, and 150 ml of water. To this, with stirring, is added rapidly a solution of 4.88 g (0.020 mol) of m-chlorosulfonylbenzoyl chloride and 150 ml of methylene chloride, and stirring is continued for 30 min. The polymeric solid is collected, washed with water, and dried to yield 85% (4.85 g) of polymer; inherent viscosity 0.45 dl/g in NMP at a concentration of 0.5 g/100 ml.

4.1-9 Poly(oxamides)

The use of oxalic acid esters in the formation of amides (specifically oxamides) improves considerably the physical and mechanical properties over ordinary polyamides [31]. These advantages include increased solubility, modulus, melt temperature, and stability. These poly(oxamides—which could be called nylons X2, where X is 6, 8, 10, or 12—are prepared by bulk condensation of the aliphatic diamines with either butyl or ethyl oxalate. The dipiperidinyl or bridged dipiperidinyl amine function is incorporated in a solution process with oxalyl chloride rather than the ester [28].

$$RO-\overset{O}{\overset{\|}{C}}-\overset{O}{\overset{\|}{C}}-OR + NH_2-(CH_2)_x-NH_2 \longrightarrow -\!\!\!\left[\overset{O}{\overset{\|}{C}}-\overset{O}{\overset{\|}{C}}-NH-(CH_2)_x-NH\right]_n$$

R is : C_2H_5 or $n-C_4H_9$ X is : 6, 8, 10, 12

also :

$$Cl-\overset{O}{\overset{\|}{C}}-\overset{O}{\overset{\|}{C}}-Cl + HN\langle\rangle-(CH_2)_y-\langle\rangle NH \longrightarrow \left[\overset{O}{\overset{\|}{C}}-\overset{O}{\overset{\|}{C}}-N\langle\rangle-(CH_2)_y-\langle\rangle N\right]_n$$

y is : 0,2 or 3

The polymers are generally about 50% crystalline and soluble in common solvents. As would be expected, the Tm of a material is inversely related to the size of the amine hydrocarbon, ranging from 195°C for the dipiperidinyl to 320°C for Nylon 62; i.e., with larger aliphatic groups, less hydrogen bonding per unit length is possible and the polymer approaches polyethylene in character. The melting point of the dipiperidinyl polymer is higher than expected even at that low value, since it has no opportunity to form hydrogen bonds. This deficiency is evidently compensated for by the ring stiffness of the piperidinyl system.

All poly(oxamides) demonstrate similar thermal stabilities in nitrogen by showing catastrophic weight loss at 400°C by TGA due to homolytic cleavage at the CO-CO weak link.

> *Method* [31] To a stirred solution of the diamine in toluene at room temperature an equimolar amount of the dialkyl oxalate is added rapidly in one portion. The mixture is stirred at room temperature for 1-5 hr, whereupon the polymer is isolated under anhydrous conditions and dried in vacuo at 70°C. The polymer is then postcured at 240-260°C for 1-5 hr under vacuum. The product, in quantitative yield, is then ground and dried at 80°C in vacuo; reduced viscosities are from about 1.0 to 3.4.

4.1-10 Poly(4,4'-dipiperidylamides)

Piperidinyl functions can be included in amide backbones by the usual interfacial or solution polymerization techniques [28]. If imide formation were possible, the reaction would terminate with trimer formation or, at best, a very low molecular weight polymer.

$$\left[-N\langle\ \rangle-(CH_2)_n-\langle\ \rangle N-\overset{O}{\overset{\|}{C}}-R-\overset{O}{\overset{\|}{C}}- \right]_m$$

R is: $(CH_2)_x$ where x is 0,1,2,4 or 8

o-m-p- C_6H_4

n is: 0, 2 or 3

One advantage of the dipiperidinyl amines is that, being secondary amines, they cannot form imides with succinyl or phthaloyl chlorides. Thermal stabilities under nitrogen are quite good for these systems. The range is from 400 to 465°C, which approaches that of the aromatic polyamides. However, in air the stability is limited by oxidation of the piperidinyl ring which begins at 200°C.

In view of the fact that no hydrogen bonding is possible with the piperidinyl group, one might expect a low PMT; but surprisingly high temperatures are observed (160-455°C). It is postulated that the deficiency in intermolecular secondary bonding is compensated for by ring, and therefore backbone, stiffness. The polymers are readily soluble in common solvents (chloroform, 90-95% formic acid, m-cresol, and conc. H_2SO_4) and can be cast into colorless films or wet-spun into fibers.

Method [28] In a blendor with vigorous stirring are mixed 200 ml of ice and water, 0.03 mol of the diamine, 6.4 g of Na_2CO_3, 10 ml of chloroform, and 0.25 g of sodium lauryl sulfate (duPont Dupanol ME). Added to this at one time is a solution of 0.03 mol of the acid chloride in 40 ml of chloroform, and the resulting emulsion is stirred for 15 min. The mixture is added to 200 ml of hot water; the chloroform removed by evaporation; the polymer isolated, washed with water, and dried to give an essentially quantitative yield. (Polyamides from oxalyl chloride are prepared by the solution method in chloroform due to the hydrolytic instability of this acid chloride.) (See also Section 4.1-9.)

4.1-11 Ordered Azoaromatic Polyamides

Azo and amide functions have been incorporated with aromatic groups in a backbone to give yellow- or orange-colored polymers [5]. The synthetic method involves first the oxidative coupling of an asymmetric aromatic diamine, i.e., asymmetric in the sense that the amino groups have different basicities. Careful control of reaction conditions insures that the more basic amino group will oxidatively couple to form an azo dimer:

$$NH_2\text{-}Ar-X-Ar'-NH_2 \xrightarrow[\text{DMAc, pyridine}]{O_2,\ CuCl} NH_2\text{-}Ar-X-Ar'-N=N-Ar'-X-Ar-NH_2$$

Where X is an amide or benzoxazole group.

The azodiamine products are then:

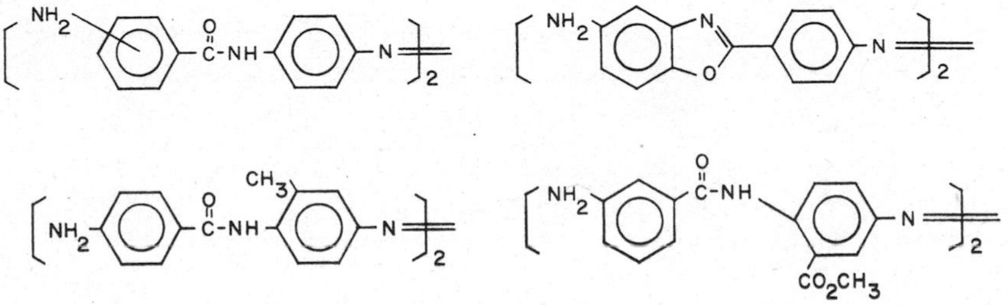

If the reaction temperature is raised to 75°C, these diamines will couple further to form azoamide block copolymers [3].

These simple dimeric azoaromatic block diamines can also be condensed in a low-temperature solution process with iso- or terephthaloyl chloride to provide additional amide backbone functions.

The poly(azoamides) can be solution-cast to tough yellow or orange films. These films are stable to 360°C for 263 hr if an isophthalamide link is used and for 479 hr for terephthalamide catenation. However, these stabilities are measured by embrittlement onset and not by TGA or tensile strength and are therefore lower than values expected from the latter methods.

Method [5] *Dimerization* A catalyst solution is prepared by bubbling oxygen (59 ml at 25°C) into a solution of 1.0 g of CuCl, 80 ml of DMAc, and 20 ml of pyridine. To this is added 9.08 g of 2'-methyl-4,4'-diaminobenzanilide, and oxygen flow is maintained for 1 hr until absorption ceases (445 ml O_2 at 25°C). The slurry is poured into 5% aq. ammonia and filtered to isolate the azodiamine.

Polymerization The phthaloyl chloride is added with stirring and cooling to an equimolar amount of diamine dissolved in DMAc containing 5% LiCl. The mixture is added to an excess of water, and the polymer is isolated by filtration, washed with water, and dried. Inherent viscosities range from 0.2 to 7.8 dl/g in DMAc/5% LiCl solvent at a concentration of 0.1 g/100 ml.

4.2 Poly(diphenylamine)

Although amines are generally avoided in high-temperature systems due to their facile oxidation and discoloration, poly(diphenylamine) has been prepared and found to be stable at 465°C by TGA. This synthesis was accomplished by the use of ferric chloride as a Friedel-Crafts catalyst [6]. The structure of the product is not known and may vary considerably with solvent used since benzene yields a dark red material and acetone gives a pale green one which has low thermal stability and softening point but a high solubility. The product softens at 80-100°C, is soluble, and can be crosslinked to a thermally stable material with p-xylylene dichloride and stannic chloride, obviously another Friedel-Crafts condensation to pendant or backbone aromatic functions.

Method [6] To a solution of 16.9 g (0.1 mol) of diphenylamine in 88 ml of benzene, the catalyst ($FeCl_3$, 0.1 mol, 163 g) is added slowly with stirring while the temperature is maintained below 35°C. The mixture is then heated at 35°C for 20 min and cooled in ice. The solid product is isolated by filtration, dried at 50°C, ground, and triturated with boiling 20% HCl until the filtrate is colorless. It is then washed with water repeatedly and dried at 80°C to yield 42% of a pale green powder which is soluble in DMF, o-dichlorobenzene, and dichloroethane.

4.3 Polyimines or Polyazomethines

Another linear nitrogenous backbone which displays thermal stability (sustained exposure at 400°C) is that in which the nitrogen is double-bonded to carbon, an imine or azomethine function. These imines, or Schiff base moieties, are condensed from aromatic aldehydes and amines at high temperatures with water off-gas [9]. The product is black, which could be predicted based on its degree of conjugation. More conjugation is possible here than with polyphenylene since in this case there is no steric hindrance to coplanarity of rings.

$$NH_2-C_6H_4-NH_2 + CHO-C_6H_4-CHO \xrightarrow[-2H_2O]{250\text{-}300^\circ} \left[=N-C_6H_4-N=CH-C_6H_4-CH= \right]_n$$

The degree of polymerization is low (approximately 5), owing to the inclusion of monofunctional aldehydes and amines as chain-terminating agents to provide tractibility in the final product. Aromatic nuclei other than benzene can be used also; however, loss of backbone conjugation significantly decreases its thermal stability.

> *Method* [9] Equimolar amounts (0.004 mol) of benzaldehyde, aniline, p-phenylenediamine, and terephthalaladehyde are mixed in a reaction vessel to a yellow paste which solidifies. The vessel is fitted with a distillation apparatus and gas inlet and the system is heated under nitrogen to 220°C at a rate of 10-30°C/hr and held there for 2 hr. The black amorphous polymer is the residue, and benzylidene aniline is in the distillation receiver. A vacuum-heating process (1 torr at 300°C) may serve to complete the reaction with other monomers. The polymerization can be run initially as a 10% solution in toluene; however, a secondary heating cycle without solvent is necessary to complete the reaction. Yields are above theoretical values due to retention of byproducts.

An aromatic azoether polymer which might be considered analogous to the above polyimine has been the subject of a patent [4] diaminodiphenylether was oxidatively coupled with a cuprous chloride-amide complex and oxygen. The polymer is soluble in sulfuric acid; and from such a solution films can be cast.

$$NH_2-C_6H_4-O-C_6H_4-NH_2 \xrightarrow[\text{pyridine}]{Cu_2Cl_2,\ O_2} \left[=N-C_6H_4-O-C_6H_4-N= \right]_n$$

Limited reliable work has been done on nonaromatic conjugated backbones. This area has been extensively reviewed [16], and references are given for these types of backbones and their electrically conductive properties. However, due to the lack of confirmatory evidence and experimental detail, most of these must be considered as only possible at best.

Good evidence is available for a system which is considered at least partly imine-like and conjugated, polypropiolamide [18]. This can be prepared either by the tertiary amine catalysis of molten propiolamide [18] or by sodium cyanide catalysis of the amide in DMF [2]. A resonance structure to conjugated double bonds was proposed to explain observed conductivity data.

$$CH{\equiv}C{-}\overset{O}{\overset{\|}{C}}{-}NH_2 \xrightarrow[\text{or molten } Et_3N \text{ for 1 hr}]{\text{DMF, NaCN, 135°, 14 days}} \left[CH{=}CH{-}\overset{O}{\overset{\|}{C}}{-}NH \right]_n \longleftrightarrow \left[CH{=}CH{-}\overset{OH}{\overset{|}{C}}{=}N \right]_n$$

The brown polymers are soluble in water or cold aq. sodium hydroxide, and from the latter solution are precipitated by acid. Yields and inherent viscosities are low (24% and 0.07, respectively).

The thermal stability by TGA gives a value of 310°C for initial weight loss, with gradual loss continuing to 60% at 1000°C.

Method *Melt process* [18] To 3.0 g of molten propiolamide a total of 0.1 g of triethylamine is added in 0.02 g portions at 3- to 5-min intervals. The reaction is quite exothermic at the end. The black, sticky solid is extracted (Soxhlet) with methyl alcohol. The residue is then triturated consecutively with alcohol, 1:1 methanol-ether, and ether to leave a brown solid.

Solution process [2] A 4% solution of propiolamide in DMF (25 ml total) containing 0.2 g of sodium cyanide is heated at 135°C for 14 days. The solution is poured into 200 ml of a 3:1 benzene-methanol mixture to give a brown precipitate. The solid is reprecipitated from formic acid solution and is then freeze-dried from a benzene suspension.

A similar but even more stable polymer which is nearly semiconducting is that prepared from cyanoacetylene (propiolnitrile) [18]. This substance, the structure of which has not been elucidated, is stable to 600°C and, of course, is similar to pyrolyzed polyacrylonitrile ("black Orlon") by IR spectroscopy and in color (black). It is also brittle and intractable and, unlike black Orlon, has no processable intermediate.

Method [18] To 2 g of cyanoacetylene at 6°C under nitrogen, 0.1 g of triethylamine is added dropwise to produce a violent, exothermic reaction. The black, sticky solid is washed with acetone and dried.

4.4 Recent Developments

4.4-1 Aromatic Polyamides

Owing to the continued promise and use of high-modulus fibers, research on aramids has continued apace. There is perhaps more work being done with this specific class of material than with any other polymer which could be called thermally stable.

Perhaps three synthetic methods can be isolated as the most important recent information. One of these is by phosphorylation. A high yield of high molecular weight aromatic polyamides is obtained by the use of pyridine with diphenylphosphite [61]. Several AB, AA-BB, and ordered copolyamides and amide hydrazides are also possible [63]. Although solubility limits the molecular weight (inherent viscosity 1.1-1.6 dl/g), high tensile strengths (8-12 gpd) and moduli (435-475 gpd) are observed.

Second, direct polycondensation has also resulted in aramids. p-Aminobenzoic acid yields poly(p-benzamide) when reacted at 80-105°C, for 6 hr with phenol, pyridine, and phosphorus trichloride in NMP with 4% lithium chloride [48]. Also diacids and diamines will yield AA-BB backbones in this way. In this case triphenylphosphite is formed in situ.

Third, polyamide synthesis was found possible by way of a formamidinium salt [65]:

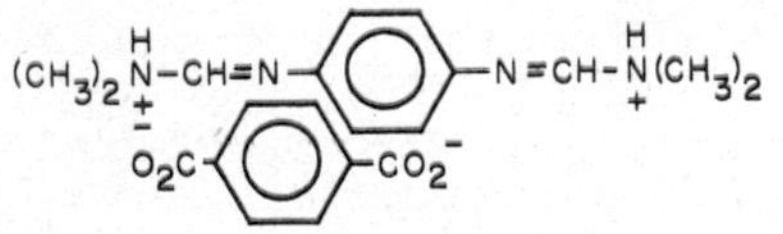

$$(CH_3)_2\overset{H}{\underset{+}{N}}-CH=N-C_6H_4-N=CH-\overset{H}{\underset{+}{N}}(CH_3)_2$$

$$^{-}O_2C-C_6H_4-CO_2^{-}$$

By heating this salt at 200-225°C in bulk or suspension, one receives the aramide with elimination of DMF.

Aramids prepared from diamines with pendant methyl groups are more soluble but less thermally stable than wholly aromatic systems [67,68]. One methyl substituent decreases the stability by 10-70°C and two substituents by 40-70°C, depending on whether the iso- or terephthalic acid is used and on the position of the pendant groups.

A catalyst of 1-hydroxybenzotriazole has been found to provide yields of 99% with inherent viscosities to 1.32 dl/g in the condensation of isophthaloyl acid derivatives with 4,4'-diaminodiphenylether [72].

The recent volume of literature on aromatic polyamides is condensed into Table 4.2.

4.4-2 Polyamines

Aromatic and aliphatic amine backbones have been synthesized and found to exhibit surprising thermal stability for such easily oxidizable functions.

Table 4.2 Aromatic Polyamide Literature

Topic	Ref.
Polyamides from pyridine-2,6-dicarboxylic acid	40
Polyamides from linear bisanhydrides [terephthalic bis(mesitoic anhydride)]	70
Extended chain polyamides	60
Polyamides from 2,2'-p-phenylene-5-oxazolone (low thermal stability)	71
Aramids containing ether sulfone units (for lower Tg and increased flexibility	47
High-Tg, low-Tm polyamides (for improved processing)	57
Optically active polyamides (stable to 450°C in nitrogen)	54
2,6-Diaminobiphenylene units as crosslinking sites in polyamides (softening temperature >360°C after cure)	66
Photoaddition of fluoroolefins on aromatic polyamides (surface treatment to increase flame resistance)	69
Sequence distribution of copolyamides	62
Synthesis, characterization, rheological and fiber formation studies of para-linked aramids (liquid crystal studies at high concentrations)	38
Liquid crystal solutions of polyhydrazides and poly(amide hydrazides) (para catenation gives optical anisotropy in H_2SO_4)	59
Ultrahigh-modulus fibers from rigid and semirigid aramids, poly(p-benzamide) and poly(terephthalamide) of p-aminobenzhydrazide, "X-500")	37
Synthesis, wet-spinning, and fiber thermal characteristics of aromatic polyamides from diaminooxanilide and terephthaloyl chloride (no weight loss or thermal transition <400°C)	64
Photodegradation of aramids in the absence of oxygen (Fries rearrangement to 2-aminobenzophenone, free radical mechanism)	41
Photodegradation of aramids with oxygen (carboxylic acids formed)	42
Rheology of an aromatic poly(amide hydrazide), "X-500" (to explain ultrahigh modulus)	73
Thermal behavior of aryl-aliphatic polyamides (end) group-initiated degradation	74
High-modulus fibers from hydrazide polymers and copolymers (terephthaloyl and oxaloyl hydrazides)	49
Thermal decomposition of aramids and chlorinated aramids (pyrolysis at 550°C and with flaming and nonflaming oxidation)	45,46
Rheological and optical properties of aramids (Kevlar, Nomex)	39

A poly(arylamine sulfone) was produced by the nucleophilic substitution of an aromatic diamine onto a bis(chloroaryl)sulfone [53]:

$$\left[(NO_2)C_6H_3-SO_2-C_6H_3(NO_2)-NH-R-NH \right]_n$$

R is : p,p'-C_6H_4-O-C_6H_4–
p,p'-C_6H_4-CH_2-C_6H_4–
$(CH_2)_6$–
m-CH_2-C_6H_4-CH_2–

The yields were 99% after 1 day and 100°C in DMSO or NMP. Thermal stabilities by TGA were nearly identical in air or nitrogen at 295-350°C. The products are soluble, transparent, orange film formers. The film is tough, flexible, and amorphous.

Polyethyleneimine was stabilized by blocking the amino group with a tin substituent [43,44]. Adding di- or trialkyl or aryl tin halide to polyethyleneimine makes the product soluble with the trialkyl or aryl substituent but insoluble for the disubstituted adduct. This is due to crosslinking rendered by the latter. Films or fibers cannot be formed, but the polymers are antifungal agents, and they do show some stability to 225-360°C in air or nitrogen.

A fluorinated aromatic amine polymer was found to be soluble in sulfuric acid, to melt at 300°C, and to be stable to 340°C in air by TGA. The Tg of the gray powder was 240°C [55]:

$$F-\left[C_6F_4-N(CH_3) \right]_n-C_6F_5$$

The polymerization was brought about by first treating the simple amine monomer with lithium amide and then reacting the intermediate with C_6F_6 in THF for 1 hr at 0°C and 1.5 hr at room temperature.

4.4-3 Doubly Bonded Nitrogen Backbones

A new synthesis of aromatic azo polymers was found by the reaction of a dinitroaromatic compound [50]:

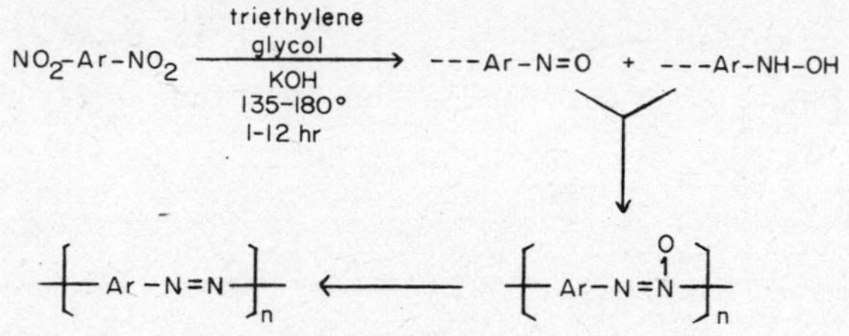

$$NO_2\text{-}Ar\text{-}NO_2 \xrightarrow[\substack{KOH \\ 135\text{-}180° \\ 1\text{-}12\ hr}]{\text{triethylene glycol}} ---Ar\text{-}N{=}O + ---Ar\text{-}NH\text{-}OH$$

$$\left[Ar-N{=}N \right]_n \longleftarrow \left[Ar-N{=}\overset{O}{\overset{\uparrow}{N}} \right]_n$$

Ar is : $m\text{-}C_6H_4\text{-}$ and $p,p\text{-}C_6H_4\text{-}X\text{-}C_6H_4\text{-}$

where X is S, O, SO_2

The product, obtained in 95% yield, had an inherent viscosity of 0.38 dl/g and thermal stability to 400°C.

Polycarbodiimides are possible by the condensation 4,4'-diphenylmethylene diisocyanate in 1-phenyl-3-methyl-2-phospholene oxide and benzene or xylene at reflux [36]:

$$C_6H_5\text{-}N{=}C{=}N\left[C_6H_4\text{-}CH_2\text{-}C_6H_4\text{-}N{=}C{=}N \right]_n C_6H_5$$

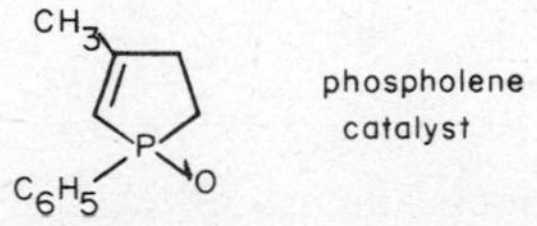

phospholene catalyst

Some phenylisocyanate is added as a chain-terminating agent. Yields range up to 98.5%, and thermal stabilities show a 10% weight loss by TGA at 425-550°C.

Polyamidines are a new class of potential thermally stable materials. They are synthesized from diamines reacting with bisketenimines [56] or triethylorthoformate [58]:

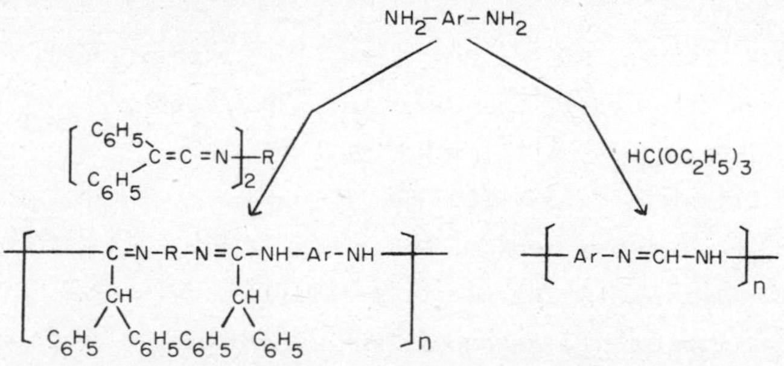

$$NH_2\text{-}Ar\text{-}NH_2$$

$$\left[\frac{C_6H_5}{C_6H_5} {>} C{=}C{=}N \right]_2 R \qquad HC(OC_2H_5)_3$$

$$\left[\underset{\substack{| \\ CH(C_6H_5)_2}}{C}{=}N\text{-}R\text{-}N{=}\underset{\substack{| \\ CH(C_6H_5)_2}}{C}\text{-}NH\text{-}Ar\text{-}NH \right]_n \qquad \left[Ar\text{-}N{=}CH\text{-}NH \right]_n$$

The former product gives initial decomposition by TGA at 230-350°C and, not unexpectedly, provides a faster reaction with aliphatic amines compared to aromatic [56].

The latter polymer is received in 98% yield, is white to yellow, and shows melting points from 180 to >360°C depending on structure [58].

4.4-4 Aromatic Polysulfonamide

A mild condensation of aromatic diamines with 4,4'-bis(1-benzotriazolyl)diphenylether gives a wholly aromatic polysulfonamide in high yield (99%) and moderate molecular weight (inherent viscosity up to 0.55 dl/g [52]:

tertiary amine
25°, 48 hr

Thermal decomposition of the polymer begins at 320°C with a 10% weight loss occurring at 360° and 380°C in air and nitrogen, respectively.

High molecular weight (inherent viscosity up to 1.2 dl/g) polysulfonamides have also been prepared in 99% yield by the condensation of the bis(sulfonyl chloride) with diamines [51]. These polymers are widely soluble, have Tgs from 120 to 155°C, and show slightly better thermal stability than those described above.

References

1. J. M. Adduci and R. Brunea, *Polym. Preprints, 15*(1), 682 (1974).
2. J. P. Allison and R. E. Michel, *Chem. Commun.*, 762 (1966).
3. H. C. Bach, *Polym. Preprints, 8*(1), 610 (1967).
4. H. C. Bach, U.S. Patent 3,637,534 (1972).
5. H. C. Bach and H. E. Hinderer, *Polym. Preprints, 11*(1), 334 (1970).
6. A. Bingham and B. Ellis, *J. Polym. Sci., A-1, 7*(11), 3229-3244 (1969).
7. W. B. Black and J. Preston, *High-Modulus Wholly Aromatic Fibers*, Marcel Dekker, New York, 1973.
8. G. L. Brode, G. T. Kwiatkowski, and J. H. Kawakami, *Polym. Preprints, 15*(1), 761 (1974).

9. G. F. D'Alelio, U.S. Patent 3,516,970 (June 23, 1970).

10. R. C. DeSelms, C. J. Fox, and R. C. Riordan, *Tetrahedron Lett.*, 781 (1970).

11. F. Dobinson and J. Preston, *J. Polym. Sci., A-1, 4*, 2093 (1966).

12. N. Doddi, *J. Polym. Sci.: Polym. Chem. Ed., 12*, 761 (1974).

13. N. Doddi, *J. Polym. Sci.: Polym. Chem. Ed., 13*, 2407 (1975).

14. L. F. Fieser and M. Fieser, *Reagents for Organic Synthesis*, John Wiley and Sons, New York, 1967, p. 485.

15. Y. Imai and H. Okunoyama, *J. Polym. Sci., A-1, 10*, 2257 (1972).

16. J. I. Jones, *J. Macromol. Sci.: Rev. Macromol. Chem., C2*(2), 303 (1968).

17. E. P. Krasnov, V. P. Aksenova, and S. N. Khar'kov, *Vysokomol. Soedin.*, Series A, *11*(9), 1930 (1969) [Chem. Abstr. 22116 (1970)].

18. B. J. MacNulty, *Polymer, 7*, 275 (1966).

19. E. E. Magat and R. E. Morrison, *J. Polym. Sci.: Polym. Symp.*, No. 51, 203 (1975).

20. G. Manecke and J. Gauger, *J. Makromol. Chem., 125*, 231 (1969).

21. G. Montaudo, G. Bruni, P. Maravigna, and F. Bottino, *J. Polym. Sci.: Polym. Chem. Ed., 12*(2), 2881 (1974).

22. P. W. Morgan, *Condensation Polymers: By Interfacial and Solution Methods*, Interscience, New York, 1965.

23. P. W. Morgan, *Polym. Preprints, 17*(1), 47 (1976).

24. E. W. Neuse, B. R. Green, and R. Holm, *Macromolecules, 8*(6), 730-733 (1965).

25. E. W. Neuse and B. R. Green, *Polymer, 15*(6), 339 (1974).

26. E. W. Neuse and B. R. Green, *J. Amer. Chem. Soc., 97*, 3987 (1975).

27. J. R. Pratt and N. J. Johnston, *Polym. Eng. Sci., 16*(5), 309 (1967).

28. J. Preston, *Polym. Preprints, 11*(1), 347 (1970).

29. J. Preston, *Org. Coat. Plast. Preprints, 35*(2), 160 (1975).

30. J. Preston, *Polym. Eng. Sci., 16*(5), 298 (1976).

31. S. W. Shalaby, E. M. Pearce, R. J. Fredericks, and E. A. Turi, *J. Polym. Sci.: Polym. Phys., 11*, 1 (1973).

32. C. W. Stephens, U.S. Patent 3,472,819 (1969).

33. J. K. Stille, P. E. Cassidy, and L. Plummer, *J. Amer. Chem. Soc., 85*, 1318 (1963).

34. M. Veda, M. Ohkura, and Y. Imai, *J. Polym. Sci.: Polym. Chem. Ed., 12*, 719 (1974).

35. N. Yamazaki, M. Matsumoto, and F. Higashi, *J. Polym. Sci.: Polym. Chem. Ed., 13*, 1373 (1975).

36. L. M. Alberino, W. J. Farrissey, and A. A. R. Sayigh, *J. Appl. Polym. Sci., 21*, 1999 (1977).

37. G. C. Alfonso, E. Bianchi, A. Ciferri, S. Russo, F. Salaris, and B. Valenti, *J. Polym. Sci.: Polym. Symp.*, No. 65, 213 (1978).

38. H. Aoki, D. R. Coffin, T. A. Hancock, D. Harwood, R. S. Lenk, J. F. Fellers, and J. L. White, *J. Polym. Sci.: Polym. Symp.*, No. 65, 29 (1978).

39. H. Aoki, J. L. White, and J. F. Fellers, *J. Appl. Polym. Sci.*, *23*, 2293 (1979).

40. A. Banihashemi and M. Eghbali, *J. Polym. Sci.: Polym. Chem. Ed.*, *14*, 2659 (1976).

41. D. J. Carlsson, L. H. Gan, and D. M. Wiles, *J. Polym. Sci.: Polym. Chem. Ed.*, *16*, 2353 (1978).

42. D. J. Carlsson, L. H. Gan, and D. M. Wiles, *J. Polym. Sci.: Polym. Chem. Ed.*, *16*, 2365 (1978).

43. C. E. Carraher, D. J. Giron, W. K. Woelk, J. A. Schroeder, and M. F. Feddersen, *J. Appl. Polym. Sci.*, *23*, 1501 (1979).

44. C. Carraher and M. Feddersen, *Angew Makromol. Chem.*, *54*, 119 (1976).

45. D. A. Chatfield, I. N. Einhorn, R. W. Mickelson, and J. H. Futrell, *J. Polym. Sci.: Polym. Chem. Ed.*, *17*, 1353 (1979).

46. D. A. Chatfield, I. N. Einhorn, R. W. Mickelson, and J. H. Futrell, *J. Polym. Sci.: Polym. Chem. Ed.*, *17*, 1367 (1979).

47. C. Chiriac and J. K. Stille, *Macromolecules*, *10*(3), 712 (1977).

48. C. Chiriac and J. K. Stille, *Macromolecules*, *10*(3), 710 (1977).

49. F. Dobinson, C. A. Pelezo, W. B. Black, K. R. Lea, and J. H. Saunders, *J. Appl. Polym. Sci.*, *23*, 2189 (1979).

50. A. de Souza Gomes and T. D. R. De Oliveira Cavalcanti, *J. Polym. Sci.: Polym. Chem. Ed.*, *16*, 2671 (1978).

51. Y. Imai, M. Ueda, and T. Iizawa, *J. Polym. Sci.: Polym. Chem. Ed.*, *17*, 1483 (1979).

52. Y. Imai, M. Ueda, and T. Iizawa, *J. Polym. Sci.: Polym. Lett.*, *15*(4), 207 (1977).

53. Y. Imai, M. Ueda, and K. Otaira, *J. Polym. Sci.: Polym. Chem. Ed.*, *15*, 1457 (1977).

54. V. Jarm and Z. Janovic, *J. Polym. Sci.: Polym. Chem. Ed.*, *16*, 3007 (1978).

55. R. Koppang, *J. Polym. Sci.: Polym. Chem. Ed.*, *14*, 2225 (1976).

56. K. Kurita, Y. Kusayama, and Y. Iwakura, *J. Polym. Sci.: Polym. Chem. Ed.*, *15*, 2163 (1977).

57. L. T. C. Lee, *J. Polym Sci.: Polym. Chem. Ed.*, *16*, 2025 (1978).

58. L. Mathias and C. G. Overberger, *J. Polym. Sci.: Polym. Chem. Ed.*, *17*, 1287 (1979).

59. P. W. Morgan, *J. Polym. Sci.: Polym. Symp.*, No. 65, 1 (1978).

60. P. W. Morgan, *Macromolecules*, *10*(6), 1381 (1977).

61. N. Ogata and M. Harada, *J. Polym. Sci.: Polym. Lett. Ed.*, *15*, 551 (1977).

62. N. Ogata, K. Sanui, and S. Kamiyama, *J. Polym. Sci.: Polym. Chem. Ed.*, *16*, 1991 (1978).

63. J. Preston and W. L. Hofferbert, Jr., *J. Polym. Sci.: Polym. Symp.*, No. 65, 13 (1978).

64. F. M. Silver and F. Dobinson, *J. Polym. Sci.: Polym. Chem. Ed.*, *16*, 2141 (1978).

65. J. Spiewak, *J. Polym. Sci.: Polym. Chem. Ed.*, *16*, 2303 (1978).

66. R. J. Swedo and C. S. Marvel, *J. Polym. Sci.: Polym. Chem. Ed.*, *16*, 2711 (1978).

67. R. Takatsuka, K. Uno, F. Toda, and Y. Iwakura, *J. Polym. Sci.: Polym. Chem. Ed.*, *15*(8), 1905 (1977).

68. R. Takatsuka, K. Uno, F. Toda, and Y. Iwakura, *J. Polym. Sci.: Polym. Chem. Ed.*, *15*(12), 2997 (1977).

69. M. S. Toy, R. S. Stringham, and F. S. Dawn, *J. Appl. Polym. Sci.*, *21*, 2583 (1977).

70. M. Ueda, O. Hara, A. Sato, and Y. Imai, *J. Polym. Sci.: Polym. Chem. Ed.*, *17*, 769 (1979).

71. M. Ueda, K. Kino, K. Yamaki, and Y. Imai, *J. Polym. Sci.: Polym. Chem. Ed.*, *16*, 155 (1978).

72. M. Ueda, A. Sato, and Y. Imai, *J. Polym. Sci.: Polym. Chem. Ed.*, *17*, 783 (1979).

73. B. Valenti and A. Ciferri, *J. Polym. Sci.: Polym. Lett. Ed.*, *16*, 657 (1978).

74. I. K. Varma, V. S. Sandari, and D. S. Varma, *J. Appl. Polym. Sci.*, *22*, 2857 (1978).

Chapter 5

Polyimides and Analogs Thereof

5.1 Polyimides

The earliest of the thermally stable polymers, and now one of the most common commercial materials, is polyimide, which was initially synthesized by duPont and marketed under the name H-Film and later Kapton as a yellow film. Some other polymers which contain imides or derivatives thereof are P 13N (TRW and duPont) [28], QX-13 (ICI) [28], and Kinel or Kerimid 601 (Rhodia, Rhone-Poulene) [12]. Further, an amide-imide copolymer is available as Amoco-AI (American Oil Company).

The available forms, chemistry, and uses have expanded considerably; e.g., polyimides come in precured films and fibers, curable enamels, adhesives, and resins for composite materials (with quartz, glass, boron, and graphite). Most recently, they have been found to be excellent hyperfiltration or reverse osmosis membranes after the incorporation of a polar function (methoxyl, carboxyl, or hydroxyl) onto the backbone [60]. The resistance of polyimides to acidic hydrolysis and bacterial attack makes them good candidates for reverse osmosis systems [57]. As discussed earlier, this particular use of high-temperature polymers is gaining popularity and is an unexpected advantage of stable backbone systems. From 1964 to 1969 syntheses and uses of wire enamels of polyimides were investigated and/or patented and discussed in a comprehensive review [53].

The polyimides can be molded (prior to final curing) and film brittleness tests have shown them to have quite good thermal stabilities. The physical properties of films are retained over a long exposure according to the following:

Temperature [°F (°C)]	Time (hr)
365 (185)	20,000
450 (235)	10,000

Table 5.1 **Polyimides and Analogs**

Name	Structure	Properties
Polyimides	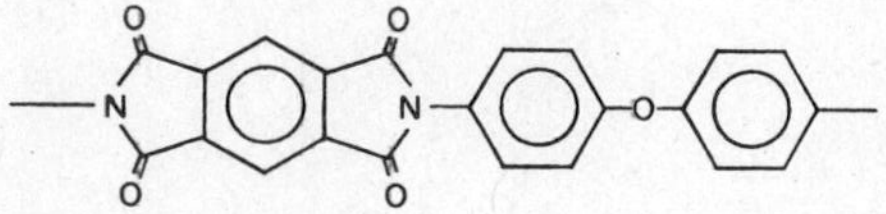	Stable to 500°C (isothermal) in N_2. Tg 300°C and higher.
	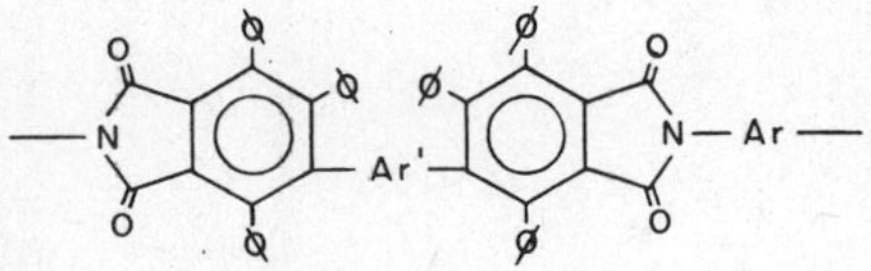	Stable to 530°C. Soluble.
	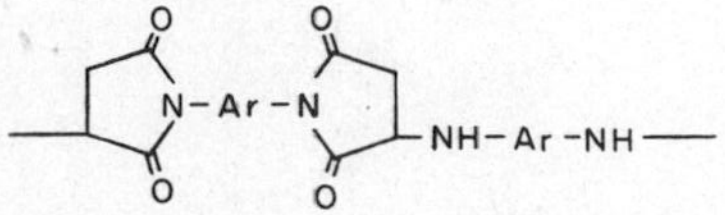	Soluble. Yellow films. Amorphous. Stable to 350°C.
	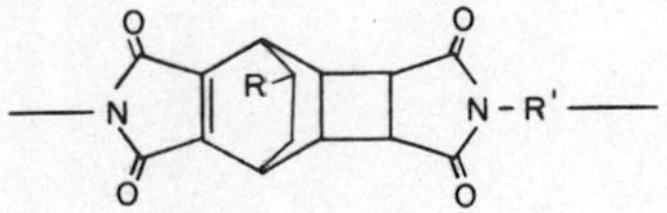	Stable at 360-485°C. White-yellow. Poor solubility.
	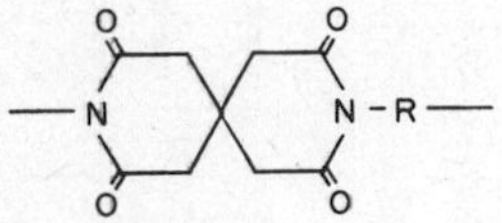	Tg 78-204°C. Tough, flexible films. Soluble. Stable to 450°C in air.
	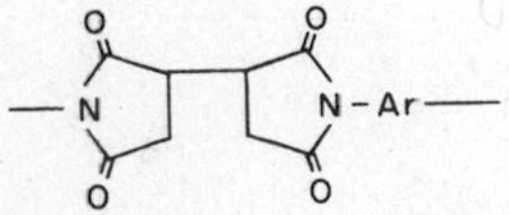	Flexible films. Tensile strength 97 MPa (14,000 psi). Tensile modulus 0.69-3.4 GPa (100,000-500,000 psi). Stable to 400-430°C.

Table 5.1 (Continued)

Name	Structure	Properties
[Polyimides]	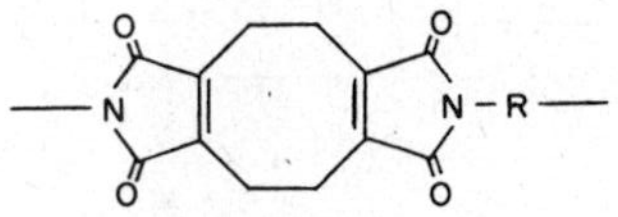	Stable to 370°C in air. Tg 140-320°C.
	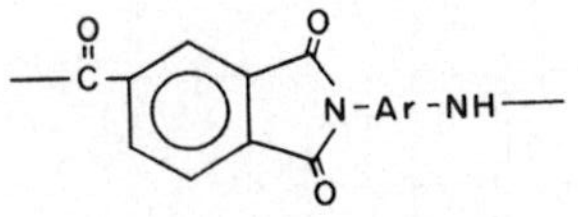	Laminating resins. Coatings. High strength and modulus. Stable to 450°C.
	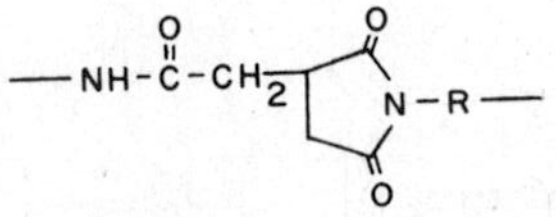	Tg 71-200°C. Stable to 417-445°C. Melt-spun fibers.
	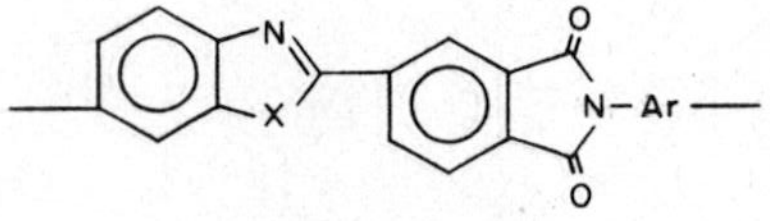	Soluble. Stable to 530-605°C. High tensile strength and modulus. Film former.
	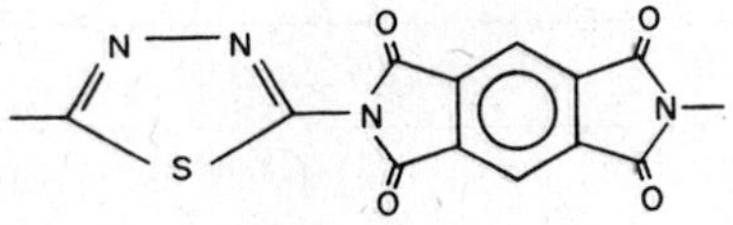	Red, clear films. Stable to 450°C in air. Zero strength at 590°C.

Polymer	Structure	Properties
	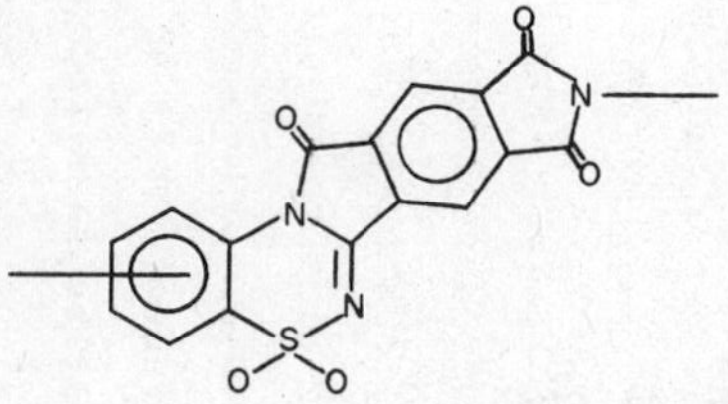	Self-extinguishing.
	 —[Polybutadiene Block]—[Polyimide Block]—[Polybutadiene Block]—	Curable. Flexural strength 48 MPa (7000 psi). Flexural modulus 3.1 GPa (450,000 psi). Stable to 450°C.
	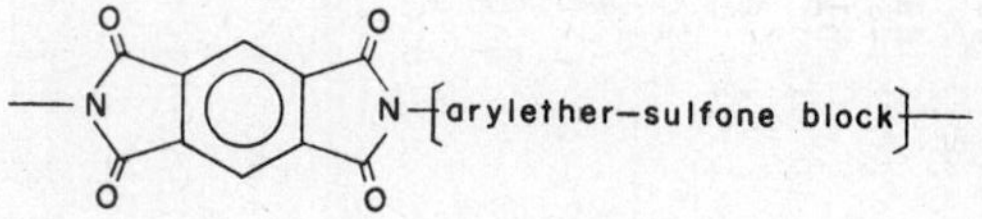 —N—[arylether–sulfone block]—	Film formers. Good tensile strength and modulus. Stable at 220°C for 5000 hr.
	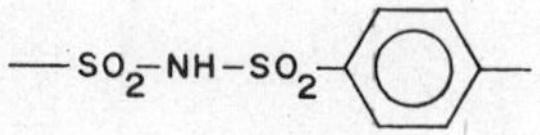 $—SO_2—NH—SO_2—C_6H_4—$	Stable to 350-410°C. Some stability.
Polypyromellitimidines	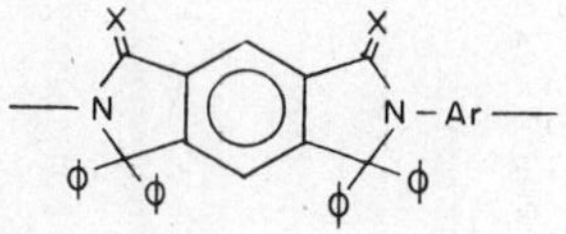 —N ... N—Ar—	Soluble. Stable to 340-650°C in N_2.
Polybenzodipyrrolediones	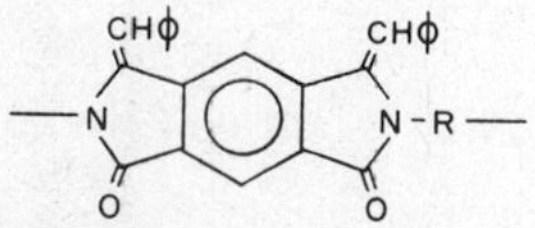 —N ... N—R—	Soluble. Tough, transparent, yellow films. Stable to 400-500°C.

Further, they demonstrate a heat deflection temperature (HDT) of 660°F (350°C).

The general method of synthesis is a two-step condensation of a tetra-acid dianhydride with a diamine:

$$\text{Ar dianhydride} + NH_2\text{-R-}NH_2 \xrightarrow{-H_2O} \left[\text{R-NH-}\overset{O}{\overset{\|}{C}}\text{-Ar}(CO_2H)(HO_2C)\text{-}\underset{O}{\underset{\|}{C}}\text{-NH}\right]_n \xrightarrow{-H_2O} \left[\text{R-N}\langle\text{Ar (diimide)}\rangle\text{N}\right]_n$$

(Amic Acid)

The extensive research to date has covered nearly every conceivable Ar and R function for this system. The most common dianhydrides are those from pyromellitic acid and benzophenone tetracarboxylic acid. Among the diamines are aliphatics, p-phenylene, and bridged biphenylene and bipyridyl [35] units (with methylene, oxygen, and sulfone bridges to provide product flexibility due to chain mobility). Both steps for synthesis involve loss of water as an off-gas, which, of course, can lead to problems in the use of polyimides as an adhesive between metal substrates or as a composite matrix. The first step is the production of the intermediate amic acid. This form is soluble and processable, and it is at this point that the final product conformation is developed (film, fiber, etc.). Then the postcure process takes place at an elevated temperature to complete the dehydration ring closure resulting in an insoluble final product.

As with most commercial systems, a compromise or tradeoff must be made with the polyimides; i.e., some desirable properties must be sacrificed partially in order to gain practicability. For example, one must frequently design a decrease in thermal stability to achieve processability, comfort, acceptable cost, flexibility, solubility, strength, etc.; of these, processing is the most elusive and important criterion, without which the polymer is merely a laboratory curiosity. From the engineer's point of view there is no value in stability if the polymer cannot be formed into an object below 400°F (205°C to the chemist) and at moderate pressures. It is noteworthy, however, that in response to these new materials new processing techniques are being developed.

For polyimides this tradeoff is accomplished by admitting to the backbone flexibilizing moieties such as bipyridyl [35], ether, sulfone, and alkylene, or blocks of these functional groups. The structure-solubility relationships in aromatic polyimides have been recently reviewed [22].

The compromise is made in another way while simultaneously solving another problem, that of off-gas. Water (or acetic acid as with QX-13) as gas creates voids and severe weaknesses in finished products, unless of course a film is produced while open to the atmosphere. To circumvent the off-gas generation and simultaneously soften the product, again with the sacrifice of thermal stability, the second step in the cure is an addition reaction rather than condensation. To accomplish this an oligomer (polyimide or other type) is capped with maleimide functions which are then polymerized by an hydrogen addition mechanism to provide polyaminobismaleoximides [9,12, 28,35,44]:

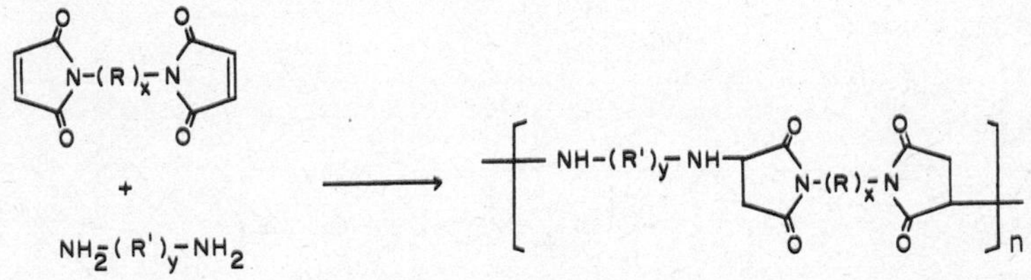

N-(R)$_x$-N

+

NH_2-(R')$_y$-NH_2

NH-(R')$_y$-NH N-(R)$_x$-N n

Another reaction which can take place with maleimide-terminated oligomers is self-addition. This is a heat-induced reaction and leads to cross-linking, which is, of course, desirable for an industrial product to impart insolubility, hardness, modulus, and strength. This portion of the curing process can be shown as follows [28]:

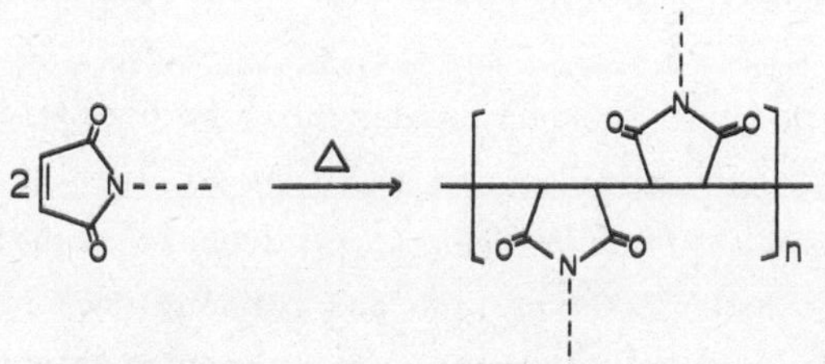

2 N---- $\xrightarrow{\Delta}$ n

These commercial materials are processed by compression, transfer, and injection molding techniques at 350-400°F (175-205°C) in 5-20 min at 21-105 MPa (3000-15,000 psi). Their processing as composite materials, especially glass-filled, has been investigated [18].

It is interesting that at very high temperatures (400°C/750°F) the oxidative stability of polyimides is sensitive to moisture; this parameter must be controlled in performing comparative tests [13]. However, at lower temperatures of 150°C or less, polyimide composites are quite resistant to water and have low absorption compared to other resin composites and for this reason are strongly favored as engineering laminates in applications such as found in supersonic aircraft (radomes and jet engine noise suppression devices) [28].

5.1-1 Common Polyimides

As mentioned above, numerous polyimides have been synthesized and characterized. What will be given here is a typical example of prepolymerizing to the amic acid, processing to a film, and postcuring to the imide for one of the most common types using pyromellitic dianhydride.

$$\xrightarrow{\Delta} \left[HO_2C, \; C(=O)\text{-}NH \; / \; NH\text{-}C(=O), \; CO_2H \right]_n$$

$$NH_2\text{-}C_6H_4\text{-}O\text{-}C_6H_4\text{-}NH_2 \xrightarrow{\Delta} \left[N \; \ldots \; N\text{-}C_6H_4\text{-}O\text{-}C_6H_4 \right]_n$$

The thermal stability of this polymer by isothermal (static) testing shows it to be excellent in an inert atmosphere up to 500°C, i.e., 5% weight loss in 15 hr, but suffering severe decomposition at 550°C. The same behavior is noted in air but to only 425°C [54]. In an excellent review of this and other systems by Arnold, it was pointed out that polythiazoles and some copoly(imide heterocycles) displayed a greater stability than commercial polyimides [3].

The thermal stability (by TGA) of the polyimides from pyromellitic dianhydride ranged from 530 to 580°C depending on the type of diamine incorporated. Naturally, those with methylene bridges (p,p'-diphenylmethylene) were lower than wholly aromatic systems (phenylene and biphenylene).

Again the glass transition temperature varied with the rigidity of the backbone, both anhydride and amine portions [17]. With pyromellitic dianhydride, any aromatic or bridged aromatic diamine gave a Tg above 300°C; so the rigidity of this anhydride seemed to mask (within the limits of the equipment) any possible effect of aromatic amine. However, other dianhydrides gave lower Tg's which can be affected by amine structure. With the diphenylether diamine, the following results are obtained:

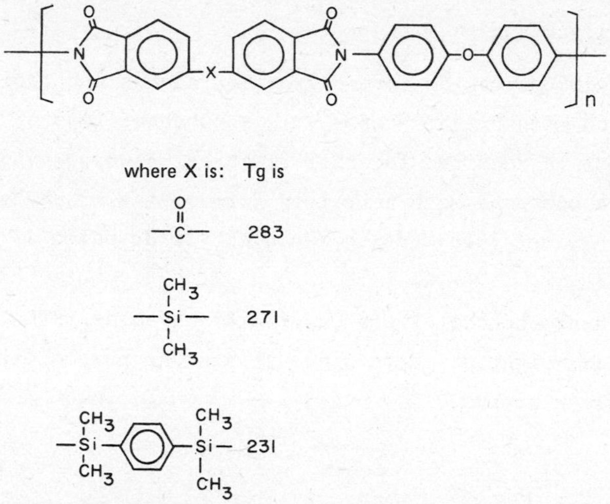

where X is: Tg is

$-\overset{O}{\overset{\|}{C}}-$ 283

$-Si(CH_3)_2-$ 271

$-Si(CH_3)_2-C_6H_4-Si(CH_3)_2-$ 231

In general, inherent viscosities are seen to follow the same pattern.

If one then considers various diamines with the benzophenone dianhydride, certain ones are seen to push the Tg to the maximum (above 300°C), i.e., the wholly aromatic or rigid fused ring types. All other para-bridged aromatics show amazingly similar Tg values, 271-285°C, while the meta-oriented bridges cause Tg to run from 232 to 257°C with the methylene bridge lowest.

Identical polyimides were prepared from pyromellitic thioanhydride with the advantages of quantitative yields and a one-step process [32]. However, only low molecular weights (inherent viscosity 0.3 dl/g in concentrated H_2SO_4) were obtained owing to precipitation of the polymer. Further, the off-gas is H_2S in this case, which is not desirable, to say the least.

Method [51,54] *Polymerization* Into a dry 500-ml flask equipped with stirrer, drying tube, nitrogen blanket, and an adapter which allows the addition of a solid under an inert atmosphere are placed 10.0 g of bis(4-aminophenyl)ether and 160 g of dry, freshly vacuum-distilled dimethylacetamide (DMAc). To the vigorously stirred solution is added over 2-3 min 10.9 g of pyromellitic dianhydride through the solid addition adapter. After the exotherm subsides (400°C), the solution (10% solids) is stored cold (-15°C) in sealed bottles.

Film preparation The above polyamic acid solution is spread into a thin (0.25-0.64 mm or 10-25 mils) layer on a glass plate and dried at 80°C for 20 min and then in a vacuum.

Postcuring The films are removed from the substrate and further heat-treated by warming to 300°C over 45 min then holding them at 300°C for an additional hour. The reaction can be followed by means of IR spectroscopy by observing the loss of the N-H/OH stretching frequency at 3400-3600 cm^{-1} and the appearance of the imide bands at 730-1800 cm^{-1}.

5.1-2 Polyimides from Diisocyanates

It is now known that polyimides can be formed from tetraacids and their anhydrides by condensation with diisocyanates with subsequent loss of CO_2 and H_2O [1,2,7,46]. Water is required for this reaction, however, and is effectively included when one uses some tetraacid form of the dianhydride. The best flexible films are obtained when the acid:anhydride ratio is between 1:1 and 1:4.

It is curious that benzophenonetetracarboxylic dianhydride (BTDA) does not provide high molecular weight polymers since it and the pyromellitic system are usually equally reactive.

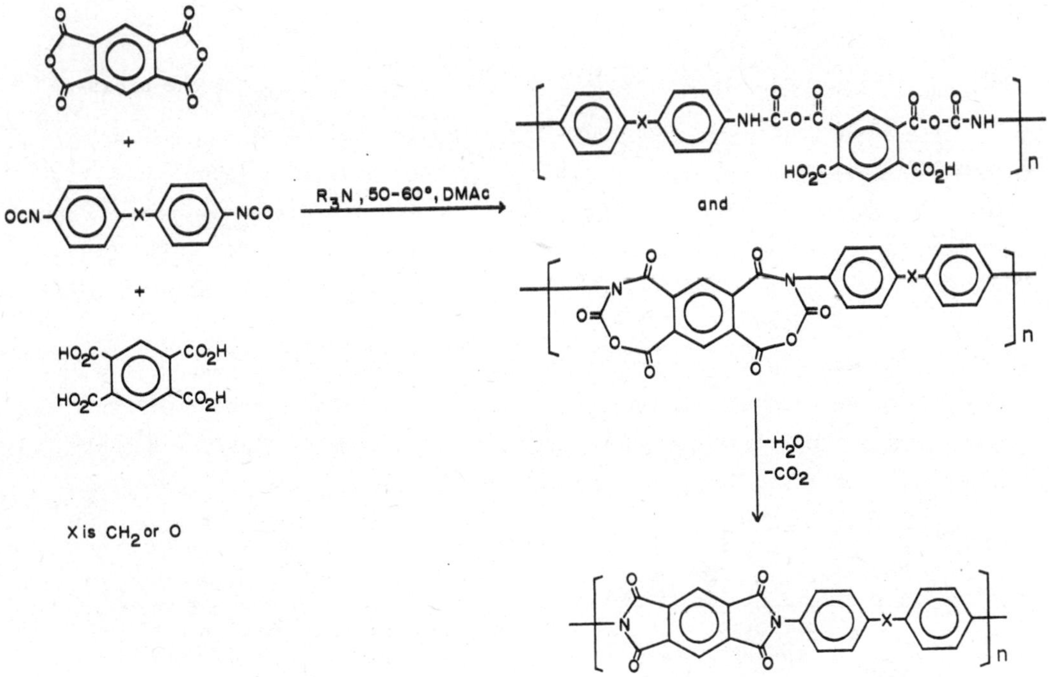

O O
O O
O O
+
OCN- -X- -NCO
+
HO_2C CO_2H
HO_2C CO_2H

R_3N, 50-60°, DMAc

-NH-C(=O)-O-C(=O)- HO_2C CO_2H -C(=O)-O-C(=O)-NH-]n

and

N O O N O O -X-]n

$-H_2O$
$-CO_2$

N O O N -X-]n

X is CH_2 or O

Inherent viscosities of the products range from 0.5 to 1.2 dl/g and thermal stabilities by TGA are near 550°C (10% weight loss) in air. Tensile strengths are above 69 MPa (10,000 psi) and elongation is approximately 8%. These materials have been shown to be useful as molding powders and as tough, flexible film formers.

> *Method* [2] To a mixture of 6.54 g (0.03 mol) of pyromellitic dianhydride and 5.08 g (0.02 mol) of pyromellitic acid in 75 g of DMAc (dry) at 50-60°C are added 5 drops of benzyldimethylamine followed by incremental portions of 12.6 g (0.05 mol) of diphenylether-4,4'-diisocyanate. The temperature is held at 50-60°C until CO_2 evolution ceases

and the viscosity increase seems to stabilize. To cast films the polymer solution (diluted if necessary) is poured onto the substrate and dried at 150-200°C. The films are then removed and postcured at 300°C.

5.1-3 Polyimides by a Diels-Alder Reaction

A unique application of the Diels-Alder reaction has been made for the preparation of polyimides as it has for polyphenylenes. Again, the reaction scheme lends itself well to the synthesis of phenylated backbones, which are more soluble than the unsubstituted polymers. (The only difficulty here is being able to visualize the rather complex intermediate and its decomposition to the desired product.)

Two approaches can be made, the first being where the imide function is preformed in the monomer and polymerization takes place via the Diels-Alder mechanism followed by dehydrogenation [23,55]:

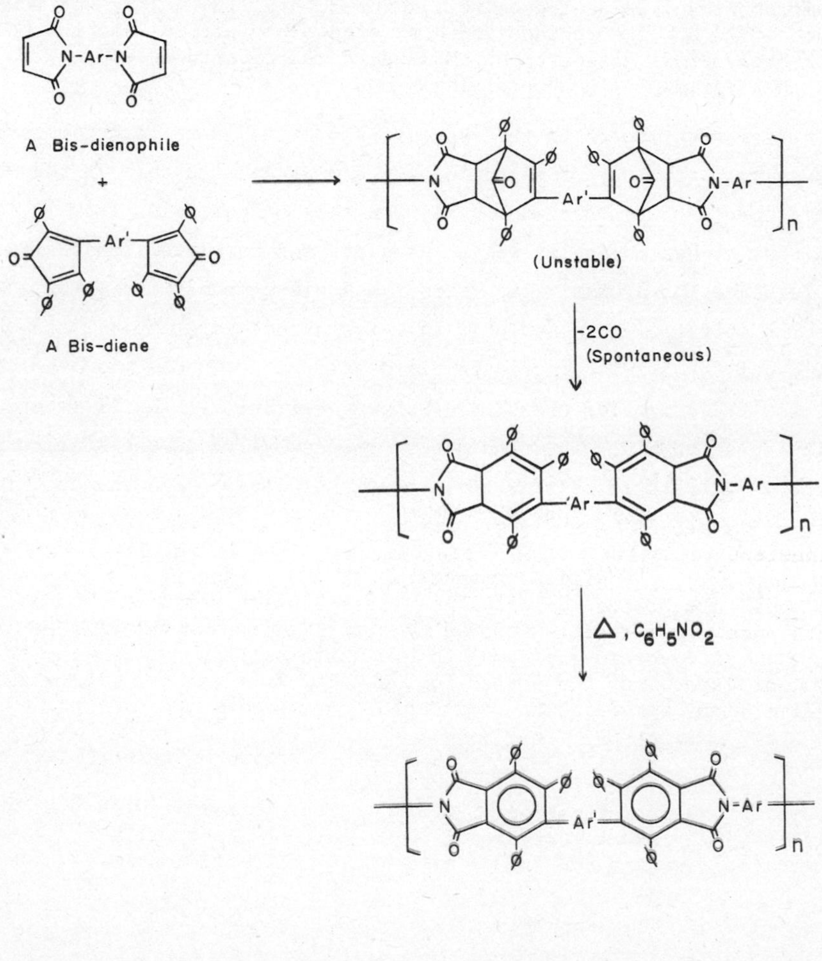

A Bis-dienophile

+

A Bis-diene

(Unstable)

-2CO (Spontaneous)

Δ, $C_6H_5NO_2$

The problems here are low molecular weights and, of course, the lack of quantitative dehydrogenation. Furthermore, a decrease in viscosity was experienced, indicating a loss in molecular weight due to thermal decomposition of the intermediate with prolonged heating.

TGA for the dihydro material shows a weight loss beginning at 300°C and becoming drastic before 500°C and going to completion in either air or nitrogen. The aromatized polyimide demonstrated a slight loss at 300°C, indicating lack of complete aromaticity, then a 7% weight loss near 530°C, dropping to a maximum loss of 30% at 600°C which is held to 800°C.

Method [23] *Polymerization* Equimolar amounts (0.002 mol) of the dimaleimide and bistetradienone are stirred in 10 ml of α-chloronaphthalene at reflux under nitrogen for 1-3 hr. Half the solvent is removed by vacuum distillation and the residue is poured into ethanol. The polymeric precipitate is collected, purified by dissolving in chloroform and reprecipitating into ethanol, and dried to give a quantitative yield of light yellow product.

Aromatization A solution of the dihydro polymer (0.5 g in 10 mil of nitrobenzene) is heated to reflux for 12 hr. The volume was reduced to one-third and the residue added to ethanol to precipitate the polyimide. The polymer is dissolved in DMF and reprecipitated in ethanol to yield a dark yellow solid upon drying.

The second approach to the above polymer type is to form a phenylated nucleus first with anhydride termination groups and use this in the more standard reaction with diamines [21]. In this manner solubility is maintained, but higher molecular weight products and complete aromaticity are possible since the aromatization step takes place prior to polymerization [20]. The white polymers, soluble in chlorinated hydrocarbons, are obtained in yields of 93-95% with intrinsic viscosities of 0.70 dl/g for the phenylene analog and 2.80 dl/g for the diphenylether derivative. The Tg's were in the opposite relationship, 413°C and 360°C, respectively, as one would expect from the relative stiffness of the aromatic linkages. TGA showed weight loss to begin at 530°C in air or nitrogen.

Method [21] Equimolar amounts of the biscyclopentadienone and the diamine are heated to reflux in m-cresol which contains isoquinoline. The water byproduct is removed by distillation and the residue is poured into absolute ethanol to precipitate the polymer. The solids are isolated and then heated to 250°C for 4 hr to complete cyclization of any remaining polyamic acid to polyimide.

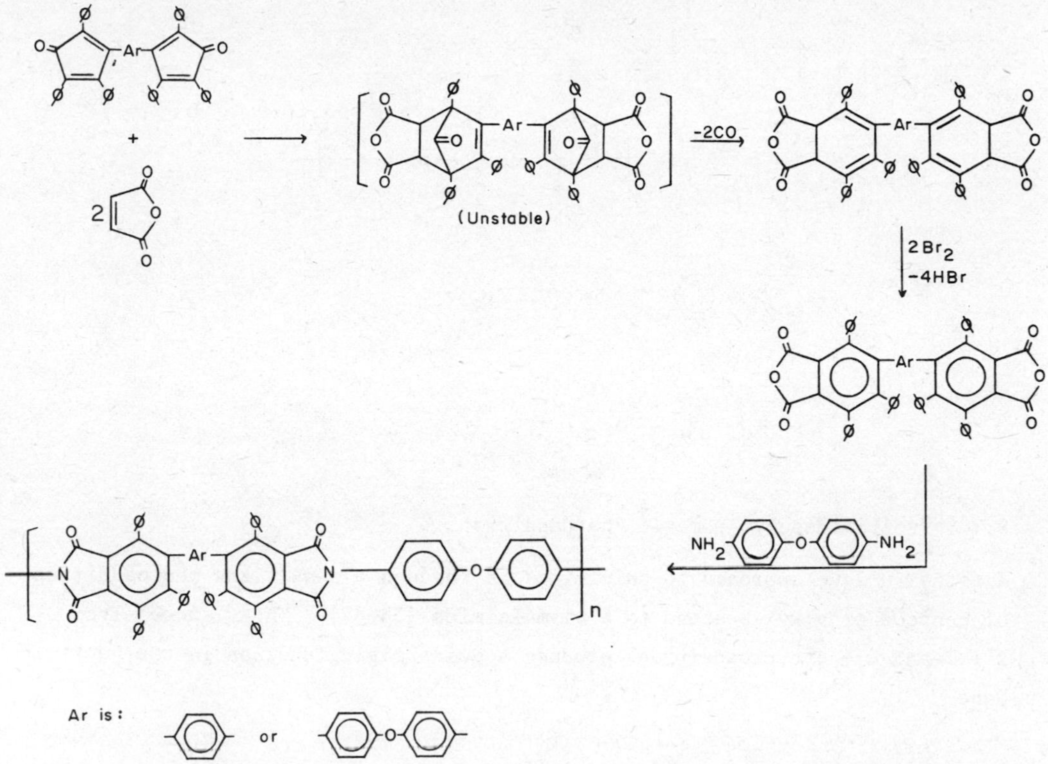

(Unstable)

$-2CO$

$2Br_2$
$-4HBr$

Ar is: or

5.1-4 Polyaspartimides

Polyaspartimides are one commercial type mentioned earlier as polyaminobismaleimides and occur from an acid-catalyzed hydrogen addition mechanism between a diamine and bismaleimide [10]. The bismaleimides are prepared by a known condensation between maleic anhydride and commercially available diamines [52].

The amorphous yellow polymers are soluble in phenolic and amide solvents from which films can be cast (intrinsic viscosities in DMF 0.25-0.66 dl/g).

Method [52] Under a nitrogen blanket equimolar amounts (0.05 mol each) of the bismaleimide and diamine are heated at 100-110°C for 3 days in 100 ml of cresol and 1 ml glacial acetic acid. The viscous mixture is poured into methanol to produce a yellow, fibrous solid which is chopped in alcohol in a blendor, extracted with hot ethanol in a Soxhlet, and dried for 1 day at 60°C in vacuo.

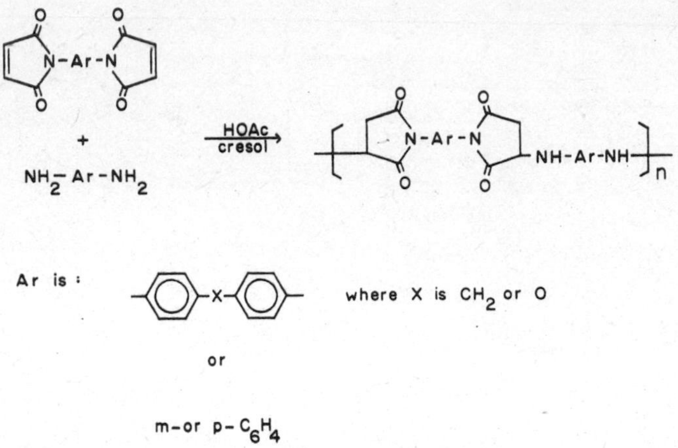

N-Ar-N (bismaleimide) + NH_2-Ar-NH_2 $\xrightarrow[\text{cresol}]{\text{HOAc}}$ [N-Ar-N ... NH-Ar-NH]$_n$

Ar is : —C₆H₄—X—C₆H₄— where X is CH_2 or O

or

m-or p-C_6H_4

5.1-5 Polyimides by Aromatic Photoaddition

A rather unique approach to polyimides is through a sensitized photoaddition of benzene or alkyl benzene to a bismaleimide [34,47]. These successive 2 + 2 and 2 + 4 cycloadditions produce a quite rigid function in the backbone.

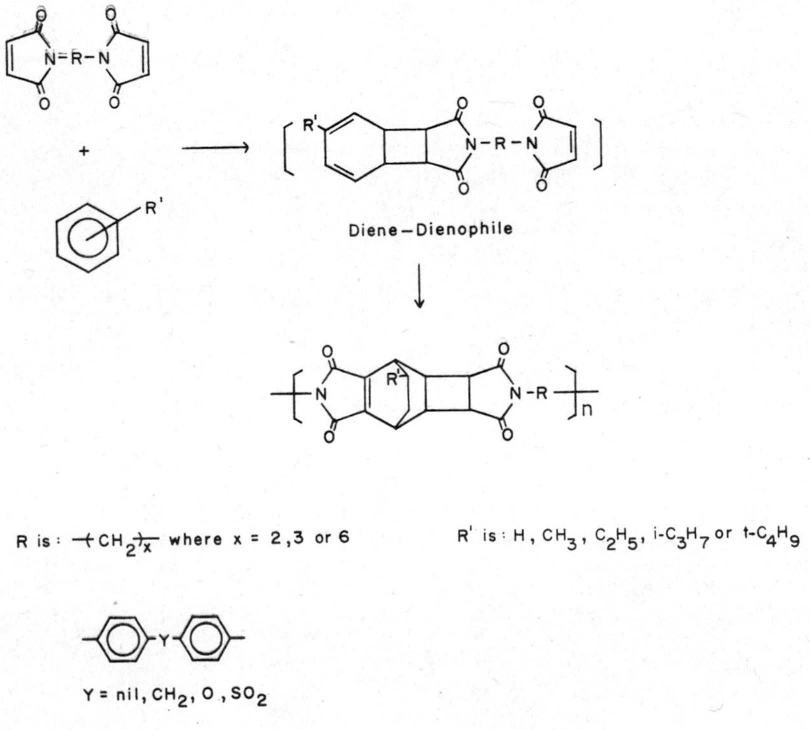

N=R-N (bismaleimide) + benzene-R' ⟶ [R' ... N-R-N]

Diene–Dienophile

↓

[N ... R' ... N-R]$_n$

R is : $\text{+}CH_2\text{+}_x$ where x = 2,3 or 6

R' is : H, CH_3, C_2H_5, i-C_3H_7 or t-C_4H_9

—C₆H₄—Y—C₆H₄—

Y = nil, CH_2, O, SO_2

The pale yellow or white polymers were obtained as powders or brittle films from the reaction and had poor solubility. The yields varied considerably (0-95%) with structure but generally seem somewhat better when Y is $-SO_2-$ and when R' is H, CH_3-, or t-butyl.

The thermal stability of these materials was determined visually in melting point capillaries, and the benzene adducts showed a slightly higher initial decomposition temperature (410-485°C) than the arene products (360-485°C). Severe degradation occurs at 480-510°C. The lack of correlation of structure to stability suggests that decomposition is due to decyclization of the photoadduct portion of the backbone. Indeed this is confirmed by the identification of benzene or arene as a major product on thermal decomposition.

> *Method* [47] The bismaleimide [52] (0.02 mol) is irradiated with a 450-watt UV light for 5-40 hr in an excess of benzene (100 ml) with 2.5 ml of acetophenone sensitizer and sufficient acetone to dissolve the monomer. The precipitated polymers are washed with acetone and dried.

5.1-6 Novel Polyimides from Less Common Dianhydrides

To this point all imide systems were derived from maleic or phthalic anhydride analogs. Other cyclic anhydrides can also be condensed with diamines to provide imides.

5.1.6.1 Polyspiroimides Good thermooxidative stability can be realized in nonaromatic polyimides [40,41]; however, the aromatic systems are still superior. One such material is synthesized from methane tetraacetic acid and diamines by condensing a 1:1 molar salt or by first preparing the dainhydride and subsequently forming the imide in the well-known fashion.

High molecular weight polymers (inherent viscosities up to 1.03), whose Tg's are inversely proportional to the length of the aliphatic R group, are obtained: Tg 205°C for $-C_2H_4-$ and 78°C for $-n-C_{10}H_{20}-$. Tough flexible films can be cast from these soluble (in hot m-cresol or trifluoroethanol) polyimides. The exception is the diphenylether derivative, the film of which is brittle.

The thermal stability by TGA of the aliphatic polyspiroimides shows 7% weight loss at 450°C in air and 485°C in nitrogen with drastic decomposition in both atmospheres by 500°C. Isothermal aging studies show at

$C(CH_2CO_2H)_4$ $\xrightarrow{2CH_3COCl}$ (spiro dianhydride) $\xrightarrow[\Delta]{NH_2\text{-}R\text{-}NH_2}$ $[HO_2CCH_2, CH_2C(O)\text{-}NH\text{-}R\text{-}; \text{-}NH\text{-}C(O)\text{-}CH_2, CH_2CO_2H]_n$

$C(CH_2CO_2H)_4$ $\xrightarrow[NH_2\text{-}R\text{-}NH_2]{DMF}$ 1:1 salt $\xrightarrow{\Delta}$ (polyamic acid) $\xrightarrow{\Delta}$ (spiroimide polymer, N-R)$_n$

R is: $-(CH_2)_x-$ where x is: 2, 6 or 10

or

$-C_6H_4-O-C_6H_4-$

100 hr a 1.8 loss at 200°C and 7.0% at 230°C. Mechanical properties reveal a strong inelastic material with a tensile strength of nearly 56 MPa (8000 psi) and an elongation-to-break of 3.6%.

Method [40] *From anhydride* A solution of 4.0 mmol of 1,6-hexanediamine (4.64 g) in 40 ml DMAc is added to a stirred solution of 4.0 mmol (8.48 g) of methanetetraacetic dianhydride under nitrogen at 10-15°C, and the clear mixture becomes cloudy on stirring at 25°C for 2 hr. The reaction temperature is then raised gradually to 165°C where it is held at 250°C and 133 Pa (1.0 torr) for 1 hr and then at 300°C and 40 Pa (0.3 torr) for 0.5 hr. To separate the solid from the vessel, liquid nitrogen cooling is utilized. The product is finely ground, washed with acetone and then water, and dried to yield 10.5 g (85%); reduced viscosity in m-cresol 0.98.

From salt A solution of 1.41 g (1.21 mmol) in 15 ml of absolute ethanol is added to a solution of 3.00 g (1.21 mmol) in 100 ml of absolute ethanol. After about 15 hr the salt is isolated by filtration in a 99% yield (4.38 g), mp 225°C. The salt is placed in a glass tube which is sealed under vacuum and then heated at 245°C for 3.5 hr and then at 260°C for 0.5 hr. The product demonstrates a reduced viscosity of 1.03 and a Tg of 125°C.

Another nonaromatic polyimide system is very similar in structure to the above spiroimide. It is derived from 1,2,3,4-butanetetracarboxylic acid dianhydride and various aromatic diamines [41] and, of course, is not truly a thermally stable material because of the alicyclic portions of the backbone.

By using both cis and trans forms of the original diacid, the meso and dl stereoisomers of the tetraacid are formed. The only discernable difference between the dl and meso polymers is in tensile modulus; for the bridged

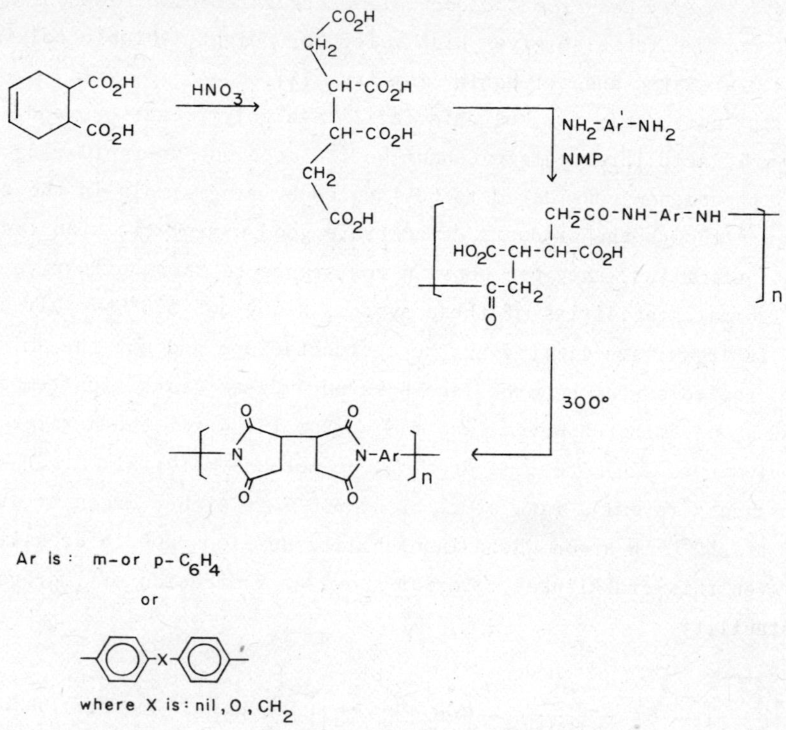

Ar is: m-or p-C_6H_4

or

where X is: nil, O, CH_2

diphenyl diamines, it is four times greater for meso forms; but for m-phenylene diamine, the value for the meso isomer is nearly half that of the dl.

Flexible films can be cast whose tensile strengths run around 90-97 MPa (13,000-14,000 psi), moduli from 0.69 to 3.4 GPa (100,000-500,000 psi), and elongations from 2 to 11% at room temperature. All values are decreased to about half at 200°C.

Isothermal weight loss studies show a 20% weight loss at 200°C and 1600 hr or at 400 hr at 240°C. A 10% loss by TGA occurs at 400-430°C in air or nitrogen, depending on the type of diamine used, methylene bridges being lower in thermal stability. Polymer melt temperatures are above 400°C.

Method [41] *Anhydride monomer* A 2:1 weight ratio of acetic anhydride and 1,2,3,4-butanetetracarboxylic acid [42] is heated to reflux for 30 min, cooled to room temperature, and filtered to give a 60% yield of meso-dianhydride (mp 248°C dec.) or a 34% yield of dl (mp 170°C dec.).

Polymerization A 20-30% solution of equimolar amounts of monomers in NMP is stirred under nitrogen and allowed to stand approximately 15 hr. Intrinsic viscosities range up to 0.73 dl/g. To make film, solutions are cast onto aluminum foil and heated for 1 hr each at 100°, 200°, and 240°C and for 30 min at 300°C. The aluminum is removed by dissolving it in HCl.

5.1.6.2 Polyimides from Cyclooctadienetetracarboxylic Dianhydride. This novel dianhydride also gives high molecular weight, soluble polyimides with bridged aromatic and aliphatic diamines [15].

Interestingly, only two solvents facilitate polymerization, and the usual polyamic acid intermediate cannot be isolated due to rapid ring closure, the latter phenomenon considered to be induced by ring strain in the cyclooctadiene. Although the products demonstrate good hydrolytic stability and mechanical properties, they are poor in resistance to thermooxidative degradation. Thermal stabilities of these systems being 365-370°C in air and 440-450°C in argon, are limited by the cyclooctadiene and not the amine structure, regardless of diamine incorporated. Glass transition temperatures are dependent on amine, however, and range from 140°C for the hexamethylene diamine polymer to 320°C for the diphenylene methane material. Isothermal weight loss data reveal a rate of 2%/hr at 300°C in either argon or air but a 0% rate at 250°C in argon where insolubility develops due to crosslinking. However, even this crosslinked material shows no indication of improved thermal stability.

110° m-cresol

R is: $-(CH_2)_6-$

or

where X is: O or CH_2

Method [15] The diamine (10 mmol) is dissolved in 50 ml of m-cresol under argon and 10 mmol (2.48 g) of 1,5-cyclooctadiene-1,2,5,6-tetracarboxylic dianhydride [19] and 10 mmol (1.26 g) of quinoline are added, and the mixture is stirred at 110°C for 28 hr. The viscous solution is poured into 300 ml of stirred methanol to precipitate the pale yellow, fibrous polymer which is subsequently washed repeatedly in a blendor with methanol and dried in vacuo at 100°C. Inherent viscosity in m-cresol at 0.5 g/100 ml is 1.9 dl/g.

5.1-7 Copolyimides

5.1.7.1 Amide Imides Commercial polymers are available which are copolymers of imides. For example, by incorporating the amide linkage a more processable, soluble, moldable, comfortable material is possible

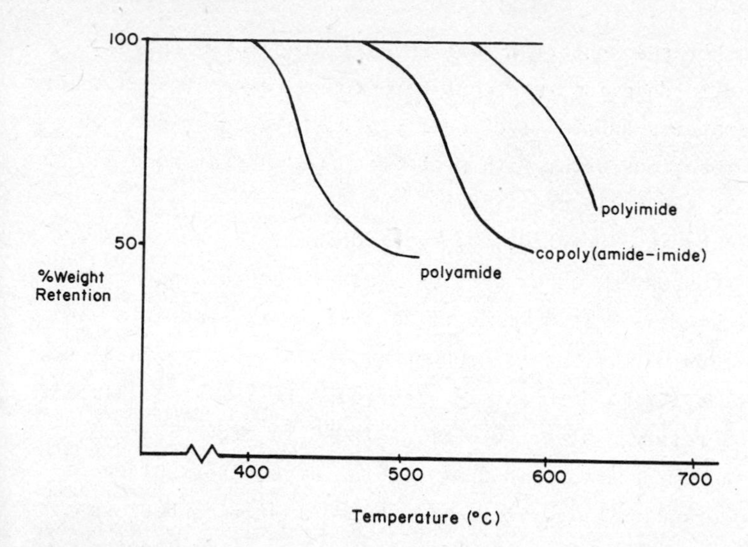

Figure 5.1 TGA curves showing the effect of amide and imide backbone composition.

(Amoco AI). To be sure, it involves a trade-off again with a decrease in thermal stability as is amply demonstrated by the TGA curves shown in Figure 1. Even with this sacrifice, however, the product can withstand use at 290°C (550°F) for 2000 hr as a laminating resin or wire coating.

Almost all the amide imide copolymers are synthesized from trimellitic acid or its derivatives, commonly the anhydride acid chloride, and the multitude of diamines available, including some fused heterocyclic sulfones [29]. The polymer structure is shown head to tail but also exists head to head and tail to tail where the R group is flanked by either two imide or two amide groups.

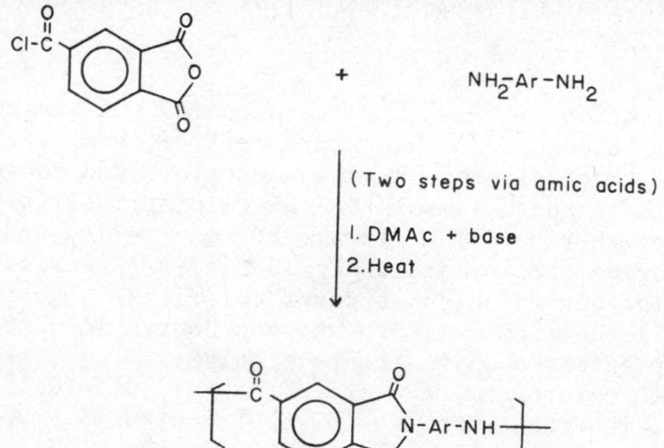

Cl-C(=O)-(trimellitic anhydride) + NH_2-Ar-NH_2

(Two steps via amic acids)
1. DMAc + base
2. Heat

$\left[-C(=O)-\text{(phthalimide)}\text{N-Ar-NH}- \right]_n$

Leaving groups other than chloride were investigated but with less success. The acid and imidazolide functions were found to react at a lower rate and higher temperatures and to give lower yields than chloride. Of course, again, these reactions occur with production of the intermediate amic acid which must be thermally cyclized.

Sulfone ether blocks incorporated into an amide imide backbone impart impact strength as discussed for other systems (sulfone-containing polyamides and polysulfone ether imides). This amide imide sulfone is tractable with a Tg of 270°C, tensile modulus 2.6 GPa (370,000 psi), tensile strength 86 MPa (12,500 psi), elongation 15%, and impact 14 J/cm^3 (171 ft-lb./in^3). Impact strength is nearly 60% retained after 5000 hr at 220°C in air. The additional advantage of the amide imides is that with a large ether sulfone block they can be processed under conventional thermoplastic conditions (325-335°C) and remain uncrosslinked. Good strength retention is demonstrated by these amide-imide composites with glass or carbon fibers after 175°C for 1000 hr in air.

The most immediately obvious advantage of the amide group is the solubility which it imparts to the final product. One example of this phenomenon is where bipyridyl units are incorporated via the diamine [37]. In the above reaction Ar is [36]:

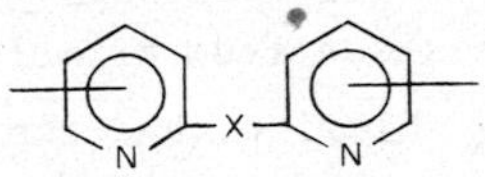

Where X is: nil, S, SO, SO_2, NH, NCH_3, C_2H_4 or $NCOCH_3$

The bipyridyl polyamide imides are generally soluble in DMAc, NMP, m-cresol, dichloroacetic acid, and H_2SO_4.

TGA of these materials shows that the sulfide linkage is weaker in thermooxidation than the methylamino. The former shows a 10% weight loss at 350°C and the latter at 440°C.

Method [37] *Polymerization* A solution is prepared containing 0.218 g (1 mmol) of 5,5'-diamino-2,2'-bipyridyl sulfide, 4 ml of DMAc, and 0.14 ml (1 mmol) of triethylamine, as acid acceptor, and cooled in an ice bath. To this is added 1 mmol (0.2105 g) of trimellitic anhydride acid chloride, whereupon the mixture becomes yellow and the triethylamine hydrochloride precipitates. In 15 min it is warmed to 25°C and stirred for 7 hr during which it turns colorless. The reaction mixture is then diluted with 2 ml of DMAc and poured into 500 ml of chloroform to precipitate the white, fibrous polymer. The solid is isolated, washed with chloroform, and dried to yield 97% (0.380 g) of polyamide amic acid; inherent viscosity 0.76 dl/g at 0.25 g/100 ml DMAc.

Cyclization Imidization is brought about by heating the above synthesized amide amic acid at 200°C in vacuo for 2.5 hr to result in yellow polyamide imides.

Another approach to polyamide imides is to use a trimellitic acid derivative which utilizes an imidazolide function as a leaving group and a preformed imide function so that postcuring is unnecessary [24,61]. (This reaction is, incidentally, the organic chemist's delight since yields are reported to be 100-105%.)

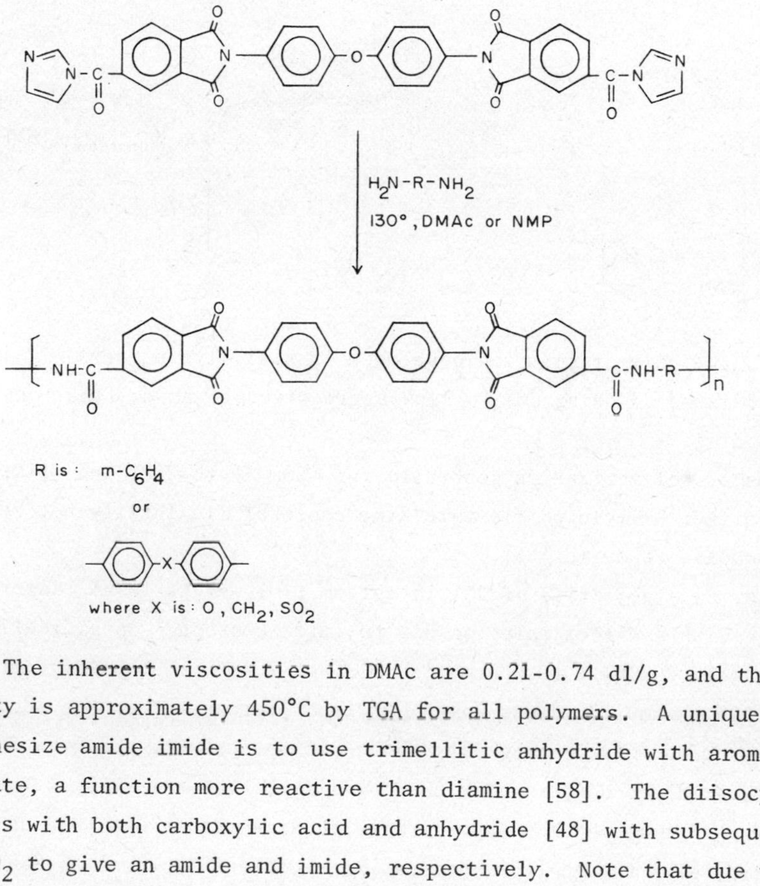

$H_2N-R-NH_2$

130°, DMAc or NMP

R is: m-C_6H_4

or

where X is: O, CH_2, SO_2

The inherent viscosities in DMAc are 0.21-0.74 dl/g, and thermal stability is approximately 450°C by TGA for all polymers. A unique method to synthesize amide imide is to use trimellitic anhydride with aromatic diisocyanate, a function more reactive than diamine [58]. The diisocyanate reacts with both carboxylic acid and anhydride [48] with subsequent loss of CO_2 to give an amide and imide, respectively. Note that due to the reactivity of the isocyanate, no second curing step is needed for imide formation.

By the partial replacement of the triacid with an aromatic diacid or tetraacid dianhydride, the amide or imide content respectively could be increased. It is no surprise that increased imide function leads to a better thermal stability. However, upon deviation from the 50:50 functional group molar ratio, the polymer solution developed instability in terms of viscosity. This is postulated as being due to slow aggregation of amide functions by hydrogen bonding.

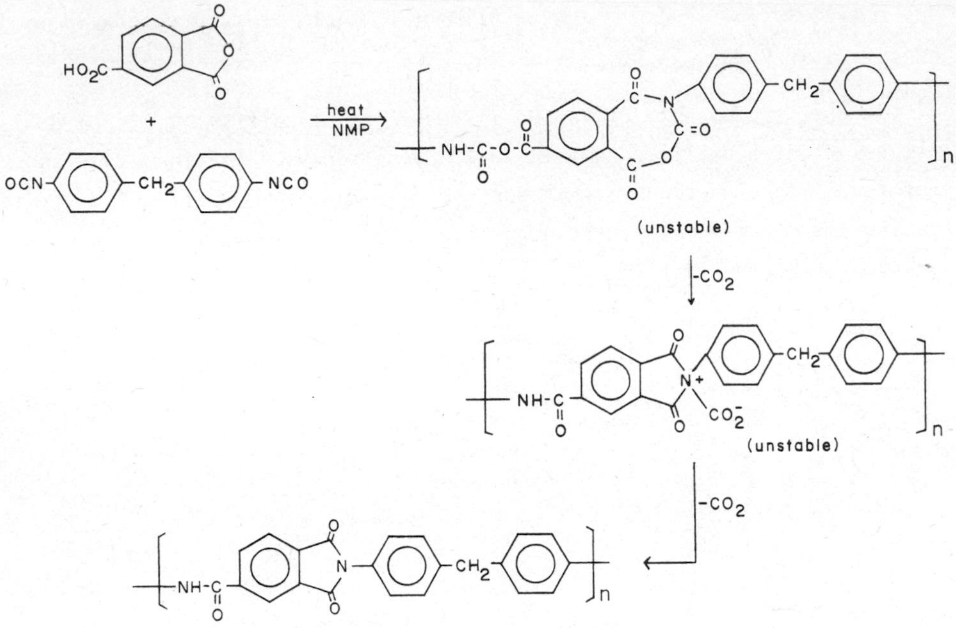

Degrees of polymerization generally run from 50 to 200, and films can be cast as either brittle or flexible, the cause of flexibility not being obvious from data reported [58].

The thermal stabilities of TGA in helium (10% weight loss) range from 485 to 565°C with a direct relationship to imide content. Specifically, the polymer with 20 mol % terephthalic acid was least in stability, while that with 100% pyromellitic was most. It is of interest that the ideal, i.e., 50:50, amide imide, being only about 5°C more stable than the 20% terephthalic acid product, is near the low side of the stability. The increased use of cyclopentane tetracarboxylic dianhydride in place of trimellitic acid, even though it increased imide content, lowered the thermal stability owing to its cycloaliphatic oxidative weak link [6].

Method [58] Equimolar amounts (0.1 mol) of monomers are weighed into a reaction system equipped for stirring, reflux, and gas blanketing; and enough NMP with xylene (80:20) is added to make a 25% solid solution. The mixture is heated slowly (over 10 hr) to the reflux temperature of NMP (177°C) until a maximum viscosity is reached. This solution is cooled to room temperature and used to cast films.

A fourth type of polyamide imide is that which incorporates aliphatic and cycloaliphatic repeating units [6] which have been noted to reduce thermal stability [58]. However, due to the lower Tg values for these systems (50-200°C), one can obtain fibers (melt spun) and plastics with a high-impact strength but still with TGA stability values of 417-445°C.

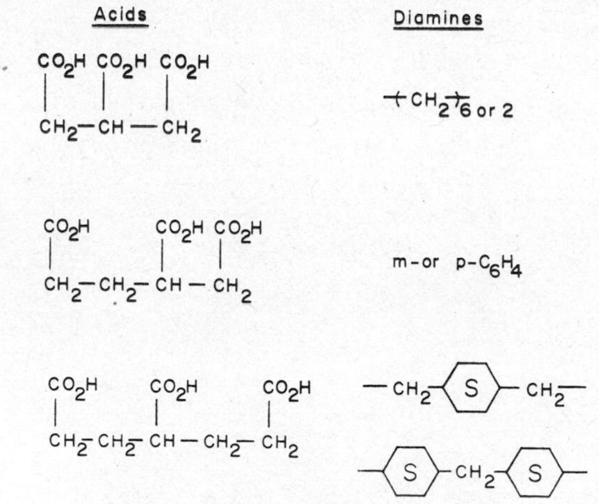

Acids	Diamines
CO_2H CO_2H CO_2H on $CH_2-CH-CH_2$	$-(CH_2)_{6\ or\ 2}-$
CO_2H CO_2H CO_2H on $CH_2-CH_2-CH-CH_2$	m-or p-C_6H_4
CO_2H CO_2H CO_2H on $CH_2-CH_2-CH-CH_2-CH_2$	$-CH_2-$(S)$-CH_2-$ $-$(S)$-CH_2-$(S)$-$

It is interesting that even for the pentane tricarboxylic acid, predominately amide imide polymers are formed rather than branched amides. Further, the diamines are arranged in order of increasing Tg (from 71 to 200°C) and PMT (polymer melt temperature) (from 120 to 305°C) for the propane triacid. The butane triacid gives thermal analytic data lower by 25-70°C.

Monoacids included in the reaction act as capping agents, and inherent viscosities are obtained up to 1.2 dl/g in m-cresol.

> *Method* Equimolar amounts (10-50 mmol) of triacid and diamine are heated under inert atmosphere pressure to 200-300°C for several hours and finally under a gas flush under vacuum to a near-quantitative yield.

Finally, a poly(amide imide) has been synthesized to include a terephthaloyl function in an effort to provide tractibility while maintaining thermal stability [56]. The acid derivative was

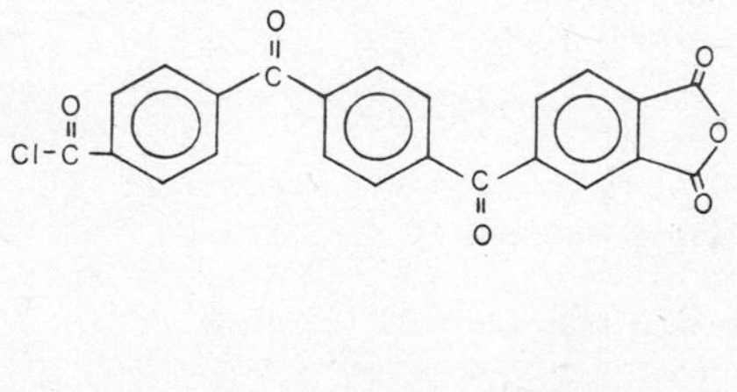

Cl-C(=O)-C_6H_4-C(=O)-C_6H_4-C(=O)-(phthalic anhydride)

which was condensed with diamines. However, the final polymers were insoluble and were stable to 400-500°C, about the same or lower than other poly(amide imides). They also had Tg values ranging from 200 to 350°C.

5.1.7.2 Poly(heterocyclic imides) In a successful effort to improve mechanical properties, thermal stability, and processability of polyimides coupled with increased modulus, Tg, and crystallinity, various heterocyclic functions have been incorporated on a regular, highly ordered basis in an imide backbone [14]. The approach was to synthesize a dianhydride of a tetraacid which forms heterocyclic functional groups during the usual imidization postcure.

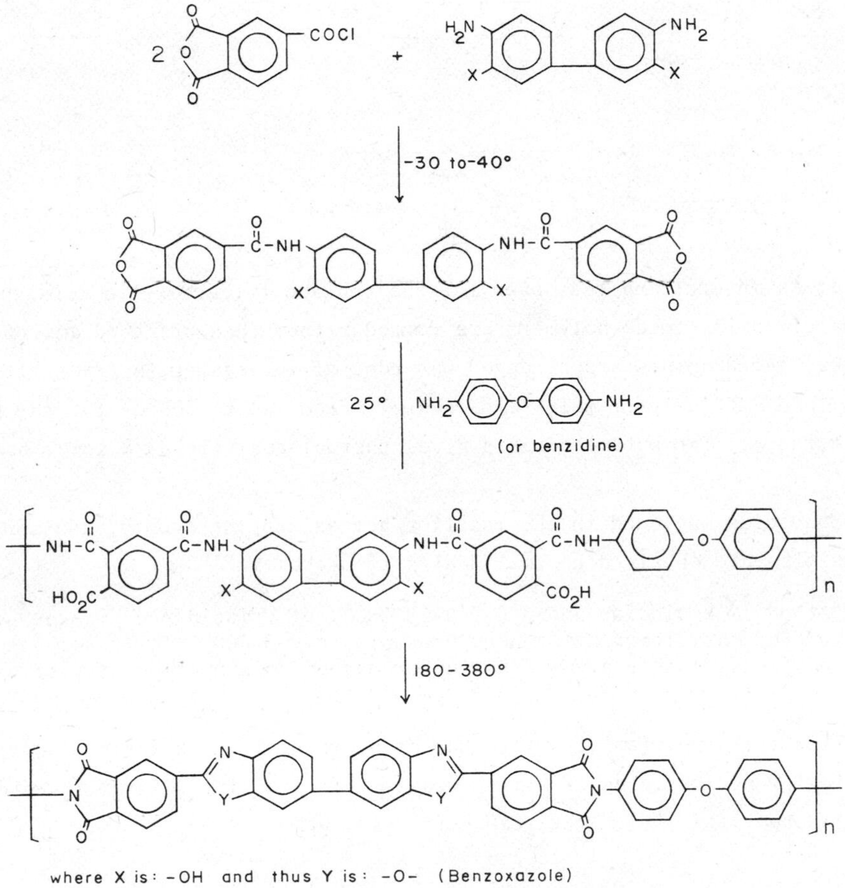

where X is:	and thus Y is:
-OH	-O- (Benzoxazole)
$-CO_2H$	-C(=O)-O- (Benzoxazinone)
$-NH_2$	-NH- (Benzimidazole)
-SH	-S- (Benzothiazole)

High molecular weights are possible because of high solubilities of the products; i.e., the polymers are not precipitated prematurely (inherent viscosities to 1.87 dl/g). Solution stabilities at 25°C (as measured by inherent viscosity in NMP) are better for the heterocyclic copolymer than the simple polyimide from pyromellitic acid.

Thermal stabilities are comparable to analogous homopolymers but in the order of oxazole (530°C), oxazinone (510°C), and thiazole (430°C), as determined in air by TGA (10% loss).

Mechanical properties of a 50-μm film include tensile strength 166 MPa (17 kg/mm^2 or 23,700 psi), elongation 21%, tensile modulus 3.44 GPa (353 kg/mm^2 or 491,000 psi), and 30,000 cycles folding endurance (MIT). Electrical properties are also reported [14].

Method [14] *Monomer preparation* A solution of 1.36 g (0.005 mol) of 3,3'-benzidinedicarboxylic acid [35] in 20 ml of NMP is added dropwise to a solution of 2.10 g (0.01 mol) of 4-chloroformylphthalic anhydride at -30°C, and the mixture is stirred at -30°C for 3 hr and then at 25°C for 1.5 hr. The precipitate is isolated by filtration, washed with 50 ml of benzene, and dried 3 hr at 180°C under vacuum for a 77% yield (2.46 g).

Polymerization A powder of the above dianhydride (2.07 g, 0.003 mol) is added to a stirred solution of 0.615 g (0.003 mol) of benzidine in 20 ml of NMP with 0.4 g of LiCl at 5°C under N_2. After being stirred for 4.5 hr at 50°C, the viscous solution is poured into acetone; and the polymer is isolated, washed with water and acetone and dried at 80°C in vacuo to yield 97% of polyamic acid. Films can be cast from a 15% solution to a glass plate with drying at 120°C for 20 min.

Cyclization The above produced prepolymer powder is heated at 180-380°C at 0.1 mmHg. Depending on structure, dehydration begins at 180-200°C and is completed by 360-380°C.

A slightly different approach to this same general method can be taken which results in a less ordered benzheterocyclic imide backbone [50]. This synthesis is accomplished by using an unsymmetrical diamine which contains preformed heterocyclic groups (benzimidazole, benzoxazole, benzothiazole, and quinoxaline). The limited order comes by virtue of the direction of coupling of the diamine in the anhydride. A structure different from the head-to-tail shown may be one where the anhydride is flanked by like amino groups (head-to-head or tail-to-tail). This limited order probably is the cause of poor fiber properties but good film properties (good flexibility and low embrittlement) because of "diluted" crystallinity.

Regarding thermal properties, it is no surprise to find the melting or softening points of the benzophenone adducts below those of pyromellitic anhydride. Thermal stabilities are not affected as much as physical properties by the keto function. The respective melting points are 305-575

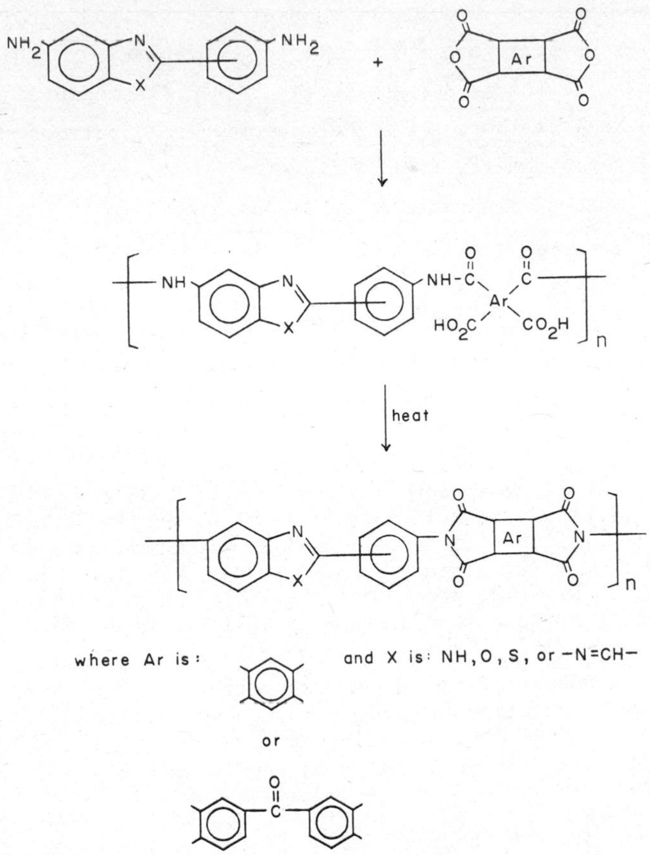

heat

where Ar is: and X is: NH, O, S, or −N=CH−

or

and 540-575°C and the TGA stabilities in nitrogen 525-575 and 550-605°C (by inflection point in the weight vs temperature curve). In contrast to the data for the analogous amide systems (vide infra), the decomposition temperatures are quite similar for the various polymers. This fact would indicate that stability is dependent on something other than the heterocyclic group in the backbone.

Most interesting are the results showing a benzoxazopyromellitide polymer to be more stable than an ordinary diphenylether-pyromellitide commercial system when subjected to isothermal aging. While the commercial material lost essentially all weight in 12 days at 350°C, the heterocyclic imide backbone retained 85% of its original weight.

Method [50] To a solution of 0.042 mol of diamine in 80 ml of DMAc is added 0.042 mol of the dianhydride at 0°C, and the solution is stirred for 7 hr and then at 25°C overnight. Films are cast from this mixture and are postcured at 110°C for 30 min, then at 145°C for 16 hr and finally at 300°C for 30 min in air.

Amide heterocyclic backbones have also been prepared like the above imides, but by using iso- and terephthalic acid [50] in a low-temperature process [49]. Of course, melting and decomposition points were considerably lower than the corresponding imides by more than 100°C, even in the best amides.

Still another polybenzheterocyclic imide has been prepared in an effort to remove all hydrogen from the composition and thereby improve thermooxidation resistance. This was accomplished by Hirsch [27] by condensing pyrazinetetracarboxylic dianhydride with 2,5-diamino-1,3,4-thiadiazole. Studies on oxadiazole backbones like these have also appeared [50]. The solubility of the amic acid intermediate decreases significantly with higher molecular weights. For this reason the polymerization is terminated prior to attaining maximum degrees of polymerization. This limitation in turn adversely affects film strength.

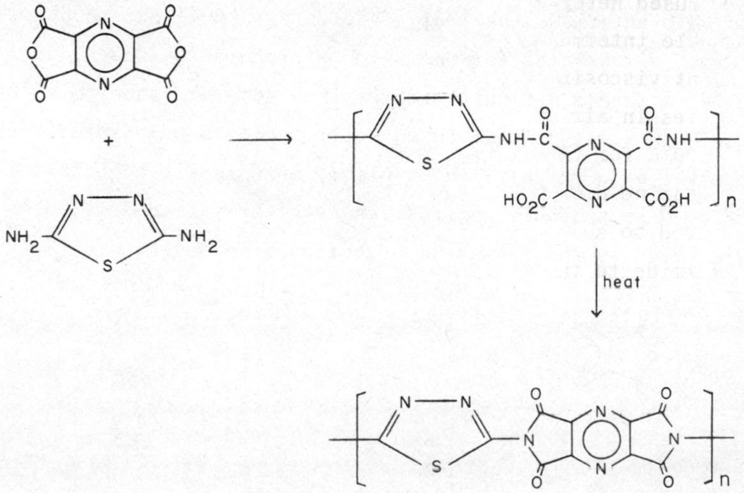

Although outstanding thermal stability is claimed for these deep red, clear films, the TGA curves show a 10% loss at 450°C in air and 410°C in N_2. (Note the apparent better stability in air. Above 600°C a N_2 atmosphere is less damaging than air, as expected.) However, a different kind of test was applied to this backbone for comparison to the polypyromellitimide from diaminodiphenylether. A 139 kPa (20 psi) stress was applied to a strip of film and the sample heated at 4°C/min to demonstrate zero strength temperatures at 580-590°C, but with the film maintaining clarity and flexibility. The standard polyimide chars at 320°C under these conditions. This study perhaps serves to point out the different data and comparative results which are available to the research chemist or engineer by choice of test method.

Method *Monomer preparation* Pyrazinetetracarboxylic acid [43] is recrystallized from HCl and 16 g of it is stirred in 400 ml of acetic anhydride under nitrogen while it is slowly heated to 85°C maximum and maintained there for 30 min. The mixture is evaporated to a slush then to dryness at 80°C. The product is purified by sublimation under vacuum to pale green crystals, dec. 180°C.

Polymerization In a 1-oz. screw cap bottle are placed 0.4624 g of the dianhydride and 10 ml of DMAc, and solution is effected at 25°C. The diamine, 2,5-diaminothiadiazole [4] (0.2440 g), is added and the mixture is shaken to effect a color change to dark and finally light yellow and then for 1.5 hr up to the point of gelation. Films are cast directly with a doctor knife and are cured by heating to 250°C at 4°C/min, then to 300°C at 1.1°C/min, holding at 300°C for 1 hr, then to 340°C at 1.1°C/min, at 350°C for 1-1/4 hr., then to 375°C at 1.1°C/min, and finally holding at 375°C for 2 hr.

Finally, a fused heterobicyclic imide copolymer has been synthesized through two soluble intermediates [26]. Yields are excellent (>85% overall) with fair inherent viscosities for the soluble precursors (0.2-0.44 dl/g). Thermal stabilities in air are quite impressive, however, showing a 10% weight loss at 500°C for the final fused heterocycle, a poly(imide oxoisoindolobenzothiadiazine dioxide). The typical, more reactive character of carboxylic compared to sulfonic derivatives is clearly demonstrated by the ability to form imide to the exclusion of sulfonamide moieties.

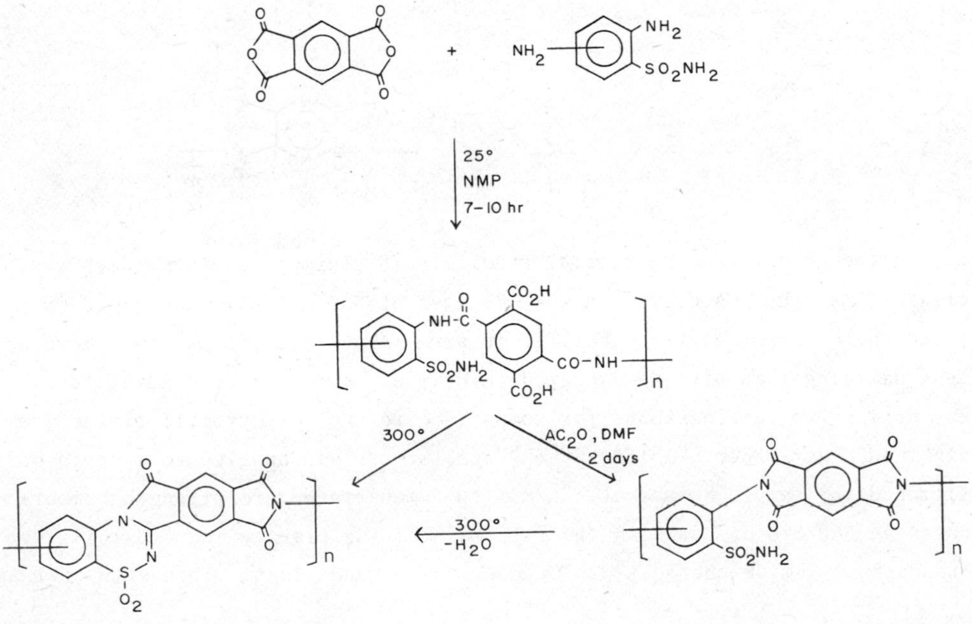

The aryl diamine was used as two different isomers--2,4 and 2,5--with the latter giving slightly higher yields and viscosities. Another dianhydride (from benzophenone tetracarboxylic acid) was also employed but again with slightly lower yields and viscosities than the pyromellitic.

Thermally, the first wholly amic acid prepolymer is least stable (200°C), but (owing to the sulfur dioxide group which acts as a free radical quencher in the flame front) all materials are self-extinguishing in a flame.

Method [31] *Polyamic acid* To a solution of 0.936 g of 2,5-diaminobenzenesulfonamide in 10 ml of NMP is added 1.101 g of pyromellitic dianhydride, and the mixture is stirred for 7 hr at 25°C to give a dark yellow solution. The viscous solution is then poured into 300 ml of methylene chloride to precipitate the polymer which is isolated, washed with methylene chloride, and dried at 50°C in vacuo to yield 2.008 g (98%).

Polyimide sulfonamide The above polyamic acid (1.06 g) is dissolved in 20 ml of acetic anhydride and 4 ml of DMF, and the solution is allowed to stand for 2 days. The precipitated polymer is collected, washed with water and dried at 100°C in vacuo to yield 0.852 g (88%).

Poly(imide oxoisoindolobenzothiadiazine dioxide) The above precursor is heated for 30 min at 300°C under nitrogen.

5.1.7.3 Polybutadiene Block Copolyimides In order to overcome the processing and intractibility problems with polyimides and to allow fabrication of thick samples and adhesives, an extreme step was taken to softening, i.e., the incorporation of polybutadiene blocks. This was done by first reacting an isocyanate-terminated polyalkene with an aromatic diamine then condensing the resulting oligomer with a dianhydride [26]. So the result is an B-I-B type block (B = polybutadiene, I = polyimide) with a rigid center and two flexible ends. The 1,2-polybutadiene blocks also offer a site for peroxide cure, which, of course, produces no off-gas.

Compositions, and therefore properties, can be varied by adjusting the size of either type of block. A dicumyl peroxide cure at 150°C (300°F) for 3 hr or 175°C (350°F) for 10 min gives the following mechanical properties:

Wt % butadiene	Mol wt (Visc. ave)	% Peroxide	Flexural Strength, MPa (psi)	Flexural Modulus GPa (psi)	HDT (°C)
40	21,700	12	52 (7500)	3.0 (430,000)	178
30	30,000	8	46 (6700)	3.14 (457,000)	168

TGA analysis revealed a surprisingly high stability in either air or nitrogen for the cured material. A 70% imide copolymer showed a 10% loss at 465°C, and a 60% imide survived to 450°C with the same loss.

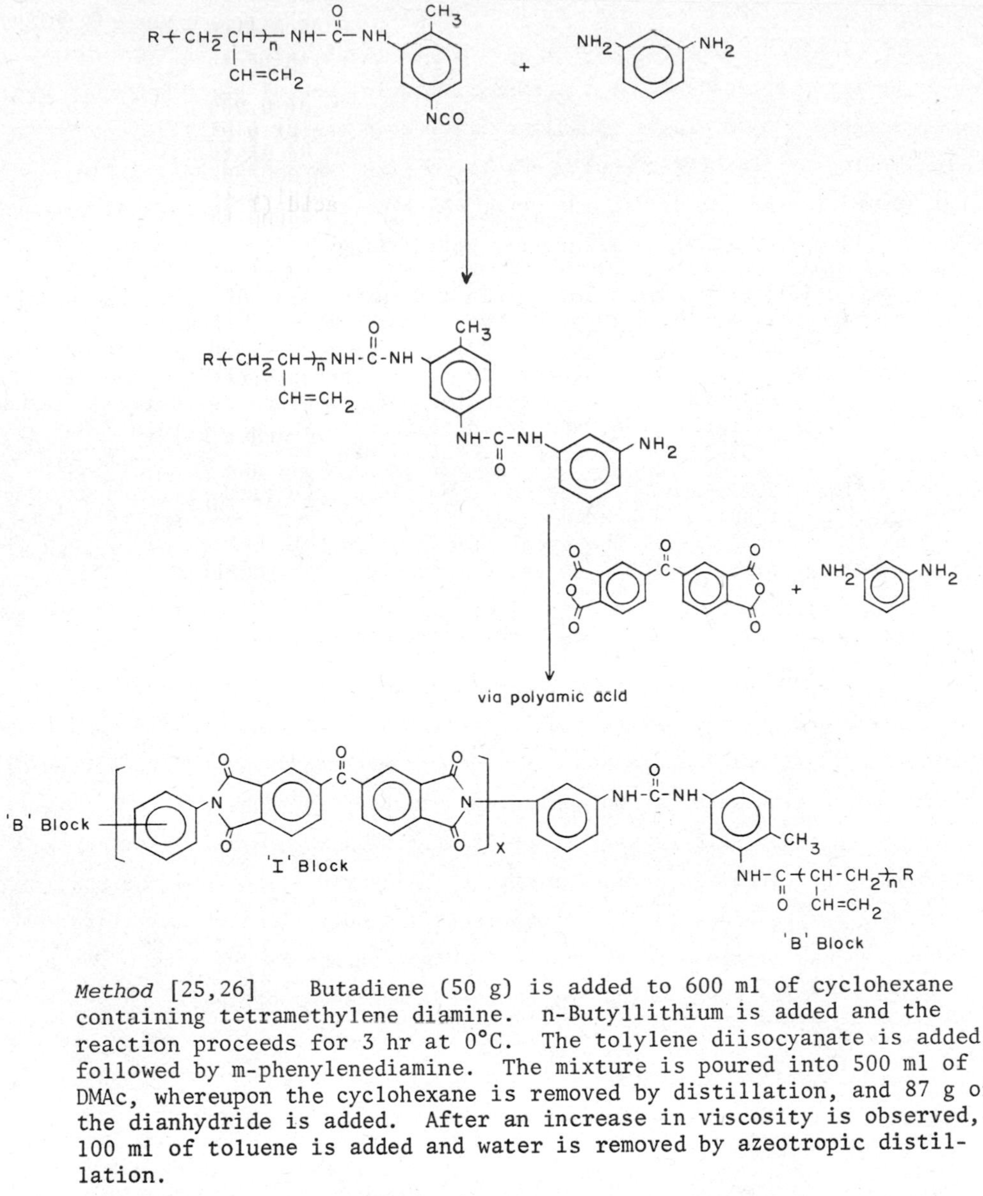

$R{+}CH_2{-}CH{+}_n NH{-}\overset{O}{\overset{\|}{C}}{-}NH$ (with $CH=CH_2$ side chain; tolyl ring bearing CH_3 and NCO) + NH_2–C_6H_4–NH_2

↓

$R{+}CH_2{-}CH{+}_n NH{-}\overset{O}{\overset{\|}{C}}{-}NH$ (with $CH=CH_2$ side chain; tolyl ring bearing CH_3) $NH{-}\overset{O}{\overset{\|}{C}}{-}NH$–$C_6H_4$–$NH_2$

↓ (benzophenone tetracarboxylic dianhydride + NH_2–C_6H_4–NH_2)

via polyamic acid

'B' Block — ['I' Block]$_x$ — $NH{-}\overset{O}{\overset{\|}{C}}{-}NH$ (CH_3) $NH{-}\overset{O}{\overset{\|}{C}}{+}CH{-}CH_2{+}_n R$ ($CH=CH_2$) 'B' Block

Method [25,26] Butadiene (50 g) is added to 600 ml of cyclohexane containing tetramethylene diamine. n-Butyllithium is added and the reaction proceeds for 3 hr at 0°C. The tolylene diisocyanate is added, followed by m-phenylenediamine. The mixture is poured into 500 ml of DMAc, whereupon the cyclohexane is removed by distillation, and 87 g of the dianhydride is added. After an increase in viscosity is observed, 100 ml of toluene is added and water is removed by azeotropic distillation.

5.1.7.4 Polyimides with Poly(ether sulfone) Blocks Not only processing but also impact strength can be improved by inclusion of flexible links such as ether and sulfone in a backbone [5]. The effect on this property is regulated by the size of the sulfone ether block included. The size of the arylether sulfone block is regulated by the amount of aryldiol which is incorporated in the first step. In the actual example of the synthetic method given below n = 0 in this block since none of the diphenolic compound is used.

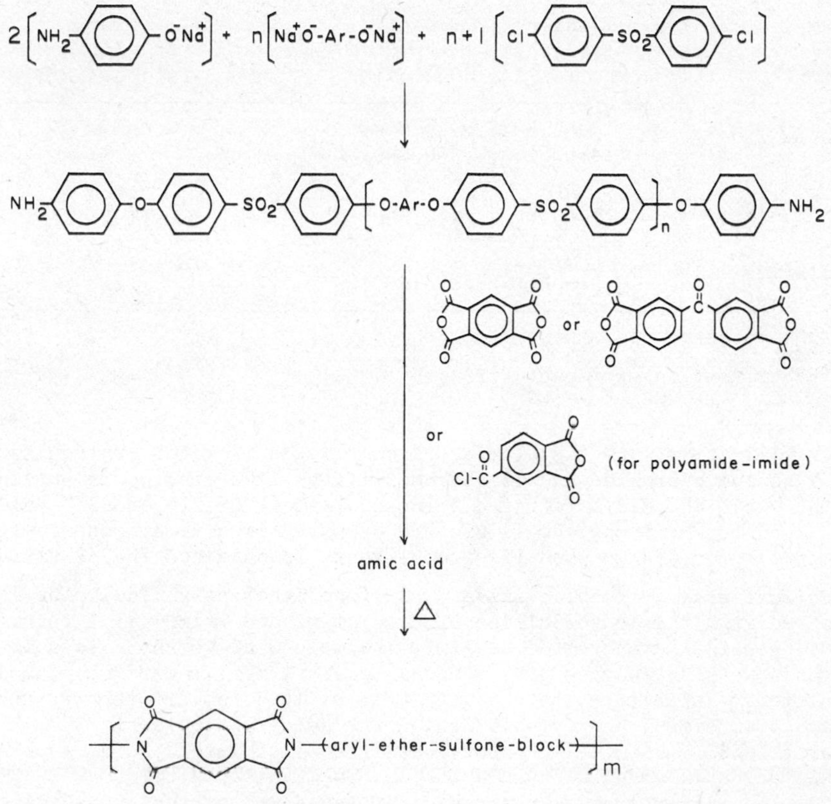

$2[NH_2-C_6H_4-O^-Na^+] + n[Na^+O^--Ar-O^-Na^+] + n+1[Cl-C_6H_4-SO_2-C_6H_4-Cl]$

↓

$NH_2-C_6H_4-O-C_6H_4-SO_2-C_6H_4-[O-Ar-O-C_6H_4-SO_2-C_6H_4]_n-O-C_6H_4-NH_2$

or

(for polyamide-imide)

↓

amic acid

↓ Δ

aryl-ether-sulfone-block

It is obvious at this point that polyamides and polyamide imides can also be synthesized by merely using the various phthaloyl chlorides or trimellitic anhydride acid chloride, respectively. The poly(amide imide) synthesis is given below since it is the most tractible of the series and has a Tg as high as one known polyimide.

The amorphous, film-forming polymers display good thermal and mechanical properties as summarized in Table 5.2.

Method [5] *Sulfone ether diamine* To a 12-liter, three-necked flask equipped with a Dean-Stark trap, condenser, nitrogen tube, and thermometer are added 1500 g of p-aminophenol, 2.8 liters of DMSO, and 3.0 liters of toluene. The solution is saturated with nitrogen and heated to 50°C, and 1082 g of 50.2% sodium hydroxide is added. After dehydration, toluene is removed by distillation and the mixture cooled to 100°C. Dichlorodiphenylsulfone, 1880 g, is added in portions to keep the temperature below 160°C; the mixture is then heated to 165°C for 2 hr. After the solution is cooled to 25°C, sodium chloride is removed

Table 5.2 Thermal and Mechanical Properties of Film-Forming Polymers

Acid type	Tg (°C)	Tensile strength, MPa (psi)	Tensile modulus, GPa (psi)	% elongation	Impact, J/cm^3 ($ft\text{-}lb./in^3$)	Film embrittlement at 220°C (hr)
Benzophenone	260	96 (14,000)	3.0 (440,000)	8	5.7 (70)	>5000[a]
Pyromellitic	320	66 (9,600)	1.9 (270,000)	15	16.2 (197)	—
Trimellitic	270	86 (12,500)	2.6 (370,000)	15	14.0 (171)	>>5000[b]

[a]20% retention of impact strength.

[b]57% retention of impact strength.

by filtration, and the diamine is coagulated in water containing about 2% sodium hydroxide and 1% sodium sulfite. The diamine is obtained in 93% yield and had a titrated amine equivalent of 216 and 189-191°C.

The same technique is used to prepare oligomer diamines; e.g., n = 1, Ar = 4,4'-diphenylisopropylidene, is obtained in 95% yield.

Sulfone ether polyamide imide Sulfone ether polyamide imide is prepared from the above sulfone ether diamine and trimellityl chloride by using phthalic anhydride to limit the molecular weight. To a 12-liter resin kettle equipped with a mechanical stirrer, thermometer, and nitrogen inlet tube are charged 900 g of the diamine prepared above and 4 liters of anhydrous DMAc and the solution cooled to 0°C. Thereafter, 435.1 g of trimellitoyl chloride mixed with 2.30 g of phthalic anhydride is added at such a rate as to maintain a temperature of 0-10°C. After 1 hr at this temperature, 420 g of anhydrous triethylamine is added, followed gradually by 3200 ml of DMAc after 30 min, and the mixture is reacted an additional 3 hr. Then 210 g of acetic anhydride are added and, after 18 hr at room temperature, the polymer is coagulated in water, filtered, washed with acetone, and dried in vacuo to afford a quantitative yield.

Another approach to including sulfone ether blocks is by end capping such blocks with maleimide functions, then polymerizing the oligomers by an addition mechanism. Again the volatile curing byproduct is eliminated [39]. The prepolymer can be transformed to a B stage at 180-220°C and this curing process is facilitated by a peroxide additive. For example, 1.5% dicumyl peroxide produces a full cure at 200°C in 5 min.

Tg's of cured materials (being inversely proportional to the length of the sulfone ether block) run from 185° to >330°C. Shear moduli are all 10-12 pPa ($1.0\text{-}1.2 \times 10^{-10}$ dyne/cm^2).

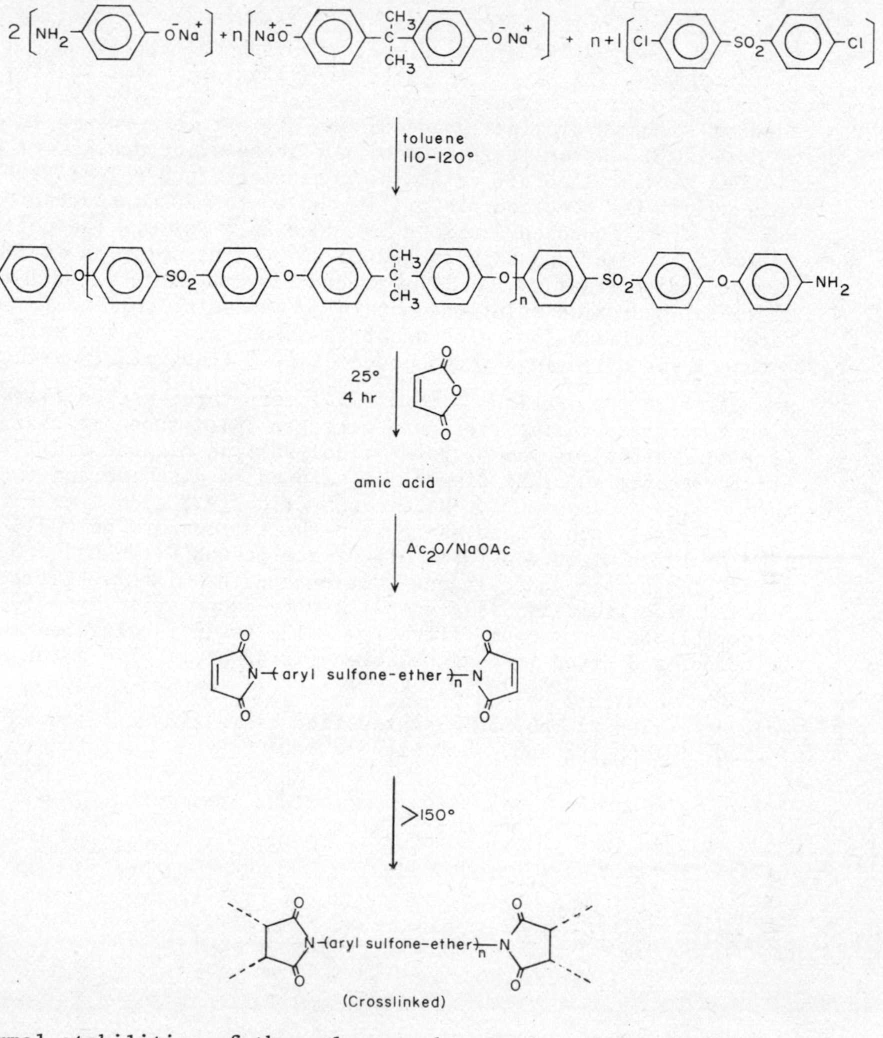

$$2\left[NH_2-C_6H_4-O^-Na^+\right] + n\left[Na^+\,^-O-C_6H_4-C(CH_3)_2-C_6H_4-O^-Na^+\right] + n+1\left[Cl-C_6H_4-SO_2-C_6H_4-Cl\right]$$

toluene
110-120°

$$NH_2-C_6H_4-O-\left[C_6H_4-SO_2-C_6H_4-O-C_6H_4-C(CH_3)_2-C_6H_4-O\right]_n-C_6H_4-SO_2-C_6H_4-O-C_6H_4-NH_2$$

25°
4 hr

amic acid

$Ac_2O/NaOAc$

N-(aryl sulfone-ether)$_n$-N

>150°

N-(aryl sulfone-ether)$_n$-N

(Crosslinked)

Thermal stabilities of the polymers show losses of 5.5-12.5% at 450°C, which are also inversely proportional to sulfone content. This correlates with the fact that the pure polyarylether sulfone (Union Carbide 1700) shows only a 2% weight loss at 450°C to 7% at 500°C.

These materials are proposed for use as adhesives and coatings and in composites. Lap shear adhesive specimens were prepared by applying a solution coating to aluminum panels and drying and bonding the coated panels at 300°C (595°F) under pressure for 1 hr. Failure data were 2.8 MPa (1865 psi) at room temperature and a still quite good 10.2 MPa (1480 psi) at 260°C (500°F).

Method [39] *Sulfone ether diamine oligomer (n = 2)* Into a 5-liter flask are charged 342.4 g (1.5 mol) of bisphenol A, 165.33 g (1.515 mol) p-aminophenol, 1.7 liters of dimethyl sulfoxide, and 1.1 liter of toluene. After a nitrogen purge, 368.16 g (4.515 mol) of a 49.06% solution of sodium hydroxide is added, and the pot temperature is brought to 110-120°C. Water is removed by a toluene azeotrope, after which the toluene is distilled off until the pot temperature reaches 150°C. At this point, the reaction mixture is cooled to 120°C, and 646.5 g (2.25 mol) of dichlorodiphenylsulfone is added as a solid. The pot is heated to 160°C for 2 hr and then cooled to 25°C. The solution is filtered to remove sodium chloride, and the product is coagulated in a blendor from 2% sodium hydroxide solution containing 1% sodium sulfite. The precipitated diamine is washed with a hot 1% solution of sodium sulfite and methanol and dried in a vacuum oven at 80°C (equivalent weight 710).

Bismaleimide preparation To a 0.5-liter, three-necked flask equipped with a stirrer, thermometer, and nitrogen inlet tube are charged 71 g (0.1 equivalent) of the oligomeric polysulfone diamine and 200 ml of dimethylacetamide. The diamine is allowed to dissolve and the solution then cooled to about 0°C. Maleic anhydride (9.8 g, 0.1 mol) is added as a solid at such a rate as to keep the temperature below 15°C. After 1 hr at 10-25°C, 10.4 g (0.1 mol) of acetic anhydride and 1.0 g of triethylamine are added in one portion and the mixture stirred for 4 hr. Bismaleimide isolation is accomplished by coagulation from 10 parts water followed by vacuum filtration. The resin is slurried in water, filtered, and dried in a vacuum oven for 48 hr to give a tan powder, 99.8 g.

Polymerization The above synthesized bismaleimide is cured at 275°C for 30 min to give a product with Tg 180°C.

5.1.7.5 Polyimides with Bridged Bipyridyl Functions Polyimides which are soluble in dichloroacetic acid, m-cresol, H_2SO_4, and DMAc are

NH_2 X N N NH_2 + O O O Ar O O O

25°

amic acid

190-200°
vacuum
2 hr

N X N N Ar N O O O O n

Ar is:

or

X is: nil, NH, NCH_3, $NCOCH_3$, S, SO, SO_2, C_2H_4

synthesized to include bipyridyl units [36]. This is very similar to the Kurita and Williams work discussed in Section 5.1.7.1 (p. 112) on amide imides [37]; therefore, this discussion will limit itself to the unique applicability to the polyimide backbone. The chemistry is common to polyimide formation, i.e., condensation of a diamine with a tetraacid dianhydride through an amic acid intermediate. Inherent viscosities of the polyimides in DMAc were as high as 2.51 dl/g but more commonly were 0.2-0.5 dl/g.

The pyridyl systems are less stable than their carbocyclic analogs, 410° and 495-505°C, respectively, for a 10% weight loss in air.

Method [36] *Polymerization* The dianhydride (1 mmol) is added to a solution of 1 mmol of the diamine in 4 ml of DMSO under nitrogen with stirring at 25°C. After 1.5 hr 3 ml of DMSO is added to the viscous solution, and the mixture is poured into 500 ml of chloroform to yield 0.415 g (95%) of white to pale yellow fibrous polyamic acid.

Imidization The above polyamic acid is heated at 200°C under vacuum for 2 hr to result in a light yellow solid.

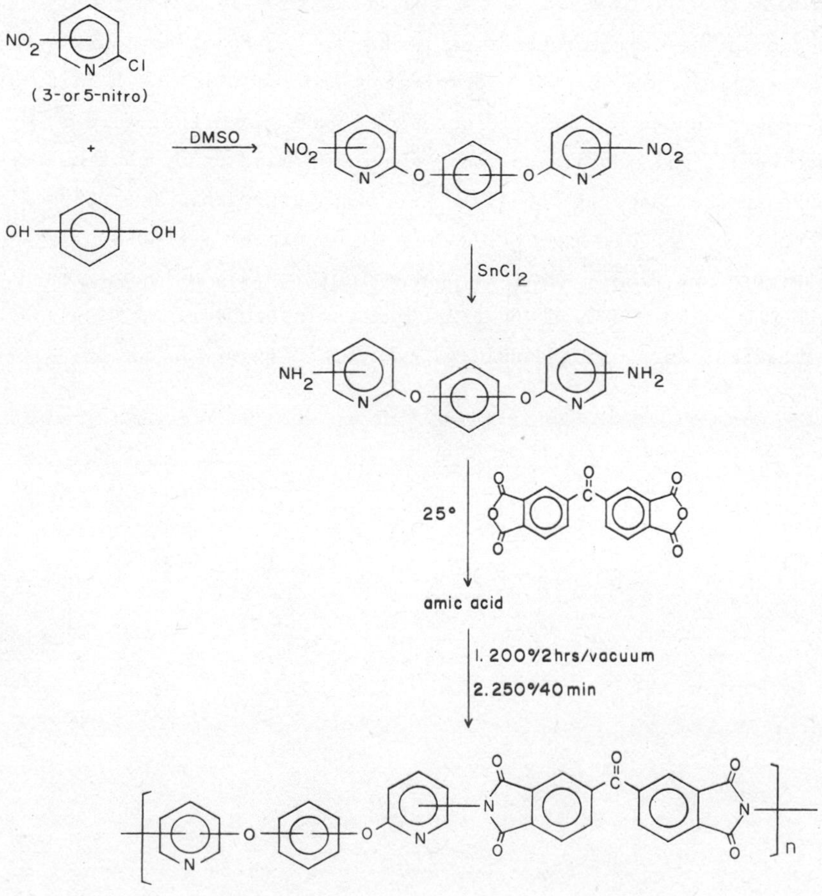

(3- or 5-nitro)

DMSO

$SnCl_2$

25°

amic acid

1. 200°/2 hrs/vacuum

2. 250°/40 min

The incorporation of both pyridyl and phenylether linkages into polyimides is possible and the resulting terpolymer exhibits good solubility [38].

Comparing the effect of catenation on the pyridine and phenylene rings shows that para orientation across the pyridyl ring allow a higher molecular weight than the ortho. In contrast, the phenylenedioxy orientation had little or no effect. Yields were commonly as high as 99% in reaction times of 3-24 hr; and products with inherent viscosities of up to 1.30 dl/g were obtained.

The polymers are soluble in DMAc, NMP, m-cresol, and dichloroacetic acid with Tg's at 230-310°C. Ranging from 390 to 440°C in this case, TGA stabilities in air (10% loss) are lower than the usual polyimides.

> *Method* [38] The process is identical to that reported above [36] involving stirring the reactants in DMSO at room temperature and precipitation into chloroform.

5.1-8 Polysulfonimides

The low molecular weights are probably due to the fact that, prior to extensive propagation, the polymer precipitates from solution as a salt. This, of course, is caused by the acid acceptor, K_2CO_3, and therefore seems to be unavoidable. Yields are commonly above 80%, and decomposition temperatures are in the range of 350-410°C by TGA in nitrogen. The polymeric salt is much better (485°C), however, as one might expect for an ionic structure. The final product displays rather poor solubility (soluble only in hot H_2SO_4, $HCONH_2$, and DMSO with 5% LiCl) and is insoluble in strong bases, the latter fact surprising considering the acidity of the imide in this situation.

$$H_2N-C_6H_4-SO_2NH_2 \xrightarrow[\text{CuCl, HOAc}]{\text{Diazotize, } SO_2} ClO_2S-C_6H_4-SO_2NH_2 \xrightarrow[10\%\ \text{LiCl},\ K_2CO_3,\ \Delta]{\text{acetone or pyridine}} \left[SO_2-\overset{K^+}{\bar{N}}-SO_2-C_6H_4 \right]_n$$

$$\xrightarrow[\text{HCl gas}]{HCONH_2} \left[SO_2-NHSO_2-C_6H_4 \right]_n$$

Several attempts were made to improve the solubility and thereby the molecular weights attainable. One was using aryldisulfonamides (or N-methyl derivatives) with aryldisulfonyl chlorides including phenylene, tolylene, and diphenylene ether systems. The N-methylsulfonamides, being secondary amines, precluded the formation of the intermediate salt, thereby improving solubility. Although these new groups afforded higher inherent viscosities (0.2 dl/g dl/g up from 0.1 dl/g) and better solubilities in hot DMSO, m-cresol, diphenyl ether, and hexamethylphosphoramide, their stabilities were on the lower end of the range for this system (350-370°C).

$$[CH_3-NH-SO_2-C_6H_4]_2O \xrightarrow[CH_3OH]{\overset{+}{Na}\overset{-}{O}Me} [CH_3-\overset{Na^+}{N^-}-SO_2-C_6H_4]_2O$$

$$\downarrow \; [Cl-SO_2-C_6H_4]_2O, \; 120°\text{, tetramethylene sulfone}$$

$$\left[-SO_2-\overset{CH_3}{\overset{|}{N}}-SO_2-C_6H_4-O-C_6H_4-\right]_n$$

Method [45] p-Chlorosulfonylbenzenesulfonamide (2.75 g, 10.7 mmol) is dissolved in 40 ml of solvent, with 4.13 g (3.0 mmol) of K_2CO_3, and the mixture is heated to reflux for 12 hr. The polymer is isolated by filtration, washed well with water, and dried at 30°C in vacuo to yield a white, free-flowing powder as the potassium salt.

The polymeric salt (0.5 g) is dissolved in 20 ml of $HCONH_2$ at 180°C, cooled to 100°C, and HCl gas is bubbled in for 2 min. The mixture is poured into water to precipitate the polymer, which is isolated and dried.

5.2 Polyimidines

5.2-1 Poly(pyromellitimidines)

A new class of polymers akin to a phenylated imide is prepared by condensing a fused lactone with diamines [9]. Pendant phenyl groups are present which certainly aid the solubility. The monomer synthesis is made possible by the peculiar rearrangement in the Friedel-Crafts reaction of benzene with pyromellitoyl chloride to give a bislactone rather than the expected tetrabenzoyl benzene. The yield of pure monomer by the route shown is quite low (6%), and only one of the two possible (cis/trans) materials is isolated. A new synthetic pathway which affords yields of 30-40% of each from pyromellitic dianhydride has been found for both the cis and trans monomers [16].

HO_2C, CO_2H (pyromellitic acid) $\xrightarrow[C_6H_3Cl_3]{PCl_5}$ $ClOC$, $COCl$ (pyromellitoyl chloride) $\xrightarrow[AlCl_3]{\phi H}$ tetraphenylpyromellitide (cis or trans) $\xrightarrow[2.\ \Delta]{1.\ NH_2\text{-}R\text{-}NH_2,\ \Delta}$ polyimidine $[-R-N\ldots N-]_n$

R is: $-(CH_2)_6-$, p-phenylene, 4,4'-diphenyl ether ($-C_6H_4-O-C_6H_4-$)

The polymerizations are conducted either in refluxing biphenyl (250°C) or neat in a sealed tube for up to 3 days. Yields are poor to fair (10-57%) of dark products (brown-black) with inherent viscosities up to 0.15 dl/g corresponding to a number average molecular weight of 16,000. The cyclization in the last step is not complete until a postcure is performed which lowers the solubility of the product. Prior to the postcure polyimidines are very soluble in chloroform and DMF.

Thermal stabilities show a classical dependence on aromatic content. The TGA stabilities (10% weight loss) in air and nitrogen, respectively, are hexamethylene polymer, 300/340°C; biphenylether analog, 460/510°C; and the phenylene analog, 420/450°C.

Method [9] *Poly(hexamethylene-3,3,5,5-tetraphenylpyromellitimidine) by solution polymerization* A 1-g portion (0.002 mol) of 3,3,5,5-tetraphenylpyromellitide [9,16], 0.25 g (0.002 mol) of 1,6-diaminohexane, and 5.0 g of biphenyl are heated to melt under vacuum and flushed with nitrogen. The mixture is then heated slowly to reflux (250°C) with stirring, and after 1 day the biphenyl is removed by distillation. The black residue is washed with 95% ethanol, dissolved in $CHCl_3$, precipitated into ethyl ether, and vacuum-dried at room temperature.

Poly(phenylene-3,3,5,5-tetraphenylpyromellitimidine) by sealed tube reaction A 10-ml tube containing 2.0 g (0.004 mol) of 3,3,5,5-tetraphenylpyromellitide and 0.44 g (0.004 mol) of p-phenylene diamine is flushed with nitrogen and sealed under vacuum. The tube is heated for 1 day at 150°C, and the temperature is increased at a rate of 50°/day to 250°C and held there for 1 day. The top of the tube is cooled to condense moisture there; then the entire tube is cooled. The navy blue polymer is purified by dissolving it in chloroform and precipitating it in petroleum ether to yield 1.0 g (38%) with inherent viscosity 0.07 dl/g in DMF (0.25 g/100 ml).

5.2-2 Poly(dithiopyromellitimidines)

As it turns out, a much better synthesis than the above, i.e., more rapid and with higher yield and complete cyclization, is available through the thio analog of tetraphenylpyromellitide [8]. The reactions are complete in 6 hr in refluxing carbon tetrachloride and readily followed by dissipation of the deep red-orange monomer color to a pale yellow. The yield of white polymer is quite good (80%), as are molecular weights (up to approximately 11,000) with inherent viscosity of 0.89 dl/g.

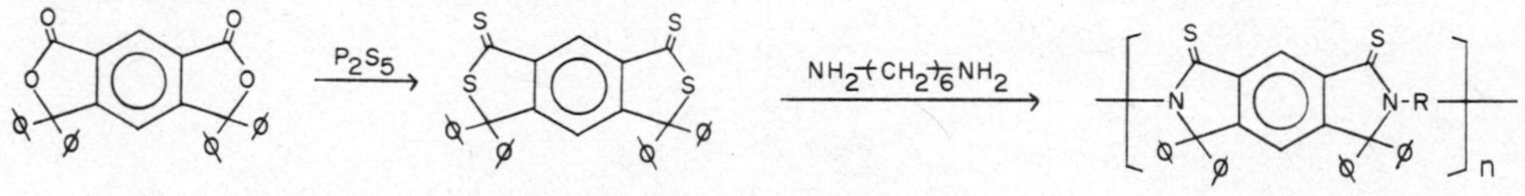

$\xrightarrow{P_2S_5}$ $\xrightarrow{NH_2(CH_2)_6NH_2}$ N-R]$_n$

The polymers are very soluble in common solvents, and slightly more thermally stable than the oxo system (10% loss at 350 and 380°C in air and nitrogen, respectively, for the hexamethylene system and up to 650°C for aromatic versions), probably owing to thermooxidative crosslinking through sulfur groups.

Method [8] *Preparation of tetraphenyltetrathiopyromellitide* A solution of 3.0 g (0.006 mol) of 3,3,5,5-tetraphenylpyromellitide [9], mp 280-282°C, and 0.89 g (0.012 mol) of phosphorus pentasulfide in 50 ml of xylene is heated under reflux for about 36 hr. The red solution is filtered hot and steam distilled until a residue precipitates. The red residue is collected by suction filtration and then dissolved in hot chloroform. Ethanol is added to the boiling chloroform solution until it turns cloudy, and it is then cooled to room temperature. The red powder, which is collected by suction filtration, melts at 338-340°C, and is received in a 90% yield.

Polymerization A solution of 0.500 g (0.0009 mol) of 3,3,5,5-tetraphenyltetrathiopyromellitide and 0.104 g (0.009 mol) of 1,6-hexanediamine in freshly distilled carbon tetrachloride is heated to reflux. The red solution gradually decolorizes in 6 hr; whereupon the solvent is removed with a rotary evaporator. The residue is dissolved in hot chloroform and purified with a small amount of decolorizing charcoal. Ethanol is slowly added to the clear filtrate to the cloud point. The cloudy solution was then cooled to room temperature and the white polymer collected from suction filtration to afford an 80% yield.

5.3 Polybenzodipyrolediones

A phthalide analog, a dibenzylidenebenzodifurandione, has proven to be more reactive in polymerizations with diamines than the phenylated pyromellitide [33,59]. These monomers (both cis and trans isomers) will react quantitatively in a week at room temperature or in a few hours at 80°C with various amines to provide polybenzodipyrolediones. Solvents used are o-cresol, DMSO, NMP, and o-phenylphenol with or without a boric acid catalyst. The process involves two steps reminiscent of polyimide preparations: first, a ring opening reaction to a polyamide; and second, a dehydration ring closure at 240°C. However, in some cases the linear intermediate is cyclized during the polymerization, especially at high reaction temperatures when a catalyst is employed.

R is: $-(CH_2)_6-$, $-CH_2-C_6H_4-CH_2-$ or $-C_6H_4-X-C_6H_4-$ where X is CH_2 or O

The polymers are soluble in hot m-cresol, NMP, hot pyridine, DMSO, H_2SO_4, or nitrobenzene and display inherent viscosities up to 1.0 dl/g. Tough, transparent, yellow films are cast from NMP. Thermal stabilities (by TGA, 10% weight loss) for the aliphatic-containing backbones are

400-430°C in nitrogen, and for the aromatic functions 460-500°C. The hexamethylene derivative shows a Tg at 150°C; but the aromatic systems give no transitions below their decomposition temperatures.

Method *Monomer synthesis* [59] In a flask is placed 39.3 g (0.17 mol) of pyromellitic dianhydride, 54.5 g (0.4 mol) of phenylacetic acid, and 2.9 g of freshly fused sodium acetate. The mixture is heated to 240°C in an oil bath to give a slurry, which is then maintained at 240-250°C for 10 hr. The solid formed is subjected to fractional recrystallization with dimethylformamide (DMF) and separated into a portion with lower solubility and one with higher solubility. The product having higher solubility (the cis isomer) is recrystallized three times more from DMF to give 13.4 g (20%) of yellow needles, mp 327°C (by DTA). The trans isomer of lower solubility is also purified by two recrystallizations from a large amount of DMF to afford 9.0 g (14%) of yellow leaflets, mp 431°C (by DTA).

Polymerization [33] A mixture of bis(4-aminophenyl)ether (0.500 g, 2.5 mmol), the cis monomeric benzodifurandione (0.916 g, 2.5 mmol), boric acid (5 mg), and solid o-phenylphenol (5 g, as the polymerization medium) is heated at 270°C under nitrogen to give a clear solution. The reaction is continued at that temperature with stirring for 24 hr. The polymer is isolated by pouring the viscous solution into methanol, washing with hot methanol and drying at 200°C, to yield 1.32 g (99%); inherent viscosity, 0.58.

5.4 Recent Developments

5.4-1 Polyimides

Polyimides are like aromatic polyamides in the sense that the commercial success of these material has maintained a steady stream of research in this field. These studies will be delineated below. Perhaps one of the least descriptive titles in polymer science has been used with polyimides: PMR, meaning polymerization of monomeric reactants.

One interesting finding is from the study of model compounds which shows that major impurities are chemically bound solvent residues (DMF or DMAc) [76]. This means that one can expect only 50-85% cyclization in polyimides, a factor which, owing to degradation mechanism effects, is important to stability.

Polyimides have resulted from the reaction of diamines with bismethylolimides in DMAc with 1% water [80]:

$HOCH_2-N$ Ar $N-CH_2OH$

The bismethylol monomers also condensed with dinitriles in PPA or sulfuric acid to give poly(amide-imides) [79].

A soluble and thermally stable polymer was received also from the reaction of either BTDA or DMDA with a cycloaliphatic diamine, 5- or 6-amino-1-(4'-aminophenyl)-1,3,3-trimethylindane [62]:

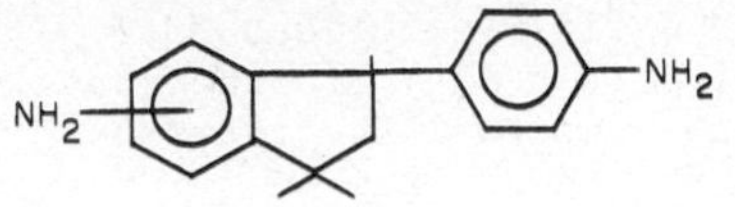

An electrochemical (anode) condensation yields (quantitatively) a polyimide from 4-amino-phthalic acid [83]. The rigid backbone gives a dark amber, brittle film with very good thermal stability (500°C in air, 540°C in nitrogen).

Polyimides from addition reactions on exocyclic vinyl [75] or N-allylic [78] groups provide thermosetting resins. The latter type uses N-allyl-phthalimide carboxylates, while the former uses a bisitaconimide:

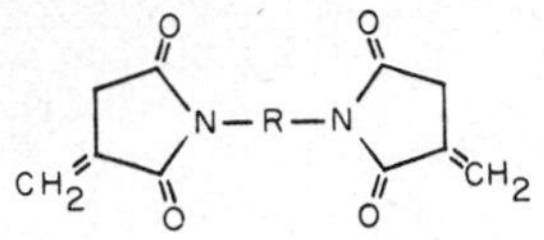

Several polyimides and poly(aroylene bisbenzimidazoles) were produced from the following dianhydrides [65]:

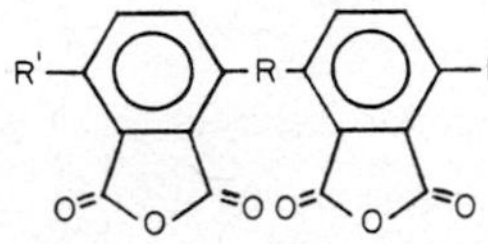

A study which incorporated multiple and single bonds between phenylene rings in analogous polyimide backbones showed their thermal stability to be triple > double > single [77].

Other studies on polyimides encompass DTA data of imidization [86], solubilities of novel polyimides [89], effluent analysis during cyclization [90], and photoreactive precursors [84]. The work on solubility showed this property to be improved by the use of bridged dianhydrides and o,p'-substituted diamines.

The effects of multifunctional amines as crosslinking agents on the properties of polyimides was investigated [71,72]. Of course, tri- and tetraamines increase the Tg's and softening temperatures, as well as the thermooxidative stability of BTDA systems. However, these amines are detrimental to the stability of PMDA-derived polyimides.

Structure-property relationships induced by isomer diamines (diamino-diphenylmethanes, diaminobenzophenones [64], and 2,5- or 2,7-diaminofluorene or 2,7-diaminofluorone [63] have been elucidated.

The use of nadimide end groups as crosslinking sites was investigated [87,88]. These oligomeric polyimides were, of course, quite soluble (up to 40% concentration) in DMF or NMP and were curable at 285°C. The prepolymer showed itself to be a good adhesive to titanium and an impregnant for graphite fiber.

A very recent publication discusses structure-property relationships of various aromatic polyimides from PMDA [91]. The viscosity, density, and thermal stability decrease with increasing meta catenation in the backbone; these are all phenomena which are expected. Another property investigated is water permeation through Kapton (duPont) [85]. The concentration of absorbed water is proportional to the relative humidity, and not to temperature or thickness of the sample. Further, the maximum water absorption is one water molecule per repeating unit.

5.4-2 Polyimidines

Several new polyimidines have been reported. All four analogs of tetraphenylpyromellitide [73], cis/trans and oxo/thio were polymerized with several amines, and the resulting polyimidines were compared [68]:

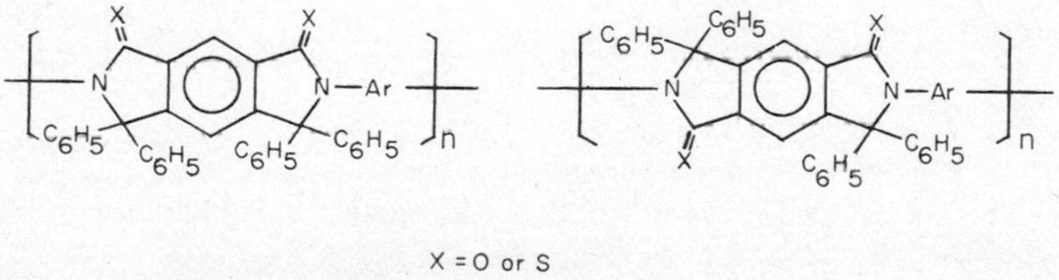

X = O or S

The thio monomers were found to be more reactive and soluble than the oxo; and they also provided more thermally stable polymers. Cis isomers were more soluble than trans, but less reactive. Thermal stabilities were excellent, ranging up to 560°C in air and 650°C in nitrogen for one aromatic backbone. The polymers were soluble in chloroform, and other solvents and brittle films could be cast from solution or melt-pressed.

Another series of polyimidines was derived from the bisphthalide monomer of BTDA [66,67,70,82]:

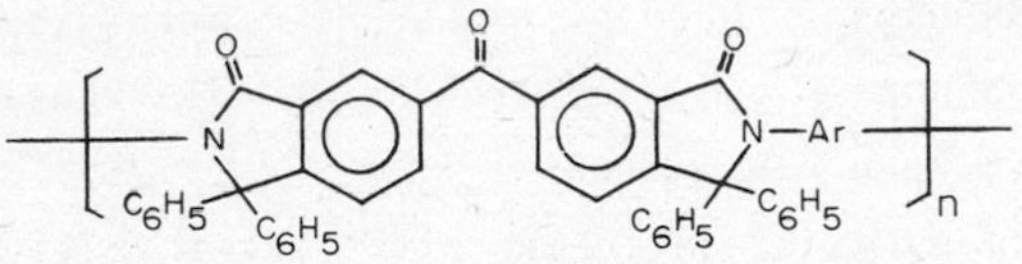

These polymers are light-colored, soluble in common organic solvents, and stable to 525°C in air and 540°C in nitrogen.

Two other polyimidines with rigid backbones have been prepared as candidates for liquid crystals [69,74,81]:

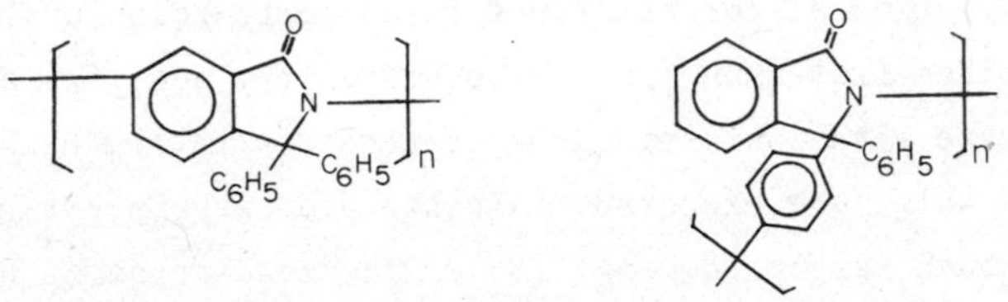

Inherent viscosities ranged up to 0.68 dl/g, but thermal stabilities were less than previous polyimidines. One interesting phenomenon was discovered during the syntheses of these rigid-backbone polyimidines. Many products from earlier work were dark owing to the extreme conditions necessary for reaction. (This was not so for the thio analogs, however.) Therefore the water byproduct was removed from the reaction by maintenance of a vacuum during the polymerization. Even with the subsequent partial loss of diamine monomer, the products were higher in molecular weight, stability, and yield and much lighter colored than when water was retained in the sealed tube.

References

1. L. M. Alberino, W. J. Farrissey, and J. S. Rose, U.S. Patent 3,708,458 (1973).
2. W. M. Alvino and L. E. Edleman, *J. Appl. Polym. Sci.*, *19*(11), 2961 (1975).
3. C. Arnold, Jr., Report SC-M-720559, Sept. 1973, Sandia Corp., Albuquerque, N.M.
4. H. Beyer, *Chem. Ber.*, *82*, 143 (1949).
5. G. L. Brode, G. T. Kwiatkowski, and J. H. Kawakami, *Polym. Preprints*, *15*(1), 761 (1974).
6. R. W. Campbell and H. W. Hill, *Macromolecules*, *8*(6), 706 (1975).
7. P. S. Carleton, W. J. Farrisey, and J. S. Rose, *J. Appl. Polym. Sci.*, *16*, 2983 (1972).
8. P. E. Cassidy and F. Lee, *J. Polym. Sci.: Polym. Chem. Ed.*, *14*, 1519 (1976).

9. P. E. Cassidy and A. Syrinek, *J. Polym. Sci.: Polym. Chem. Ed.*, *14*, 1485 (1976).
10. J. V. Crivello, *J. Polym. Sci.: Polym. Chem. Ed.*, *11*, 1185 (1973).
11. W. H. Daly and H. J. Holle, *J. Polym. Sci.*, *B*, *10*(7), 519 (1972).
12. F. Darmory, *Org. Coat. Plast. Preprints*, *34*(1), 181 (1974).
13. R. A. Dine-Hart and W. W. Wright [C.A., *73*, 56669 (1970)], *U.S. Gov't. Res. Develop. Report*, *70*(7), 109 (1970).
14. N. Dokoshi, S. Tohyama, S. Fujita, M. Kurihara, and N. Yoda, *J. Polym. Sci.*, *A-1*, *8*(8), 2197 (1970).
15. M. Dorr and M. Levy, *J. Polym. Sci.: Polym. Chem. Ed.*, *13*, 171 (1975).
16. N. C. Fawcett, P. E. Cassidy, and J.-C. Lin, *J. Org. Chem.*, *42*, 2929 (1977).
17. J. K. Gillham and H. C. Gillham, *Polym. Eng. Sci.*, *13*(6), 447 (1973).
18. D. J. Goldwasser and E. P. Otocka, *Org. Coat. Plast. Preprints*, *35*(2), 198 (1975).
19. B. S. Green, M. Lahav, and G. M. J. Schmidt, *J. Chem. Soc.*, *13*, 1552 (1971).
20. O. Grummitt, *Organic Synthesis Coll.*, Vol. 3 (E. C. Horning, Ed.), John Wiley and Sons, New York, 1955, pp. 807-808.
21. F. W. Harris, W. A. Field, and L. H. Lanier, *J. Polym. Lett.*, *13*(5), 283 (1975).
22. F. W. Harris, W. A. Field, and L. H. Lanier, *Polym. Preprints*, *17*(2), 353 (1976).
23. F. W. Harris and S. O. Norris, *J. Polym. Sci.: Polym. Chem. Ed.*, *11*, 2143 (1973).
24. F. Hayano and H. Komoto, *J. Polym. Sci.*, *A-1*, *10*(4), 1263 (1972).
25. W. L. Hergenrother and R. J. Ambrose, *J. Polym. Sci.*, *B*, *10*, 679 (1972).
26. W. L. Hergenrother and R. J. Ambrose, *J. Polym. Sci.: Polym. Lett. Ed.*, *12*(6), 343 (1974).
27. S. S. Hirsch, *J. Polym. Sci.*, *A-1*, *7*(1), 15 (1969).
28. S. S. Hirsch and S. L. Kaplan, *Org. Coat. Plast. Preprints*, *34*(1), 162 (1974).
29. F. F. Holub and J. T. Kobakc, German Patent 1,922,339, January, 1970 [C.A., *72*, 79925 (1970)].
30. Y. Imai and H. Koga, *J. Polym. Sci.: Polym. Chem. Ed.*, *11*, 289 (1973).
31. Y. Imai and H. Koga, *J. Polym. Sci.: Polym. Chem. Ed.*, *11*(10), 2623 (1973).
32. Y. Imai and K. Kojima, *J. Polym. Sci.*, *A-1*, *10*, 2091 (1972).
33. Y. Imai, M. Ueda, and T. Takahashi, *J. Polym. Sci.: Polym. Chem. Ed.*, *14*, 2391 (1976).
34. N. Kardush and M. Stevens, *J. Polym. Sci.*, *A-1*, *10*(4), 1093 (1972).
35. M. Kurihara and N. Yoda, *Bull. Chem. Soc. Japan*, *40*, 2429 (1967).

36. K. Kurita and R. L. Williams, *J. Polym. Sci.: Polym. Chem. Ed.*, *11*, 3125 (1973).

37. K. Kurita and R. L. Williams, *J. Polym. Sci.: Polym. Chem. Ed.*, *11*, 3151 (1973).

38. K. Kurita and R. L. Williams, *J. Polym. Sci.: Polym. Chem. Ed.*, *12*, 1809 (1974).

39. G. T. Kwiatkowski, L. M. Robeson, G. L. Brode, and A. W. Bedwin, *J. Polym. Sci.: Polym. Chem. Ed.*, *13*, 961 (1975).

40. L. T. C. Lee, E. M. Pearce, and S. S. Hirsch, *J. Polym. Sci.*, *A-1*, *9*, 3169 (1971).

41. D. G. Loncrini and J. M. Witzel, *J. Polym. Sci.*, *A-1*, *7*, 2185 (1969).

42. J. W. Lynn and R. L. Roberts, *J. Org. Chem.*, *26*, 4303 (1961).

43. H. I. X. Mager and W. Berends, *Rec. Trav. Chim.*, *76*, 28 (1957).

44. M. A. J. Mallet and F. Darmory, *Org. Coat. Plas. Preprints*, *34*(1), 173 (1974).

45. H. Meerwein, G. Dittmar, R. Goollner, K. Hafner, F. Mensch, and O. Steinfort, *Chem. Ber.*, *90*, 841 (1957).

46. R. A. Meyers, *J. Polym. Sci.*, *A-1*, *7*, 2757 (1969).

47. Y. Musa and M. P. Stevens, *J. Polym. Sci.*, *A-1*, *10*(2), 319 (1972).

48. L. Otras, J. Marton, and J. Meisel-Agoston, *Tetrahedron Lett.*, *2*, 15 (1960).

49. J. Preston and W. B. Black, *J. Polym. Sci.*, *B*, *4*, 267 (1966).

50. J. Preston, W. DeWinter, W. B. Black, and W. L. Hofferbert, Jr., *J. Polym. Sci.*, *A-1*, *7*(10), 3027 (1969).

51. S. R. Sandler and W. Karo, *Polymer Synthesis*, Academic Press, New York, 1974, pp. 216-224.

52. N. E. Searle, U.S. Patent 2,444,536 (1948); *Chem. Abstr.*, *42*, 1340 (1948).

53. C. E. Sroog, *Encycl. Polym. Sci. Technol.*, *11*, 247-272 (1969).

54. C. E. Sroog, A. L. Endrey, S. V. Abrano, C. E. Beer, W. M. Edwards, and K. L. Olivier, *J. Polym. Sci.*, *A*, *3*, 13973 (1965).

55. J. K. Stille, F. W. Harris, H. Mukamol, R. O. Rakutis, C. L. Schilling, G. K. Noren, and J. A. Reeds, *Adv. Chem. Series*, No. 91, 628 (1969).

56. L. H. Tagle, J. F. Neira, F. R. Diaz, and R. S. Ramirez, *J. Polym. Sci.: Polym. Chem. Ed.*, *13*, 2827 (1975).

57. H. Strathmann, W. Schumann, and P. E. Gruber, *Org. Coat. Plast. Preprints*, *34*(1), 487 (1975).

58. S. Terney, J. Keating, J. Zielenski, J. Hakala, and H. Sheffer, *J. Polym. Sci.*, *A-1*, *8*(3), 683 (1970).

59. M. Ueda, T. Takahashi, and Y. Imai, *J. Polym. Sci.: Polym. Chem. Ed.*, *14*, 591 (1976).

60. A. Walch, H. Lukas, A. Klimmer, and W. Pusch, *J. Polym. Sci.: Lett. Ed.*, *12*, 697 (1974).

61. W. Wrasidlo and J. M. Augl, *J. Polym. Sci., A-1, 7,* 321 (1969).

62. J. H. Bateman, W. Geresy, and D. S. Neiditch, *Org. Coat. Plast. Preprints, 35*(2), 72 (1975).

63. V. L. Bell, *J. Polym. Sci.: Polym. Chem. Ed., 14,* 225 (1976).

64. V. L. Bell, B. L. Stump, and H. Gager, *J. Polym. Sci.: Polym. Chem. Ed., 14,* 2275 (1976).

65. A. A. Berlin, B. I. Liogon'kii, B. I. Zapadinskii, E. A. Kazantzeva, and A. O. Stankevich, *J. Macromol. Sci. Chem., A11*(1), 1 (1977).

66. P. E. Cassidy and S. V. Doctor, *J. Polym. Sci.: Polym. Chem. Ed., 18,* 69 (1980).

67. P. E. Cassidy and A. Kutac, *J. Polym. Sci.: Polym. Chem. Ed.,* in press.

68. P. E. Cassidy, J. C. Lin, and N. C. Fawcett, *J. Polym. Sci.: Polym. Chem. Ed., 17*(5), 1309 (1979).

69. P. E. Cassidy, R. A. Lohr, and N. C. Fawcett, *Polym. Preprints, 20*(1), 467 (1979).

70. P. E. Cassidy, R. A. Lohr, A. Kutac, and S. Doctor, *Polym. Preprints, 19*(2), 46 (1978).

71. P. Delvigs, *NASA Technical Memorandum,* Lewis Research Center, Cleveland, Ohio 44135 (1976).

72. P. Delvigs, *Polym. Eng. Sci., 16*(5), 323 (1976).

73. N. C. Fawcett, P. E. Cassidy, and J. C. Lin, *J. Org. Chem., 42*(17), 2929 (1977).

74. N. C. Fawcett, R. A. Lohr, and P. E. Cassidy, *J. Polym. Sci.: Polym. Chem. Ed., 17,* 3009 (1979).

75. S. L. Hastford, S. Subramanian, and J. A. Parker, *J. Polym. Sci.: Polym. Chem. Ed., 16,* 137 (1978).

76. J. H. Hodgkin, *J. Appl. Polym. Sci., 20,* 2339 (1976).

77. K. Inove and Y. Imai, *J. Polym. Sci.: Polym. Chem. Ed., 14,* 1599 (1976).

78. K. Iwata, M. Ogasawara, and S. Hara, *Makromol. Chem., 179*(5), 1361 (1978).

79. K. Kurita, H. Itoh, and Y. Iwakura, *J. Polym. Sci.: Polym. Chem. Ed., 16,* 779 (1978).

80. K. Kurita, H. Itoh, and Y. Iwakura, *J. Polym. Sci.: Polym. Chem. Ed., 17,* 1187 (1979).

81. R. A. Lohr and P. E. Cassidy, *Makromol. Chem.,* in press.

82. R. A. Lohr, P. E. Cassidy, N. C. Fawcett, and A. Kutac, *J. Chem. Eng. Data, 24*(2), 156 (1979).

83. D. C. Phillips, S. Spewock, and W. M. Alvino, *J. Polym. Sci.: Polym. Chem Ed., 14,* 1137 (1976).

84. R. Rubner and E. Kiihn, *Org. Coat. Plast. Preprints, 37*(2), 118 (1977).

85. E. Sacher and J. R. Susko, *J. Appl. Polym. Sci., 23,* 2355 (1979).

86. Yu. N. Sazanov, L. V. Krasilnikova, and L. M. Shcherbakova, *Europ. Polym. J.*, *11*(11), 801 (1975).

87. A. K. St. Clair and T. L. St. Clair, *Org. Coat. Plast. Preprints*, *35*(2), 185 (1976).

88. A. K. St. Clair and T. L. St. Clair, *Polym. Eng. Sci.*, *16*(5), 314 (1976).

89. T. L. St. Clair, A. K. St. Clair, and E. N. Smith, *Polym. Preprints*, *17*(2), 359 (1976).

90. G. F. Syres and P. R. Young, *J. Appl. Polym. Sci.*, *21*, 2393 (1977).

91. I. K. Varma, R. N. Goel, and D. S. Varma, *J. Polym. Sci.: Polym. Chem. Ed.*, *17*, 705 (1979).

Chapter 6

Heterocyclic Polymers: Five-Member Rings

During the first 15 years of the high-temperature polymer field, much effort has been expended and much success derived in terms of polymers with heterocyclic functional groups, particularly five-member rings, in the backbone. Each polymer discussed in this chapter (which excludes polyimides) contains nitrogen as one of the heterocyclic atoms. The abundance of work on this topic is because such polymers are readily synthesized, are aromatic or nearly so, and are quite resistant to thermooxidation due to their aromatic character.

The field of thermally stable polymers became more widely known through research on polyheterocycles, particularly the polybenzimidazoles. And many successful examples (in terms of stability, tractibility, processability, and applications development) are nitrogen heterocyclic aromatic backbones: polybenzimidazoles, polypyrrones, and polyoxadiazoles.

Because of the large bulk of information available for this class of polymers, many examples have been moved to other chapters. Polyimides are grouped with nonheterocyclic polymers, polypyrrones with ladders, etc.

The most important functions remaining in this chapter, then, are oxadiazoles and benzazoles (benzimidazoles, benzoxozoles, and benzothiazoles). Much work has been done on oxadiazoles, as is evident by the large section on these materials. The chapter is organized according to the number and type of heteroatoms, beginning with a single nitrogen and progressing to nitrogen and sulfur and finally to four nitrogens. Table 6.1 gives a list of structures in the order discussed. All of the functional groups and their nomenclature for this and other chapters also appear in the appendix. Because of the large number of polybenzimidazoles which have been prepared, their structures are given separately in Table 6.2 (Section 6.10).

Table 6.1 Five-Member Ring Heterocyclic Polymers

Name	Structure	Comments
Poly(phenylenepyrrole)		Yellow. Soluble. Stable to 520°C in air.
Poly(phenylenethiophene)		Orange. Insoluble. Stable to 580°C in N_2.
Polypyrazole		White. Soluble, film former. Stable to 450°C.
Polyisoxazole		Soluble, film former. Stable to 370°C in N_2.
Polyimidazolidine		Soluble, film former. Stable to 400°C.
Polybenzoxazole		Limited tractability. Stable to 550°C in air.

Polymer	Structure	Properties
Polyoxazole		Limited solubility. Stable to 440°C in N_2.
Poly(oxazolone enamine)		Yellow, transparent films. Fluorescent solutions. Poor hydrolytic and thermal stabilities (300°C in N_2).
Poly(arylenesulfimide)		Insoluble, dark, brittle solid. Off-gassing during postcure.
Polybenzimidazoles		Much development done. Soluble intermediate, intractable final product. Stable to 500°C in N_2.
Polythiazole and polybenzothiazole		Stable to 500°C. Aliphatic: soluble, film and fiber former, crystalline.
		Aromatic: poor solubility.
Poly-1,3,4-oxadiazole		Widely studied. Soluble, film and fiber formers. Partially crystalline. Stable to 450°C.
Poly-1,2,4-oxadiazoles		Crosslinkable gums possible. Stable to 310°C.

Table 6.1 (Continued)

Name	Structure	Comments
Poly-1,2,5-oxadiazole-N-oxide (or polyfuroxan)	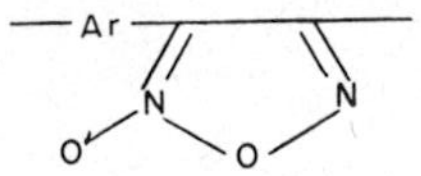	Yellow-brown. Poor thermal stability (220°C).
Poly-1,3,4-thiazidazole	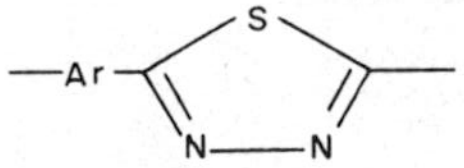	Flexible, films and fibers. Stable to 400-600°C. More tractable than oxo analog.
Poly(ester-1,2,5-thiadiazole)	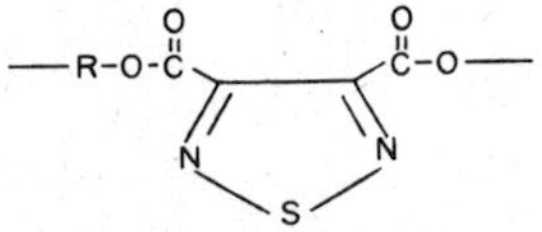	Soluble. Clear films, fibers. Stable to 400°C.
Polytriazole	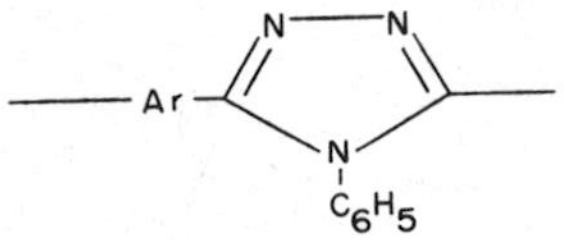	Acid-soluble. Soften at 240-310°C. Stable to 400-512°C. Film and fiber former. Crystallizable.
Polytriazoline	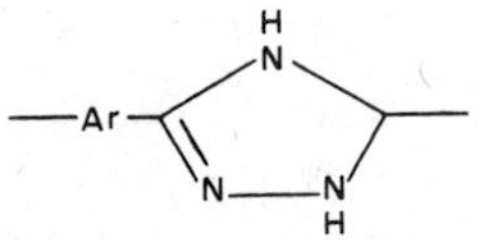	Acid-soluble. Transparent, yellow films. Poor thermal stability (310°C in air).
Polytetrazole	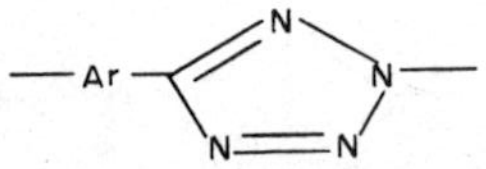	Soluble in H_2SO_4. White to yellow. Poor thermal stability (220-245°C in N_2).

6.1 Poly(phenylenepyrrole)

A simple heterocyclic polymer with a single nitrogen heterocyclic atom is made from polydiethynylbenzene. This polymer is prepared by cyclization and aromatization of the backbone with concurrent insertion of nitrogen [8].

$$HC{\equiv}C{-}C_6H_4{-}C{\equiv}CH \xrightarrow{O_2,\ Cu^+} \left[C_6H_4{-}C{\equiv}C{-}C{\equiv}C \right]_n \xrightarrow[Cu^+,\ 150^\circ]{ArNH_2} \left[\text{pyrrole}(N{-}Ar){-}C_6H_4 \right]_n$$

Ar is C_6H_5 or $CH_3C_6H_4$

In following sections it will be seen that cyclization of the polyalkyne can be accomplished in other ways. This unstable polymeric acetylene is produced to a degree of polymerization of 200 [51,170]. Its reaction with primary amines occurs in the absence of oxygen to nearly quantitative yields usually within 1 hr at 150°C. Actually, yields considerably above 100% were observed frequently. Rather than refute the first law of thermodynamics, the authors chose to explain the phenomenon as follows. Although a large amount (10%) of Cu_2Cl_2 is desirable to allow rapid reaction, it also leads to excess addition of the amine across the triple bond forming enamines, thus providing excess weight:

$$C_6H_4{-}C(NH\phi){=}CH{-}C(NH\phi){=}CH{-} \quad \text{or} \quad C_6H_4{-}C(NH\phi){=}CH{-}CH{=}C(NH\phi){-}$$

The second product, from addition to the 1 and 4 positions, can cyclize to pyrrole. However, the first product, the 1,3 adduct, cannot cyclize and therefore adds weight to the polymer above the theoretical amount.

The polypyrroles are yellow and soluble in common solvents (toluene, chlorobenzene, nitrobenzene, methylene chloride, DMF, and DMSO). This solubility may be aided by the presence of the pendant phenyl group which serves to decrease crystallinity. Intrinsic viscosities range from 0.3 to 0.5 dl/g in chlorobenzene.

Thermal stability measurements by TGA show an incipient weight loss at 490°C in air or nitrogen and a 10% loss at 520°C in air and 590°C in nitrogen. Isothermal aging at 350°C reveals a steady weight loss reaching 10% in 14 hr.

Method *Polydiethynylbenzene* [170] A mixture of the m- and p-diethynylbenzenes is used to alter backbone regularity and thereby increase solubility and also therefore the molecular weight of the product. Into a 125-ml Erlenmeyer flask heated to 90°C and equipped with stirrer, thermometer, and gas inlet are placed 48 ml of o-dichlorobenzene, 1.7 ml of pyridine, 0.23 ml (0.0015 mol) of N,N,N',N'-tetramethylethylene diamine, and 0.150 g (0.0015 mol) of cuprous chloride. Oxygen is bubbled into the solution with stirring, and to it is added a solution of 4.75 g (0.0376 mol) of m-diethynylbenzene and 0.25 g (0.0020 mol) of p-diethynylbenzene in 10 ml of o-dichlorobenzene. The temperature rises to 108°C in 2 min, and the viscosity increases. After 17 min of reaction, the solution is cooled to 70-80°C, whereupon it is poured into 500 ml of methanol which contains 10 ml of conc. HCl. The precipitate is collected and washed by stirring in another batch of methanol-HCl and finally with methanol alone. The polymer is isolated and dried under vacuum at room temperature to yield 4.92 g (99.5%) of pale yellow solid; intrinsic viscosity 1.44 dl/g in o-dichlorobenzene at 120°C. The dry polydiethynylbenzene decomposes violently with shock, spark, or heat. Furthermore, it is light-sensitive. Therefore manipulations with it should be conducted in the dark.

Polyphenylenepyrrole In a 250-ml Erlenmeyer flask equipped with a stirrer, gas inlet, and means of heating is placed 200 ml of aniline and 5 g of polydiethynylbenzene. The solution is purged with nitrogen, and 20 g of cuprous chloride is added. The flask is then capped and heated to 150°C for 30 min. Then it is poured into excess methanol which contains 10% conc. ammonia under nitrogen to precipitate the polymer. The solid is washed free of copper salts and dried to yield 101% of a yellow powder; inherent viscosity 0.5 dl/g in chlorobenzene.

6.2 Poly(phenylenethiophene)

The polyacetylene discussed above can also suffer cyclization to a thiophene backbone nucleus. This facile reaction occurs at room temperature with H_2S and basic catalysts. However, an elevated temperature is necessary to dissolve the starting materials.

$$\left[C_6H_4{-}C{\equiv}C{-}C{\equiv}C \right]_n \xrightarrow[\text{HMPA, } 50\text{-}80^\circ]{H_2S} \left[C_4H_2S{-}C_6H_4 \right]_n$$

The thiophene polymer is orange and insoluble, the latter phenomenon possibly due to crosslinking. Its thermal stability is quite good, however, showing a 10% weight loss by TGA in nitrogen at 580°C and only a 20% loss at 800°C.

Method [8] A solution of 0.5 g of polydiethynylbenzene in 25 ml of HMPA is flushed with nitrogen at 50°C, and H_2S is added slowly for 2 hr. The mixture is poured into methanol to precipitate the polymer which is isolated, washed with methanol, and dried at 180°C in vacuo.

6.3 Polypyrazoles

Again polydiethynylbenzene can condense with hydrazine or methylhydrazine to yield a nearly aromatic backbone with pyrazole units [9]. The unsubstituted polypyrazole can undergo substitution of acyl groups.

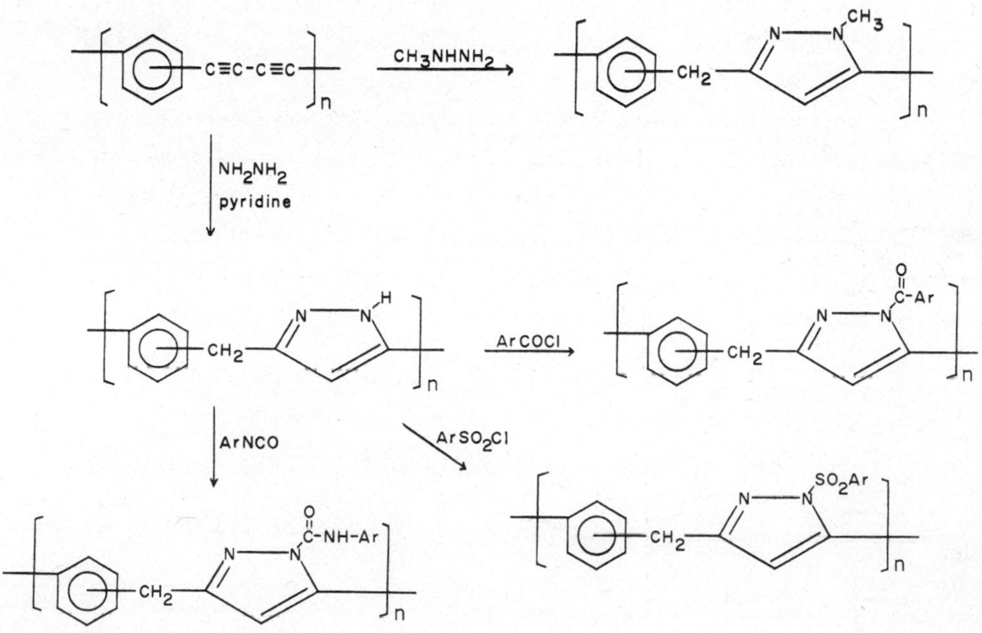

Note that a methylene unit, which undoubtedly detracts from thermooxidative stability, occurs in the backbone. Even so, these materials display good resistance to thermal degradation. Note also that the polymer consists of a random mixture of head-to-head and head-to-tail structures rather than the latter form which is pictured.

The simple polypyrazole is obtained in yields of 60-97% with inherent viscosities up to 0.8 dl/g in DMF. The white polymers are soluble in DMF, pyridine, HMPA, DMSO, and formic acid. Films can be cast which have a high tensile strength (138 MPa or 20,000 psi), which is attributed to hydrogen bonding between pyrazoles.

TGA data reveal a rather precipitous decomposition at 450°C in either air or nitrogen, a quite good stability, particularly in view of the presence of the

NH and CH_2 functions. Isothermal aging shows only a 5% loss at 300°C for 80 hr.

> *Method* [9] *Polypyrazole* In a flask with stirrer and condenser are placed 15 g of polydiethynylbenzene, 225 ml of pyridine, and 75 ml of hydrazine. The solution is heated at reflux (120°C) for 2 hr, and then 300 ml more of pyridine is added and the hydrazine pyridine azeotrope is removed. The solution is then cooled and poured into methanol to precipitate the white fibrous polymer, which is washed with methanol and dried at 50°C under vacuum. The product (16 g) had an inherent viscosity of 1.75 dl/g in DMF.

Methylhydrazine can be used to condense with the polyacetylene to yield a poly-N-methylpyrazole. However, its thermooxidative stability is poorer than the unsubstituted version.

The unsubstituted polypyrazole will react with acyl halides, sulfonyl chloride, or an aromatic isocyanate to remove the amino hydrogen and, one would expect, to improve thermal stability. Just the opposite occurs however; the N-carbamoyl substituent is worst in stability, sulfonyl next and benzoyl best, but it still undergoes decomposition near 300°C. For the latter the 10% weight loss by TGA in air is 370°C, and in nitrogen it is 400°C.

> *Method* *Poly(N-benzoylpyrazole)* [9] A solution is prepared of 1 g of the polypyrazole in 19 ml of pyridine, and to this is added at room temperature a solution of 1.35 g of benzoyl chloride in 12 ml of pyridine. The mixture is stirred at room temperature for 1 hr and at 80°C for 1 hr. The solution is poured into methanol to precipitate the white, fibrous polymer which is washed and dried. Its solubility is the same as the starting material, and its inherent viscosity is 0.75 dl/g in DMF, somewhat lower than the parent material (1.0 dl/g).

Another synthesis of polypyrazoles has been shown to give good yields (>80%) by dipolar addition reactions. The diacetylene, m- or p-diethynylbenzene, can react with the dipoles, bissydnone [158,159], or bisnitrilimine [157,158] to give poly(phenylenepyrazoles). Although both reactions are 1,3-dipolar additions, the first, with the sydnone, is more like a Diels-Alder with the elimination of CO_2, while the latter proceeds through an intermediate bisnitrilimine (see also polytetrazines and polytriazoles). One of the interesting phenomena of these reactions is that only one of two possible isomers is realized. Note that the pyrazone functions have different catenations by the two methods.

The polymers are soluble in polar, aprotic solvents (DMF, DMAc, DMSO, and HMPA) although the meta-phenylene systems are more soluble than the para-. The viscosities are moderate (0.4 dl/g at 0.25% conc.)

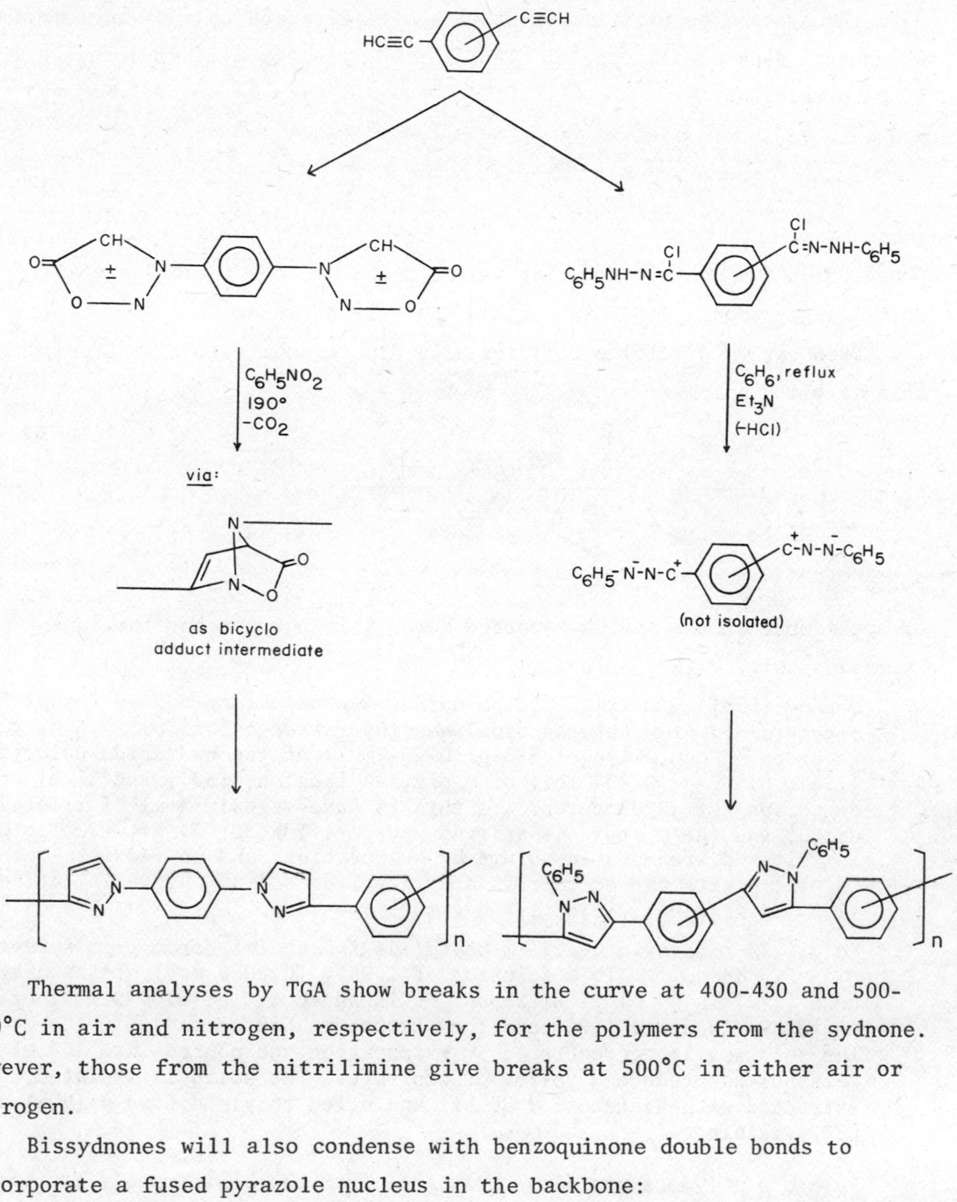
HC≡C
C≡CH
CH
N
O
N
CH
N
N
O
O
Cl
C=N-NH-C6H5
Cl
C6H5NH-N=C
C6H5NO2
190°
-CO2
via:
C6H6, reflux
Et3N
(-HCl)
N
N
O
O
as bicyclo
adduct intermediate
C6H5-N-N-C
C-N-N-C6H5
(not isolated)
N
N
N
N
n
C6H5
N
N
C6H5
N
N
n

Thermal analyses by TGA show breaks in the curve at 400-430 and 500-520°C in air and nitrogen, respectively, for the polymers from the sydnone. However, those from the nitrilimine give breaks at 500°C in either air or nitrogen.

Bissydnones will also condense with benzoquinone double bonds to incorporate a fused pyrazole nucleus in the backbone:

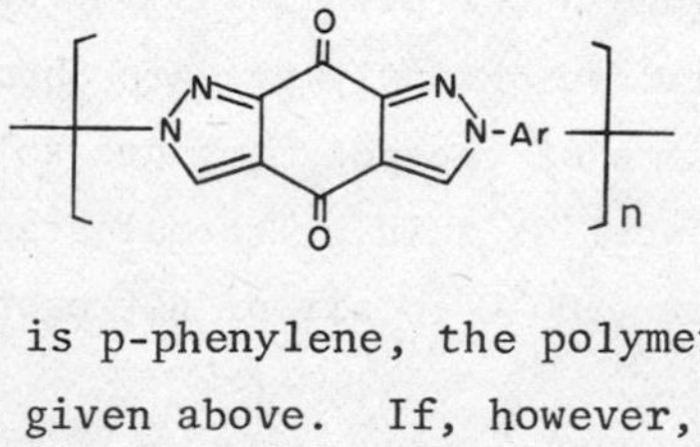
O
N
N
N
N-Ar
O
n

If Ar is p-phenylene, the polymer has thermal stabilities comparable to those given above. If, however, Ar is hexamethylene, the respective stabilities in air and nitrogen drops to 300 and 400°C, as would be expected.

The sydnone or nitrilimine dipoles will also add to a divinyl compound which acts as a dipolarophile. Thus the addition to m-divinylbenzene yields a polypyrazoline:

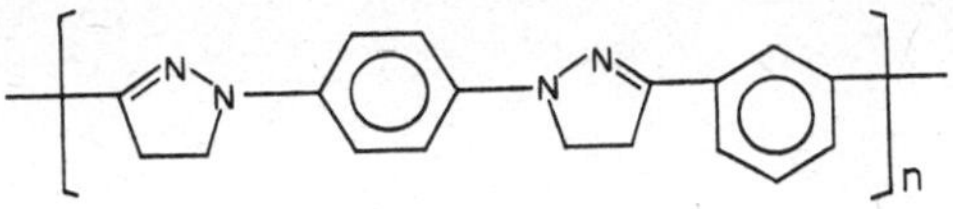

Considering the nonaromatic ring in the backbone, the thermal stability is surprisingly good: 410°C in air and 485°C in nitrogen.

Even carbon disulfide will act as a dipolarophile to give a spirobis-thiadiazoline polymer:

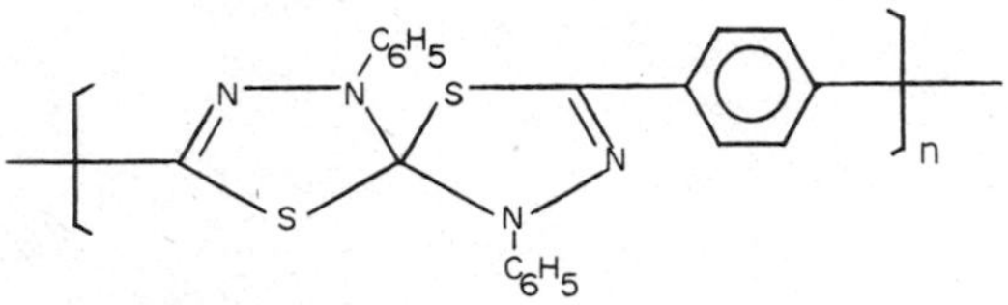

Of any member of the series reported here, this polymer has the lowest thermal stability (275°C in air).

Method [158] *Poly[1,1'-diphenyl-3,3'-p-phenylene-5,5'-m-phenylene-dipyrazole] (from terephthaloylphenylhydrazide chloride)* A mixture is prepared containing 1.533 g (0.004 mol) of the hydrazide chloride [153], 0.505 g (0.004 mol) of m-diethynylbenzene [50], and 15 ml of anhydrous tetrahydrofuran. To this is added slowly 4 ml of triethylamine, and the mixture is stirred under reflux for 24 hr. The solution is filtered, reduced in volume by evaporation, and poured into methanol to precipitate the polymer as a dark yellow solid. Reprecipitation from THF-methanol yields 1.436 g (82%).

Poly[1,1'-p-phenylene-3,3'-m-phenylenedipyrazole] (from p-phenylene-3,3'-disydnone) To a solution of 1.95 g (0.0079 mol) of the disydnone [172] in 50 ml of nitrobenzene is added 1.0 g (0.0079 mol) of m-diethynylbenzene [50], and the mixture is heated at 190°C for 24 hr. The solution is concentrated by evaporation and poured into 200 ml of methanol to produce a yellow precipitate. The solid is isolated, extracted with benzene for 20 hr, and dried to yield 0.94 g (90%) of polypyrazole.

Pyrazole polymers are also obtainable from the bistetrazole addition to diynes [156]. The tetrazole serves to generate a bisnitrilimine as discussed above, but in this case high molecular weight polymers are obtained (intrinsic viscosity 1.67). Furthermore, some of the polymers are soluble chlorobenzene or 1,2,4-trichlorobenzene as well as acid. Thermal stabilities are the same as those reported earlier (500°C in air or nitrogen).

$C_6H_5NH{-}N{=}CH{-}Ar{-}CH{=}N{-}NH{-}C_6H_5$

↓ $C_6H_5N_3$, Na

[C_6H_5–tetrazole–Ar–tetrazole–C_6H_5]

↓ heat, $-2N_2$

$C_6H_5\bar{N}{-}N{=}\overset{+}{C}{-}Ar{-}\overset{+}{C}{=}N{-}\bar{N}{-}C_6H_5$

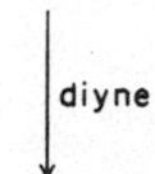

polypyrazole (see above reaction)

Method [156] *Preparation of monomer: 2,2'-diphenyl-5,5'-p-phenyl eneditetrazole* A solution is prepared from 14.4 g (0.045 mol) of terephthaloylphenylhydrazone and 10.0 g (0.430 mol) of sodium in 160 ml of 2-methoxyethanol, and to it is added 9.0 g (0.089 mol) of phenylazide. The solution is heated at reflux for 15 hr and cooled to 25°C, and the yellow precipitate is collected by filtration. The tetrazole is purified by several recrystallizations from benzene to yield 6.8 g (47%) of white solid; mp 220-225°C (dec.).

Polymerization The above tetrazole (0.366 g, 1.000 mmol) is mixed with 0.126 g (1.000 mmol) of p-diethynylbenzene and 5 ml of 1,2,4-trichlorobenzene in a polymerization tube. The system is degassed by several freeze-thaw cycles in liquid nitrogen and sealed under vacuum. The tube is placed in a reaction bomb with solvent external to the tube and is heated at 185°C for 48 hr. The contents are poured into 200 ml of methanol to precipitate the light brown polymer, which is isolated and dried in vacuo at 212°C; intrinsic viscosity 1.67 dl/g (in 98% formic acid). The yield is 89% (0.391 g).

6.4 Polyisoxazole

By condensing hydroxylamine with polydiethynylbenzene, one isolates a polyisoxazole [9]:

$$\left[C_6H_4{-}C{\equiv}C{-}C{\equiv}C \right]_n \xrightarrow[\text{picoline, } 100°,\ 18\text{ hr}]{NH_2OH} \left[C_6H_4{-}CH_2{-}\text{(isoxazole)} \right]_n$$

6.5 Poly(iminoimidazolidinediones) and Poly(1,3-imidazolidine-3,4,5-triones) (Polyparabanic Acids)

Several routes, all of which involve the cyclization of isocyanate functions with cyanide groups, are possible to poly(iminoimidazolidinediones).

Ar(NCO)$_2$ —(HCN, NaCN / NMP)→

Ar(NH–C(=O)–CN)$_2$ + Ar(NCO)$_2$ —(NaCN / NMP)→ polymer [–Ar–N(ring)N–]$_n$

OCN–Ar–NH–C(=O)–CN —(NaCN / NMP)→

Ar is : –C$_6$H$_4$–CH$_2$–C$_6$H$_4$– , –(CH$_2$)$_6$– , CH$_3$-substituted phenylene , –C$_6$H$_4$–O–C$_6$H$_4$– , CH$_3$/CH$_3$-substituted biphenylene

The second one shown was perhaps the first synthesis of this system [118]. However, more recently Patton has conducted extensive research on these polymers; and the soluble materials are being developed commercially by Exxon Chemical Company [121,122].

This rather complex-appearing reaction can be envisioned between isocyanate and carbamoyl cyanide as a four-centered cycloaddition:

—Ar–N(H)···C(=O)=N— / C(=O)—C≡N (cycloaddition transition state)

However, when an isocyanate-HCN reaction is employed, a three-step reaction takes place, the first being HCN addition to the isocyanate to form a cyanoformamide. The cyanoformamide then adds to another isocyanate to give the N-cyanoformyl urea which subsequently cyclizes [121]:

—NCO + HCN ⟶ NH–C(=O)–CN —(—NCO)→ —N(C(=O)NH—)(C(=O)CN) ⟶ —N(ring)N— (with O, O, NH)

The reactions generally proceeded in high yield (80-100%) in a rapid, exothermic fashion in a number of solvents (DMF, DMSO, NMP, or HMPA). NMP was the best solvent, and DMF and DMSO promoted crosslinking. A catalyst was necessary for the reaction, however, the nature of which depended on the types of monomers used. These catalysts included morpholine, amines, dibutyltin diacetate, tributylphosphine, trimethyl lead acetate, and analogous compounds. Tertiary amine catalysts commonly led to insoluble crosslinked products, while sodium cyanide gave high molecular weight, soluble products (in DMF).

The poly(iminoimidazolidinedione) polymers displayed inherent viscosities up to 1.36 dl/g, frequently 0.17-0.69 dl/g. It is interesting that their elemental analyses corresponded very closely to theoretical values, a condition not often seen when heterocyclic polymers are formed at high temperatures.

The thermal properties of these heterocyclic polymers are strongly dependent on the connecting function, as has been frequently seen. For example, the hexamethylene group, when included in the backbone, gives a Tg of 30°C; the diphenylmethane group has a Tg of 195°C; and the diphenylether has a Tg of 245°C. TGA data on both the diphenylmethane and diphenylether systems show a 10% weight loss at 340°C.

These imino polymers can be hydrolyzed to the trione analog, a polyparabanic acid:

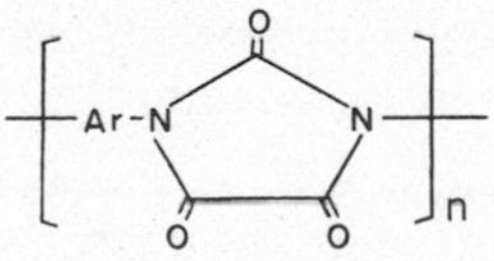

This hydrolysis is a complete one as shown by the good comparison of theoretical to found elemental analysis data.

Inherent viscosities of the polyparabanic acids are in the same range as their precursors (0.14-1.57 dl/g). They are also soluble in DMF and DMSO and from solution; films can be cast and fibers can be spun which have very promising mechanical properties: tensile strength 172 MPa (25,000 psi) at -90°C (-130°F) and 34 MPa (5000 psi) at 204°C (+400°F); elongation at yield, 10% at -90°C (-130°F) and 20% at 204°C (+400°F).

Thermal data for the parabanic acid polymers are much improved over their imino precursors. Upon hydrolysis, Tg data increase by 25-53°C, with higher Tg materials showing the largest increase. TGA stability data increase to 460°C for the diphenylmethane analog (up 120° over the imino) and to 500°C for the diphenylether system (an increase of 160° with replacement of the NH with O).

Softening points of the polyparabanic acids vary from 55° to 300°C. This value depends, of course, on the nature of the connecting group between isocyanate functions of the monomer. Any softening point can be designed into the polymer simply by mixing aliphatic and aromatic monomers.

Method [122] *Prepolymerization: poly(4,4'-diphenyleneether-5-iminoimidazolidine-1,3-diyl-2,4-dione)* Anhydrous conditions are observed for materials and equipment. In a reaction vessel equipped for cooling and maintaining a dry nitrogen atmosphere is placed a solution of 6.48 g of hydrogen cyanide in 145 ml of nitrobenzene, along with a solution of 60.48 g of 4,4'-diphenylether diisocyanate in 458 ml of nitrobenzene and 67 ml of toluene. To this is added 5.64 ml of sodium cyanide saturated NMP while the temperature is kept below 90°C by cooling the exothermic reaction. The slurry of precipitated polymer is stirred for a total reaction time of 20 min, whereupon the solid product is isolated by filtration, washed with methanol, and dried.

Hydrolysis to polyparabanic acid The prepolymer is suspended with stirring in 77.6 wt % sulfuric acid for 30 min at room temperature; isolated by filtration; washed well with water, methanol, and acetone, sequentially; and dried; inherent viscosity 1.74 dl/g in DMF.

6.6 Polybenzoxazoles

Polybenzoxazoles, also known infrequently as polyindolones, have a history similar to the polybenzimidazoles. Aliphatic backbones were patented in 1959 [29]; and in a few years the superior, all-aromatic backbones became known [70,71,76,92,108]. Only a limited amount of work has been done since the mid-1960s; and in 1969 Korshak compiled a rather extensive summary of polybenzoxazoles [134]. They are synthesized in a manner similar to that for polybenzimidazoles. In this case, a bis-o-aminophenol is condensed with a dicarboxylic acid or acid derivative.

HO, OH, NH_2, NH_2 + X–C(=O)–Ar–C(=O)–X $\xrightarrow[\text{melt or solution, heat}]{-HX}$ [–Ar–(benzoxazole)–(benzoxazole)–]$_n$

X is : OH, C_6H_5O, CH_3O, NH_2, Cl

Ar is : m-or p-C_6H_4, or $(CH_2)_8$

Another aminophenol is:

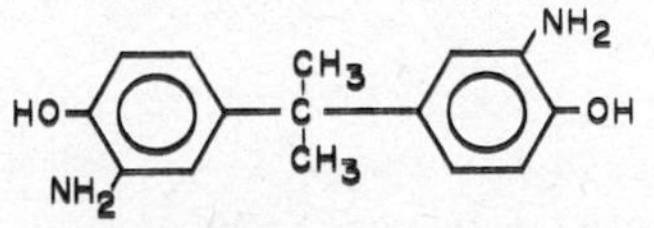

Of course the simplest polybenzoxazole is the one prepared from 3-amino-4-hydroxybenzoic acid. It is also the most rigid backbone available in this series.

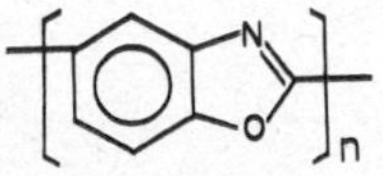

As expected, the benzoxazole ring formation is a two-step process. The first step is the reaction of the acid groups with the more basic amino group. Subsequent heating then completes cyclodehydration. It is, of course, possible to form the benzoxazole nucleus in a monomer prior to polymerization. An aromatic diamine shown below has been synthesized and then polymerized by various means:

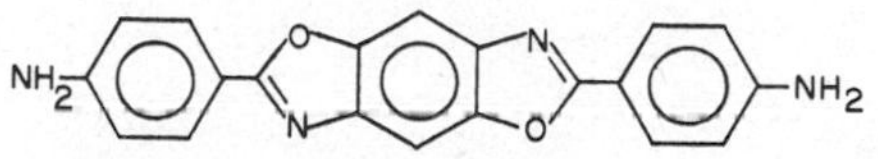

The polybenzoxazoles are not as resistant to hydrolysis as polybenzimidazoles; however, they do withstand hours of heating in KOH or H_2SO_4 solutions.

Thermostabilities of nitrogen and oxygen benzazole analogs are similar. The totally aromatic polybenzoxazole shows a 10% weight loss at 550°C in air by TGA. The aliphatic systems, not unexpectedly, begin degradation below 300°C and reach 10% loss at 350°C.

For the most part, except for some recent examples (vide supra), tractability of polybenzoxazoles is not good. They are soluble only in concentrated sulfuric acid. Applications of these polymers to commercial uses are not known.

Many syntheses using both solution and melt procedures have been published. Those given below are meant to demonstrate both types with the simple aromatic polybenzoxazoles and more complex and recent polymers containing the benzoxazole nucleus.

Method *Poly-2,2'-(m-phenylene)-6,6'-bibenzoxazole* [108,134]

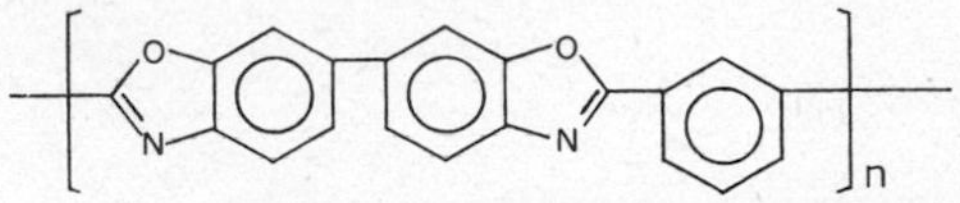

A well-powdered mixture of 2.16 g of 3,3'-dihydroxybenzidine and 3.18 g of diphenylisophthalate (0.010 mol of each) is heated under nitrogen at 370°C for 75 min, then at 300°C for 70 min, and finally at 300-320°C for 4 hr. A quantitative yield of yellow solid polymer is obtained with an inherent viscosity of 0.43 dl/g in H_2SO_4.

An interesting copolymer containing both benzoxazole and oxadiazole functions in an ordered arrangement has been synthesized [22]. The two-step process goes through a poly(amide hydrazide) intermediate, which is useful for processing since it is soluble in DMSO, NMP, DMF, and DMAc. It can be cast to films prior to cyclodehydration.

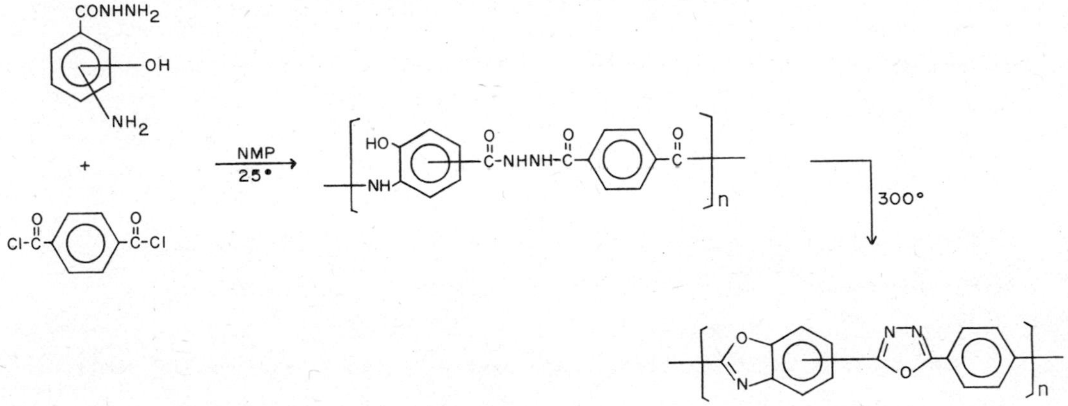

$CONHNH_2$ OH NH_2 + Cl-C(=O)-C(=O)-Cl NMP 25° HO NH C-NHNH-C C n 300° O N N-N O n

Either isomer of the poly(phenylene-1,3,4-oxadiazolylbenzoxazole) shows a 10% loss by TGA at 550°C in either air or nitrogen.

Method [22] To a solution of 8.35 g of 4-amino-3-hydroxybenzoylhydrazide in 100 ml of NMP is added 10.15 g of terephthaloyl chloride with rapid stirring at 0-20°C. After 1 hr at room temperature, the solution is cast to a film on a glass plate and dried at 90°C under vacuum for 2 hr or precipitated into water. The precipitated polymer is washed with water, then with hot ethanol, and dried at 80°C under vacuum. The white powder is obtained quantitatively with an inherent viscosity of 2.17 dl/g in DMSO.

To cyclodehydrate, the polyamide hydrazide is heated either under nitrogen or vacuum at 300°C for 1 hr.

Evers has introduced fluoroether units into a benzoxazole backbone to achieve flexibility to the point of producing a rubbery product [33]. Earlier work resulted in some crystallinity and less flexibility in the polymer [13].

HO, NH_2 – R_f' – OH, NH_2 + $CH_3O-\overset{NH}{\overset{\|}{C}}-R_f-\overset{NH}{\overset{\|}{C}}-OCH_3$ $\xrightarrow[\text{HOAc, 50°}]{\text{HFIP}}$ [benzoxazole – R_f' – benzoxazole – R_f]$_n$

R_f and R_f' are: $-(CF_2)_{3 \text{ or } 8}-$, $-(CF_2)_2O(CF_2)_5O(CF_2)_2-$

The thermal stability suffered somewhat in that weight loss began at 400-450°C in air or nitrogen by TGA. Isothermal aging at 316"C showed a 15% weight loss in 100 hr, but the polymer was still rubbery. After 200 hr, severe loss of properties was incurred. Glass transition temperatures ranging from -22 to 106°C, were also lower than with simple aromatic systems.

The polymers are soluble in hexafluoroisopropanol (HFIP) and demonstrate inherent viscosities from 0.2 to 0.9 dl/g.

Method [33] In 5 ml of HFIP are mixed 0.498 g (0.0010 mol) of dimethylperfluoro-3,6-dioxaundecanediimidate [the above monomer where R_f' is $(CF_2)_2O(CF_2)_5O(CF_2)_2$] and 0.698 g (0.0010 mol) of the bisaminophenol monomer where R_f is the same fluoroether. After a few minutes of stirring to obtain a pale yellow solution, 0.25 g of glacial acetic acid is added and the mixture is stirred at 50°C under nitrogen for 11 days. The solution is then poured into 150 ml of methanol at -78°C to precipitate the polymer. The product is collected, washed with cold methanol, reprecipitated from Freon 113 into methanol, and dried at 180°C under vacuum. The yield is 0.70 g (71%) of a light amber, rubbery polymer; inherent viscosity 0.79 dl/g in HFIP.

The fluorocarbon dibenzoxazole polymers discussed above have been studied for their thermomechanical behavior [48,136]. Torsional braid analysis (TBA) of these materials has shown a glass transition above room temperature for each, in addition to two lower temperature, glassy state secondary transitions. One of these, at -35 to -40°C, is due to submolecular motion of the R_f' group linked between the benzylene groups. The second, in the -80 to -150°C range, is due to the R_f function between the oxazole moieties.

The most recent example of modification of polybenzoxazoles is the insertion of siloxane links in the backbone [10]. Doing so results in a much improved solubility (pyridine and NMP), while good thermal stability is retained.

Inherent viscosities ranged from 0.17 to 0.38 dl/g in pyridine. A 10% weight loss by TGA occurs above 500°C in air, and isothermal aging at 350°C for 96 hr results in less than 5% loss for the worst case. Glass transition temperatures were 160 and 221°C for the meta and para catenations, respectively.

$$O\left[Si(C_6H_5)_2-C_6H_4-CO_2C_6H_5\right]_2 + \text{3,3'-dihydroxybenzidine} \xrightarrow{240-328^\circ} \left[Si(C_6H_5)_2-O-Si(C_6H_5)_2-C_6H_4-\text{benzoxazole}-\text{benzoxazole}\right]_n$$

Method [10] Pure ester monomer and careful air exclusion are necessary to obtain a soluble product. Without solvent, 0.541 g (0.00250 mol) of 3,3'-dihydroxybenzidine and 1.937 g (0.00250 mol) of the para-siloxane ester monomer are mixed and heated from 240 to 300°C in 1.6 hr while phenol is distilled from the mixture. Heating is continued at 300-305°C for 2 hr under vacuum and then at 325-328°C for 2 hr. The mass is cooled to a hard, glassy solid to yield 1.8 g (94%); inherent viscosity 0.34 dl/g in pyridine.

6.7 Polyoxazoles

Another aromatic, albeit not fused, polymer backbone contains the oxazole nucleus. Polyoxazoles are synthesized by a method reminiscent of the oxadiazole preparation, which consists of acyl hydrazidine cyclization. In this case, however, an aminoacetyl group condenses with an acyl halide to yield a β-ketoamide which cyclizes to the oxazole [150]. As with many polyamides, the intermediate polymer is best synthesized by an interfacial process. Cyclodehydration of the amide to oxazole can be effected by heating the dry amide at 300°C or by treating it at room temperature with sulfuric acid. The cyclization step is monitored by the appearance of the C=N band in the IR at 1600 cm^{-1}. The polyamide intermediates are obtained in good yield (>90%) and molecular weights. They are soluble in m-cresol, DMF with 5% LiCl, and HMPA, and have inherent viscosities of 0.45-0.63 dl/g in m-cresol (0.25%). It is not surprising in view of backbone flexibility that the order of viscosities for the different Ar groups is p-C_6H_4 > m-C_6H_4 > diphenyl ether.

The polyoxazoles are soluble only in H_2SO_4, wherein their inherent viscosities range from 0.42 to 0.56 dl/g. The thermal stability of the p-phenylene-containing polymer was measured only in nitrogen, however, so an indication of thermooxidative resistance is not possible. In nitrogen the TGA gave a 10% weight loss at 440°C.

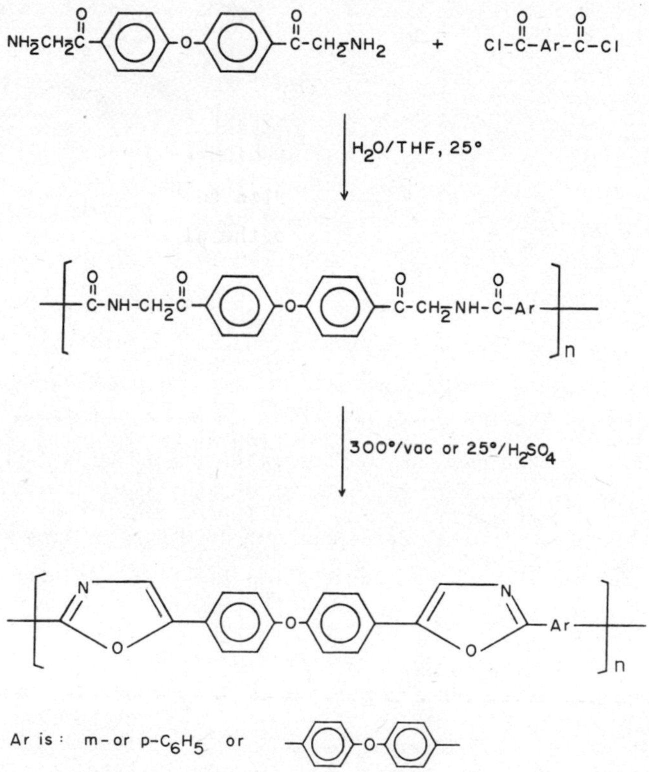

$NH_2CH_2CO-C_6H_4-O-C_6H_4-COCH_2NH_2$ + $Cl-CO-Ar-CO-Cl$

H_2O/THF, 25°

$-[-CO-NH-CH_2-CO-C_6H_4-O-C_6H_4-CO-CH_2-NH-CO-Ar-]_n-$

300°/vac or 25°/H_2SO_4

$-[-(oxazole)-C_6H_4-O-C_6H_4-(oxazole)-Ar-]_n-$

Ar is: m-or p-C_6H_5 or $-C_6H_4-O-C_6H_4-$

Method [150] *Polymerization* A solution is prepared from 1.780 g (0.0049 mol) of 4,4'-bis(aminoacetyl)diphenylether dihydrochloride, 2.120 g (0.02 mol) of sodium carbonate, and 0.2 g of sodium lauryl sulfate in 10 ml of water. The above solution is stirred at 25°C, and to it is added over 30 sec a solution of 1.024 g (0.0049 mol) of terephthaloyl chloride in 10 ml of tetrahydrofuran (THF). After an additional 5 min of stirring, the mixture is filtered to remove the polyamide, which is washed with methanol then with hexane, and dried. The yield is 1.85 g (91%) and inherent viscosity is 0.63 dl/g at a concentration of 0.25% in m-cresol.

Cyclodehydration Polyphenyleneoxazole is prepared by heating the above polyamide at 300°C under vacuum. An alternate method comprises dissolving 0.1 g of the amide in 10 ml of conc. H_2SO_4 and stirring for 1 hr; inherent viscosity 0.56 dl/g (0.25% H_2SO_4).

6.8 Poly(oxazolone enamines)

Through the condensation of bisoxazolones with primary amines, a new class of polymers has been found. These are called polyenamines by the investigators, although they also contain the oxazolone nucleus [166].

The aliphatic amines yield products which are much more soluble in polar aprotic solvents (DMAc, DMSO, NMP, HMPA) than the aromatic analogs.

$$\left[C_6H_5\text{-O-CH=}\underset{\text{O=C-O}}{\text{(oxazolone)}} \right]_2 \text{-}C_6H_4 \; + \; NH_2\text{-R-}NH_2 \xrightarrow[1/2\ hr]{25^\circ} \left[\text{-CH=(oxazolone)-}C_6H_4\text{-(oxazolone)=CH-NH-R-NH-} \right]_n$$

R is: $-(CH_2)_6-$, $-CH_2-C_6H_4-CH_2-$, $-C_6H_4-CH_2-C_6H_4-$, $-C_6H_4-O-C_6H_4-$

The solutions fluoresce (yellow-green) and can be cast to yellow, transparent films. There are, however, two serious problems: hydrolytic and thermal instabilities. Within 3 days in solution, the inherent viscosity drops to one-third of its original value (from 0.94 to 0.30 dl/g). This is said to be due to the ability of the enamine to isomerize to the imine (-CH=N-) which then hydrolyzes with a trace of moisture to the aldehyde and amine, thus resulting in chain cleavage. Thermal decomposition in air or nitrogen by TGA begins at 260°C and reaches 10% weight lo-s at 300-330°C for both aliphatic and aromatic systems.

Method [166] *Monomer synthesis* A solution of 29.8 g (0.106 mol) of N,N'-terephthaloylbisglycine (from the reaction of terephthaloyl chloride with glycine) and 36 ml (0.217 mol) of ethyl orthoformate in 66 ml of HOAc is heated with stirring at 140-150°C, with ethyl acetate being removed by distillation. After 2 hr the precipitate is collected, washed with chloroform, and dried under vacuum. Two recrystallizations from chloroform yield pale yellow leaflets (mp 273-275°C) in less than 85% yield of 2,2'-p-phenylenebis(4-ethoxymethylene-5-oxazolone).

Polymerization To a solution of 0.290 g (2.5 mmol) of hexamethylene diamine in 5 ml of DMAc is added 0.890 g (2.5 mmol) of 2,2'-p-phenylenebis(4-ethoxymethylene-5-oxazolone), and the solution is stirred at 25°C for 1 day. The viscosity reaches a maximum in 0.5 hr, and the monomer slowly dissolves. The solution is diluted with 10 ml of DMAc and poured into 500 ml of water to precipitate the fibrous polymer, which is washed with boiling acetone and dried to yield 0.94 g (99%).

6.9 Poly(arylenesulfimides)

D'Alelio has patented a sulfur-containing imide polymeric system which is said to be useful for adhesives, fiber laminates, and molding compounds [23]. This synthesis was accomplished by the self-polymerization of 6-aminosaccharin yielding some open-ring amide-sulfonic acid units as well as the sulfimide.

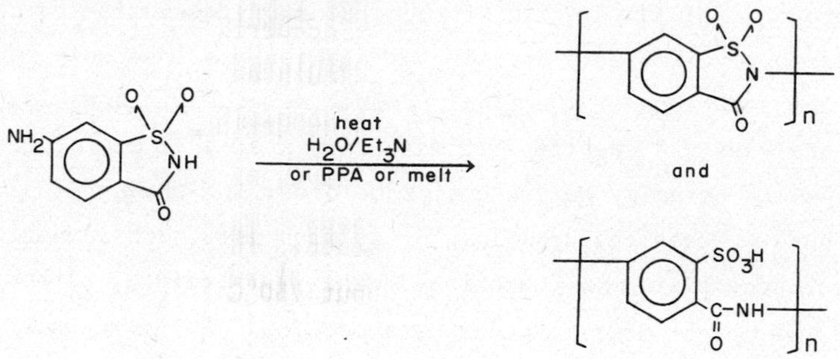

Other monomers which also undergo such a self-condensation are the 5-aminosaccharin and the naphthalene analogs shown below:

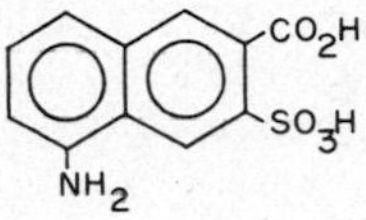

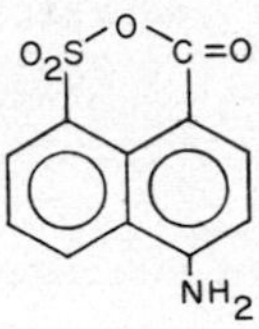

A later publication referred to these cyclized polymers as AB types of polysaccharins and reported them to have a brown poly(amic sulfonamide) intermediate which was soluble in DMAc and DMF (intrinsic viscosity 0.16 dl/g in DMF).

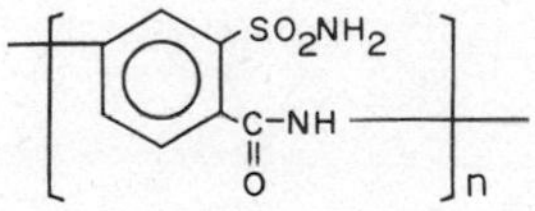

The poly(carboxylic amide) as an intermediate is more reasonable than the polysulfonamide in view of the higher reactivity of carboxylic acid derivatives compared to sulfonic acid analogs. In this case a melt procedure of the triethylamine salt of 6-aminosaccharin was preferred to a solution polymerization.

The polymerization occurs at 295°C (by DTA), just above the melting point of the monomer of 279°C. Finally, an exotherm at 390°C and an endotherm at 400°C were associated with cyclization of the intermediate. However, the triethylamine salt melts at a lower temperature (177-180°C) and gives a higher viscosity (0.41 dl/g) as a result of being in the molten state longer than the nonsalt.

The postcure or cyclodehydration of the precursors was effected at 400°C for 1-4 hr to give a black, brittle solid which was insoluble in hot DMF. However, the structure of the final product remained in question due to ill-defined analyses and spectra. It was postulated that a mixture of functional groups may actually be present, some occurring through loss of SO_2 in the postcure. Occasionally a loss of as much as 37% was experienced in this postcure, accompanied by "obnoxious" gases. Therefore, although TGA data show a 10% weight loss to occur at about 750°C in nitrogen, they are of limited value.

> *Method* [25] *Polyamide* A mixture of 1.1021 g (0.0055 mol) of 6-aminosaccharin, 0.4131 g (0.0041 mol) of triethylamine, and 2.175 g of water is degassed with nitrogen in a polymerization tube and heated at 175°C for 8 hr, during which time water distilled from the system. The viscous mass is cooled to room temperature to leave a glassy, brown solid; intrinsic viscosity 0.149 dl/g in DMF.

Yet a third publication of D'Alelio et al. describes an AA-BB type of polysaccharin derived from bisaccharin and diamines [26]. In this case, however, a solution process was preferred over melt conditions since the latter gave much ring closure and the uncyclized intermediate was not isolable. Again, the intermediate is soluble in DMF, DMAc, and aq. sodium hydroxide.

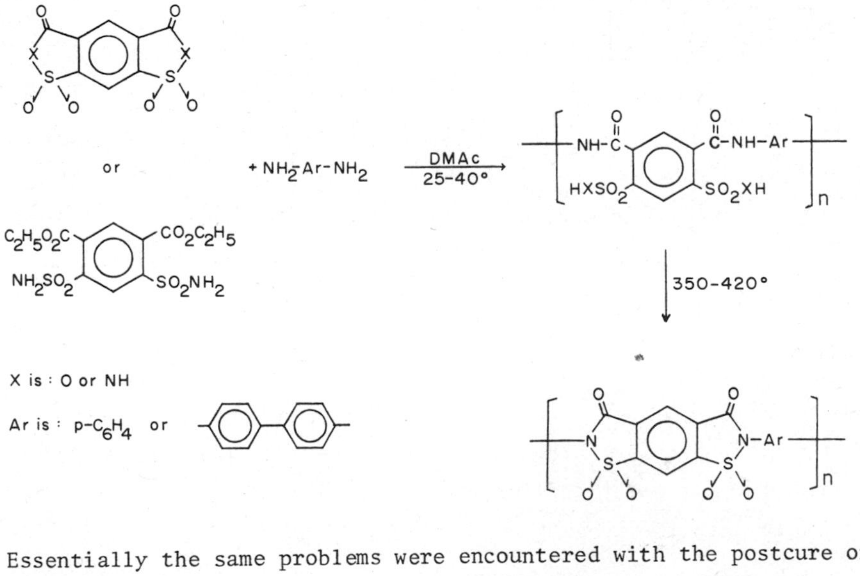

X is : O or NH

Ar is : p-C_6H_4 or

Essentially the same problems were encountered with the postcure of the AA-BB types as with the AB types discussed above. The postcure was accompanied by an odorous off-gas, and the final product was a black, porous solid; i.e., significant weight loss occurred, and the final structure is unknown. Thermal stabilities by TGA were similar also: 10% weight loss at 600°C.

Method [26] *Polyamide* To 20 ml of DMF, 2.90 g (10 mmol) of the dianhydride (cyclic monomer where X is -O-) [24] is added under nitrogen. [Evidently a chemical reaction occurs at this point since it is reported that the temperature rises to 60°C and the solution turns a light brown.] The solution is cooled again to room temperature, and 1.08 g (10 mmol) of p-phenylene diamine is added with stirring. The temperature rises to 40°C and a precipitate forms. The mixture is cooled to room temperature and stirred for 4 hr. The DMF is removed by vacuum distillation; and the brown, powdery residue is ground and dried at 50°C for 40 hr. It is washed with n-heptane and dried again at 50°C for 8 hr; intrinsic viscosity 0.126 dl/g in DMAc at 20°C.

6.10 Polybenzimidazoles

Some indication of the future of benzimidazole polymers was available in the 1959 patent of Brinker and Robinson which dealt with aliphatic systems [11]. The totally aromatic system published by Vogel and Marvel [167] in 1961, in addition to industrial patents beginning in 1958, opened the field of high-temperature polymers, especially aromatic heterocycles, to the torrent of research to follow in the next few years. The literature on the polybenzimidazoles (PBI) can only be described as voluminous.

By 1969 some 15 tetraamines and about 60 dicarboxylic acids had been condensed to PBI polymeric systems. In light of the reviews available on PBI in 1968 [78] and 1969 [38,80], particularly the comprehensive one by Levine [96], the literature prior to that time will be mentioned here only briefly and a general synthesis will be given. The work reported since 1968 will be covered in more detail. Table 6.2 gives structures of and references to most of the PBIs synthesized.

These PBI materials and others which followed caused some profound changes in associated technical areas. For example, aromatic tetraamines were essentially unavailable prior to 1960. Now, many of these can be purchased off the shelf or be obtained by custom synthesis. A second example is the development of techniques for using concentrated sulfuric acid as a processing solvent. A third effect is the appearance of heretofore exotic solvents (hexamethylphosphoramide, hexafluoroisopropanol, etc.).

Table 6.2 Examples of Polybenzimidazoles

Structure	Ref.
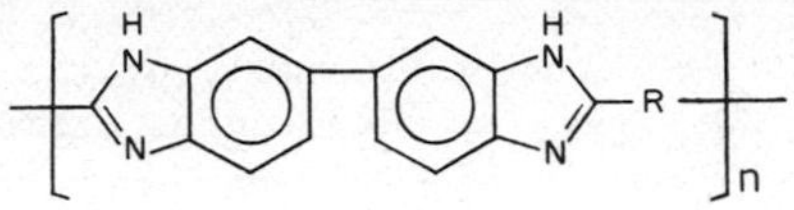 [bibenzimidazole–R]$_n$ (H, N, N, H, N, N, R, n) where R is: m-or p-C_6H_4, $-(CH_2)_4-$ o,o'- or p, p'-biphenylene, 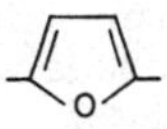 [furan-2,5-diyl, O], 1,7-napthalylene, 3,5 pyridylene, 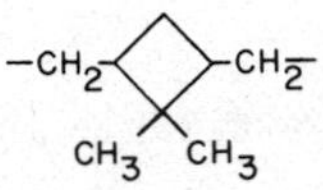 $-CH_2-$[cyclobutane]$-CH_2-$ with CH_3 CH_3	167
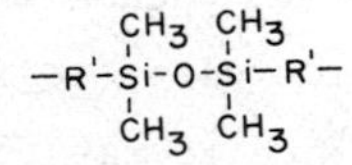 CH_3 CH_3 $-R'-Si-O-Si-R'-$ CH_3 CH_3 R' is: C_2H_5 or p-C_6H_4	111 113
 $-C_6H_4-Si(C_6H_5)(C_6H_5)-C_6H_4-$ (C_6H_5, Si, C_6H_5)	88 90
$-CH=CH-$ (trans)	165
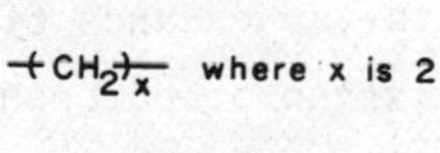 $-CH_2-C_6H_4-CH_2-$ $-(CH_2)_x-$ where x is 2, 4, 6, or 8	163
2,4- or 2,7-napthalylene	106

Table 6.2 (Continued)

Structure	Ref.
1,1' ferrocenylene	123
imidazole-4,5-diyl (N, NH ring)	123
$-(CH_2)_8-C(=O)-NH-(CH_2)_6-NH-C(=O)-(CH_2)_8-$	96 85
$-(CH_2)_8-C(=O)-O-(CH_2)_6-O-C(=O)-(CH_2)_8-$	84 96
(a) $[-B(NH)_2C_6H_3-C_6H_3(NH)_2B-R-]_n$ where R is: m- or p-C_6H_5 or 1,1' ferrocenylene	109 173
(b) $-(CH_2)_4-$	173
$[-C(NH)(N)C_6H_2(N)(NH)C-R-]_n$ R is: m- or p-C_6H_4, $-(CH_2)_4-$,	167
$-C_6H_4-SO_2-C_6H_4-$	114

Table 6.2 (Continued)

Structure	Ref.
SO_3H, SO_3H, $-(CH_2)_8-$	162
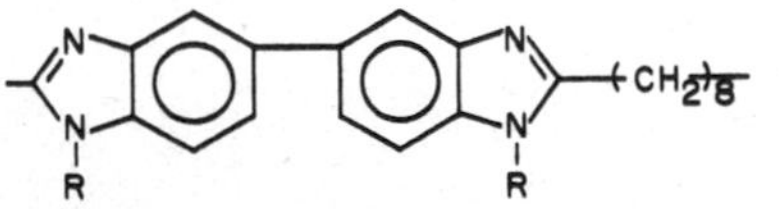 $-(CH_2)_8-$; where R is: $-CH_2CH_2OH$, $-CH_2CH_2CN$, $-CH_2CH_2CO_2H$	162
where R is: m-C_6H_4	52 37
$-C_6H_4-SO_2-C_6H_4-$	114
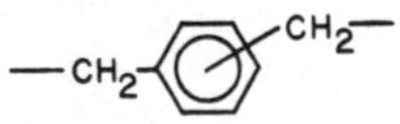 $-CH_2-C_6H_4-CH_2-$	163
$-(CH_2)_{2\ or\ 8}-$	
$-C_6H_4-P(=O)(CH_3)-C_6H_4-$	85
where R is: m-C_6H_4	52

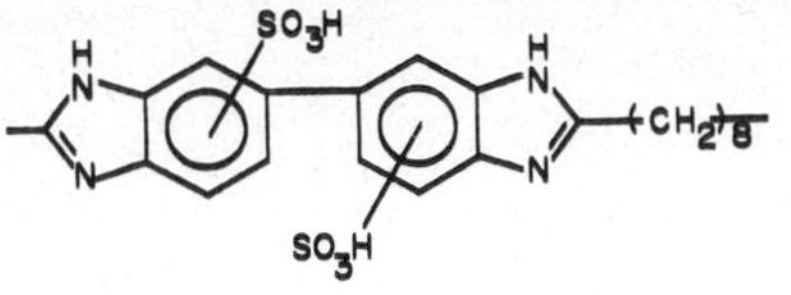

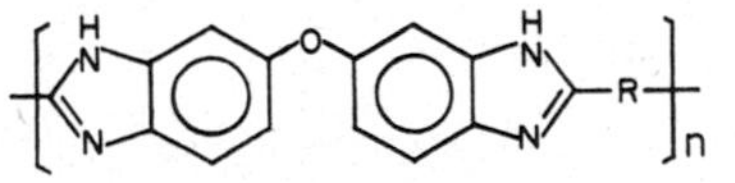

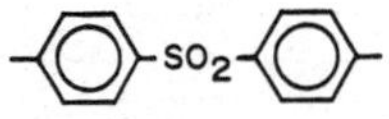

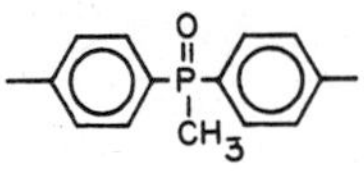

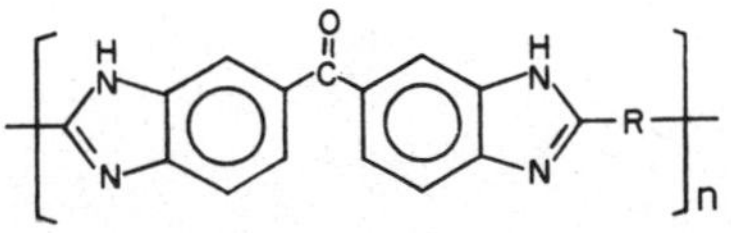

Table 6.2 (Continued)

Structure	Ref.
(structure: H, N, SO_2, H, N, R, n) where R is: m-C_6H_4 or 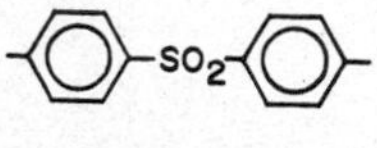 $-C_6H_4-SO_2-C_6H_4-$	114
(structure: C_6H_5, N, Ar, C_6H_5, N, R, n) Ar is: (structures; SO_2) R is: m-or p-C_6H_4, p,p'-biphenylene, p,p'-diphenylsulfone or ether, 2,6-napthalylene, 2,6-pyridinylene	82 168
(structure: H, N, CH_2, H, N, R, n) where R is: m-C_6H_4, $-(CH_2)_8-$ $-(CH_2)_4-$, p-C_6H_4, 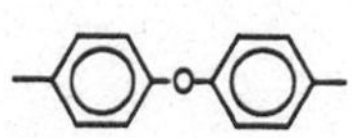 $-C_6H_4-O-C_6H_4-$	163 85
(structure: H, N, $-CH_2CH_2-$, n)	147

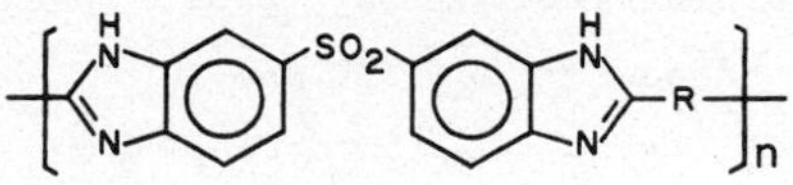

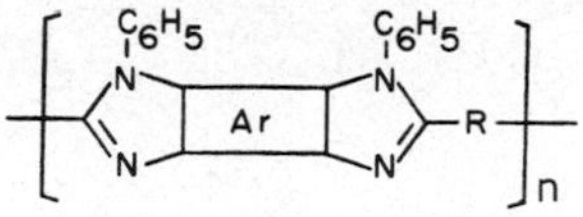

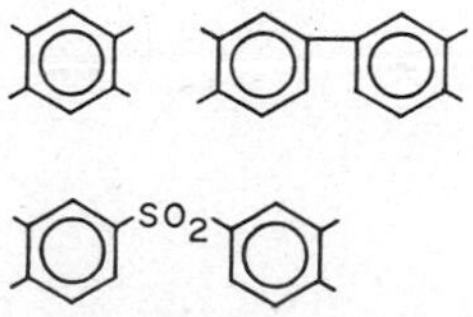

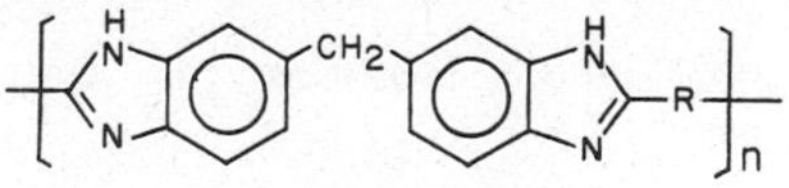

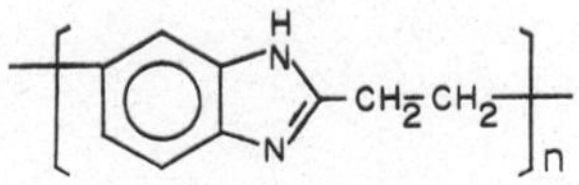

The equation showing two possible mechanistic routes for one of the simplest PBIs is given below. Both may have merit, depending on the conditions of the reaction. Now, of course, only one of three possible isomers of the polymer is shown. The others, which can be pictured by rearranging the double bond in the imidazole ring, occur as does the one shown.

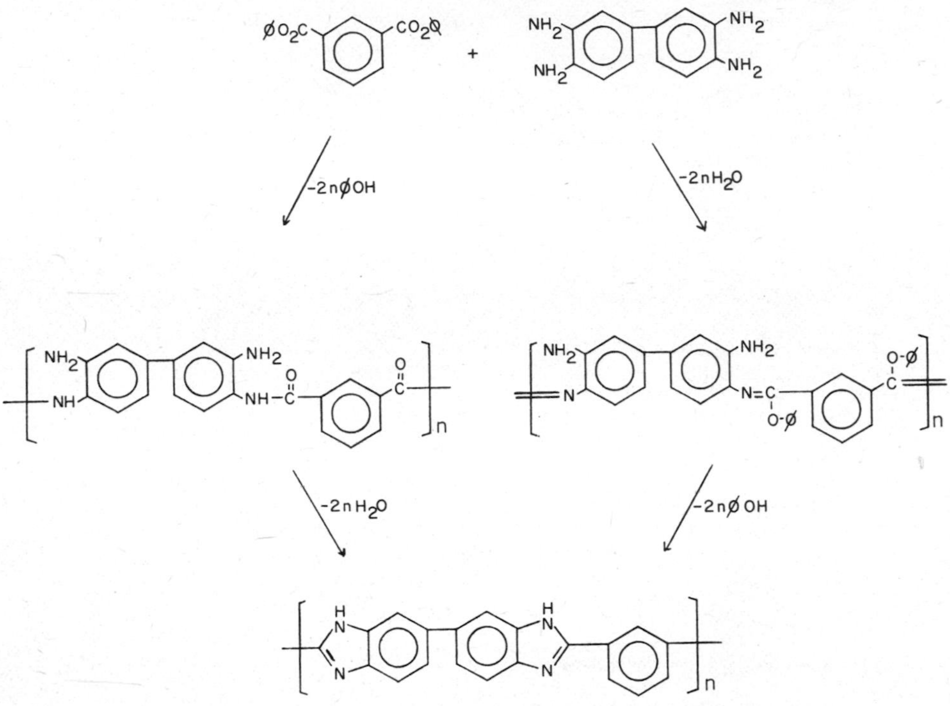

The intermediate amic acid is soluble and can be formed to films or fibers prior to the second, cyclization, step. The final product is soluble in sulfuric acid.

General precautions to follow are (1) prevent air from reaching the reaction due to the oxidizability of the tetraamine, (2) use very pure monomers, and (3) treat aromatic tetraamines as powerful carcinogens, which most of them are.

Polymerizations have been reported under melt or bulk conditions and in solutions of PPA, m-cresol, DMAc, DMF, DMSO, NMP, and phenol.

Method [19] *Poly[2,2'-(m-phenylene)-5,5'-bibenzimidazole]* In a 1-liter flask equipped with mechanical stirrer, Dean-Stark azeotrope trap, and gas inlet are placed 26.784 g (0.125 mol) of purified 3,3'-diaminobenzidine and 39.791 g (0.125 mol) of diphenylisophthalate. The system is deaerated by alternate vacuum and nitrogen purge cycles, and while under nitrogen is stirred and heated for 2 hr up to 250°C.

Then, without stirring, the temperature is raised to 290°C and held for 1.5 hr. A total of 22 ml of phenol and water collects in the trap to leave a yellow-brown friable residue; inherent viscosity 0.2-0.3 dl/g in H_2SO_4. The polyamic acid prepolymer is ground, placed in a reactor, degassed, and heated from 220 to 385°C at a rate of 1.5°C/min and held at 385°C for 3 hr. The resulting yellow-brown solid PBI has an inherent viscosity of 0.8-1.2 dl/g in H_2SO_4.

Since the late 1960s, the synthetic work on PBI encompassed mostly modifications to known polymers to improve their solubility. For example, groups were attached to the amino nitrogen; ether and carbonyl links were introduced between rings in the diaminobenzidine; and aromatic and aliphatic silanes or siloxanes were placed in the backbone. Further, two new routes to PBI were opened: from the reaction of tetraamines with dialdehydes or dinitriles. The remainder of the work reported was not synthetic in nature but rather investigated structure property relationships, processing techniques, mechanical behavior, and applications of PBI to uses as forms, adhesives, fibers, films, etc.

Hedberg and Marvel developed a single-step polymerization for the common PBI from diphenylisophthalate and 3,3'-diaminobenzidine (DAB). The ether and phenone analogs of DAB also gave polymers [52]. The key is to use refluxing sulfolane or diphenylsulfone. The method gives high yields (99%) and viscosities (0.69 dl/g in H_sSO_4) with no crosslinking. Other advantages are easy removal of byproducts, facile deaeration, and ready solvent removal. The yellow, meta polymer is soluble in DMAc and has a molecular weight (M_n) of 18,000. The para isomer proved more difficult and was successful only in diphenylsulfone to an inherent viscosity of 1.03 dl/g. Due to a higher reaction temperature in diphenylsulfone, a reaction time of only 1 hr is successful, compared to 72 hr in sulfolane for the meta isomer. The para-phenylene polymer was also more difficult to dissolve, H_2SO_4 or refluxing DMAc with 1% LiCl being necessary. To aid the monomer dissolution, an inert solvent is included at the start of the reaction; it is subsequently removed with increased temperature.

Method [52] A mixture is prepared with 50 ml of toluene, 200 g of diphenylsulfone, 5.357 g (0.0250 mol) of diphenyl terephthalate, and 7.958 g (0.0250 mol) of 3,3'-diaminobenzidine. The mixture is deaerated with nitrogen and heated with stirring to 270°C in 1 hr. The suspension turns orange and some precipitate forms. Heating at reflux is continued for 5.5 hr. The mixture is then cooled to 150°C and poured into 2.5 liters of acetone; and the yellow precipitate is collected, washed with acetone, and extracted for 1 day with ethanol and for 1 day with benzene. The PBI is dried at 120°C in vacuo to yield 7.70 g (100%) of yellow solid with inherent viscosity of 1.03 dl/g in H_2SO_4.

Silane and siloxane (aliphatic and aromatic) copolymers with benzimidazoles have been investigated since 1967. Kovacs et al. coupled a tetraphenylsilane nucleus between benzimidazoles to obtain yellow, soluble (in DMF, DMAc, DMSO, and pyridine), flexible-film formers [88,89]. Condensations were run in N,N-dimethylaniline with postcures in refluxing tetralin to give inherent viscosities up to 1.23 dl/g with molecular weight of 126,000.

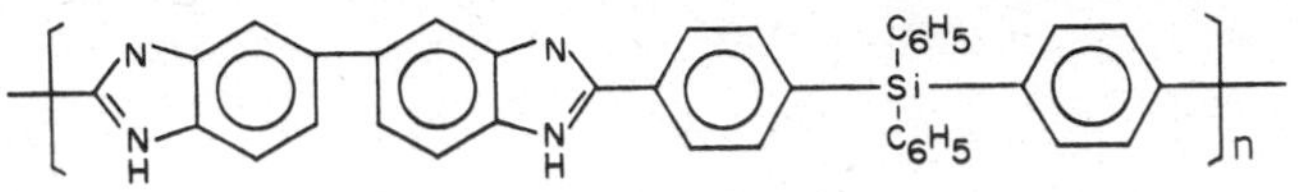

Nakajima and Marvel then published the method for preparing a rubbery aliphatic siloxane copolymer [113]. This polymer was light tan and began to decompose at 500°C; so the aliphatic portion caused some lowering of resistance to thermooxidation.

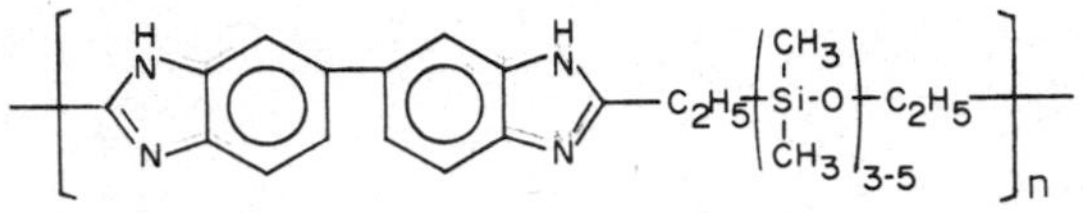

Most recently reported is an aromatic siloxane system which is not only soluble in pyridine and NMP but retains thermal stability above 500°C by TGA and shows only 3-8% weight loss after 96 hr at 350°C in air [10]. In this case then the stability has not been lessened by a gain in tractibility. Inherent viscosities in pyridine run from 0.32 to 0.59 dl/g, the para catenation giving higher values. Yields were essentially quantitative.

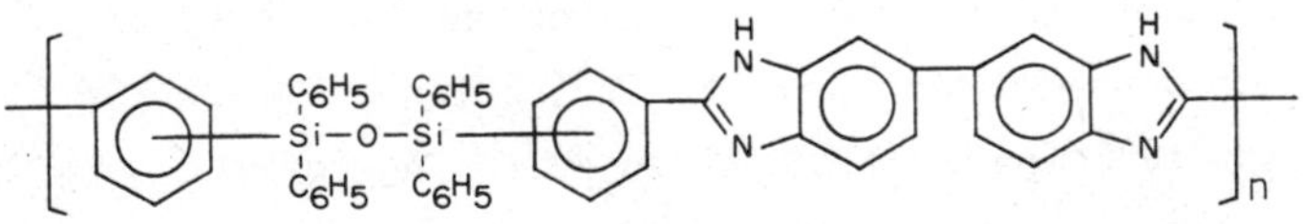

Method [10] A mixture of 0.536 g (0.00250 mol) of 3,3'-diaminobenzidine and 1.938 g (0.00250 mol) of diphenyl-4,4'-(1,1,3,3-tetraphenylsiloxanylene)dibenzoate is heated under nitrogen according to the following schedule: to 300°C in 1.4 hr, at 300°C for 2 hr, 300-350° in 0.4 hr, at 350-355° for 2 hr, and finally under vacuum at 170° for 20 hr. During the process a vacuum is applied when necessary to remove phenol and water. The yield of yellow-gold polymer is quantitative; inherent viscosity 0.47 dl/g in pyridine.

A new route to PBI was shown to be possible by the alkoxide-catalyzed addition of tetraamines to aromatic dinitriles [120]. However, due to the chain termination reactions, only low molecular weight species were obtained. Interestingly, the use of tetracyanobenzene gave structures similar to the polypyrrones, which are discussed in Chapter 9.

A more successful new route to PBI, via the reaction of various tetraamines with the bis(bisulfite adduct) of isophthalaldehyde [61], was discovered by Higgins and Marvel. Mild conditions and short reaction times give products with moderate inherent viscosities (0.3-0.5 dl/g in formic acid).

Method [61] *Monomer preparation* Into 500 ml of methanol and 75 ml of water are added with stirring 5.0 g (0.037 mol) of isophthalaldehyde and 8.80 g (0.085 mol) of sodium bisulfite. The mixture is stirred at 25°C overnight and then filtered. The solid is washed with 50 ml of warm methanol and dried in vacuo at 100°C for 8 hr to yield 10.5 g; mp 300°C.

Polymerization A solution of 0.6310 g (0.00294 mol) of 3,3'-diaminobenzidine and 1.0388 g (0.00304 mol) of the above bis(bisulfite adduct) of isophthalaldehyde in 100 ml of DMAc (or DMF, NMP, or DMSO) is heated at reflux under nitrogen for 5 hr. Then about 65 ml of DMAc is removed by vacuum distillation, and the residue is poured into 100 ml of water to precipitate a cream-colored polymer. The solid is isolated by filtration, washed well with water, and dried in vacuo at 160°C for 10 hr. The yield of poly-2,2'-(1,3-phenylene)-5,5'-bibenzimidazole is quantitative, and the inherent viscosity is 0.47 dl/g in formic acid.

The structure-property relationships of benzimidazole polymers are discussed in most of the papers published on them. After all, the purpose of most synthetic work has been to gain certain properties by modification of the backbone. The general relationships between various backbones is the same with PBI systems as other polymers. Some of these can be summarized as follows: (1) aromatic nuclei display highest stability (>500°C in nitrogen) and lowest tractibility [82,162], and para more so than meta; (2) ether and sulfone linkages increase solubility and flexibility and lower thermooxidative stability [82,163]; (3) aliphatic groups lower thermal stability [163,165] by 150-350°C in nitrogen; (4) substitution of a methyl onto the amino nitrogen lowers the softening temperature by 140°C, increases solubility by a factor of 5, and decreases thermal stability by 100°C (TGA), all due to loss of hydrogen bonding; and (5) methyl groups on the aromatic ring of the tetraamine lower softening temperature by 10-40°C, and larger groups effect an additional 10-30°C drop [82,163].

PBIs have demonstrated good adhesion as films when cast from solution onto glass plates. This quality of course leads to their use in glass composites, laminates, and filament-would structures. Fibers have been wet spun from DMAc solution, and a deep gold-color, woven cloth has been made from this fiber by Celanese. The cloth is said to be more comfortable than cotton (due to high moisture retention) and have greater flame resistance than Nomex (oxygen index of 28% for PBI compared to 17% for Nomex).

The U.S. Air Force has tested flight suits of PBI and found them superior to other materials. Other applications are in drogue parachutes and lines for military aircraft and ablative heat shields. However, owing to the cost of the monomers, the PBI has yet to replace either material. Further, the PBI fibers show promise as reverse osmosis membranes and in graphization to high-strength, high-modulus fibers for composites. The above areas of applications have been well reviewed by Leal [94]. More recently, the development of ultrafine fibers of PBI for use in fuel cell and battery separator applications has been undertaken by Celanese Research Company in Summit, N.J.

A new and interesting technique of simple precipitation has been used to process PBI polymers into films [54]. High strength molecular composite films are obtained by collecting platelets on a glass filter. Tensile strength is in the region of 137 MPa (20,000 psi) and is attributed to the coalescence of microscopic, ordered sheets of polymer. This method has also been applied to BBL and polybenzoxazole polymers. This type of innovation in processing may well open new vistas in high-temperature polymers.

The PBI polymers have also been studied as foams [87] (including syntactic types with phenolic or glassy carbon microballoons [108]). The foams have been developed to a commercial product by Whittaker Corporation [171]. Because of its ability to be fabricated as a foam and its high char yield at 800°C (76%), the PBI polymer (m-phenylene type) is superior to other materials for a low-weight, high-strength, thermally stable, machinable insulation, much needed in the aerospace industry. Foam densities of 24-80 kg/m^3 (1.5-5.0 lb/ft^3) were produced by heating the prepolymer (amic acid), which contains only 95% theoretical of the amine in a nitrogen-purged mold, at a rate of 2°/min to a final temperature of 467°C (some going to 527°C). Silane surfactants were used to control bubble uniformity. Compressive strengths at 50% compression ranged from 100-800 KPa (167-1340 psi) at room temperature, for foam densities of 20-60 kg/m^3 (1.2-3.6 lb/ft^3), respectively. values which decreased to 60% of the original at 300°C. Compressive modulus was 0.8-6.2 KPa (1.4-10 psi) for low- and high-density foams, respectively. Dimensional recovery was excellent (>90%) at 10% compression even at 500°C. At 50% compression, recovery dropped rapidly above 300°C.

Aliphatic PBIs (from suberic acid) have been modified by sulfonation, N-hydroxyethylation, N-cyanoethylation, and N-carboxyethylation, the last in the series giving a product useful as heat-sterilizable battery separators [162]. This material can absorb electrolytes and has a wet tensile strength after three 36-hr cycles at 145°C in 40% KOH of 14 MPa (2000 psi) (compared to 57 MPa [8300 psi] for the unmodified backbone). Further, it can be molded by injection, compression, or hot-melt extrusion processes when the proper stoichiometry is used in the synthesis to cap it with acids or diamines.

6.11 Polythiazoles and Polybenzothiazoles

The simple thiazole nucleus was recognized early for its ability to be formed in a polymerization reaction. In the period 1944-1946, Erlenmeyer published a series of articles dealing with this reaction, a bisthioamide with a bis-α-haloketone, but left the polymers uncharacterized [53].

In 1961 Mulvaney and Marvel were the first to incorporate this moiety in a characterized backbone [110]. Their efforts resulted in fairly low molecular weights (~6000), however, and only fair thermal stability. Later this system was expanded and characterized by Sheehan; and its properties studied extensively [148,149].

The aromatic fused thiazoles, benzothiazoles, were first reported in 1964 and 1965 almost simultaneously by Hergenrother, Wrasidlo, and Levine [59,60] at Narmco-Whittaker and by Imai et al. [70,73] at the University of Tokyo.

There are three basic approaches to polythiazoles or polybenzothiazoles, and these are given below as procedures 1, 2, and 3, each followed by one example of a synthetic method.

The syntheses designated as procedure 1 are due to Erlenmeyer and involve the condensation of a bis-α-bromoketone with a thioamide or thiourea to give the simple linear thiazole backbone.

Procedure 1

Type A

$$NH_2\overset{S}{\overset{\|}{C}}{-}Ar{-}\overset{S}{\overset{\|}{C}}{-}NH_2 \;+\; \overset{Br}{\overset{|}{CH_2}}{-}\overset{O}{\overset{\|}{C}}{-}R{-}\overset{O}{\overset{\|}{C}}{-}\overset{Br}{\overset{|}{CH_2}} \longrightarrow \left[Ar{-}(\text{thiazole: N, S}){-}R{-}(\text{thiazole: N, S}) \right]_n$$

Type B

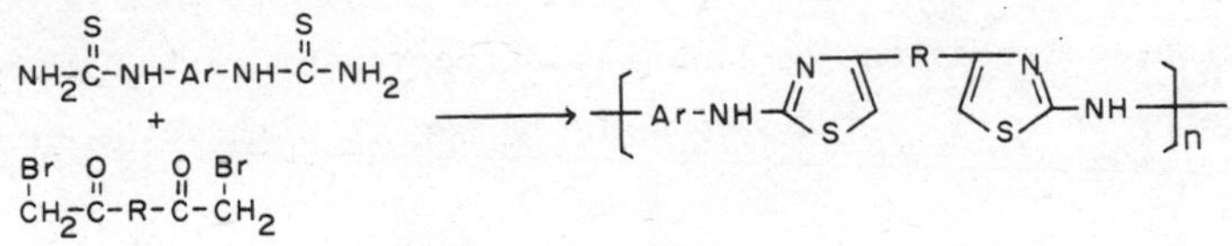

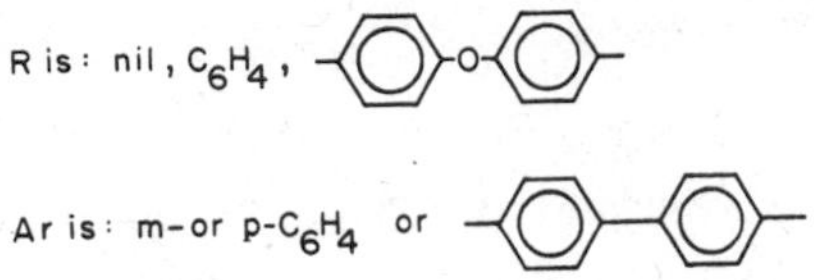

$$NH_2\text{-}\overset{S}{\overset{\|}{C}}\text{-}NH\text{-}Ar\text{-}NH\text{-}\overset{S}{\overset{\|}{C}}\text{-}NH_2 + \overset{Br}{CH_2}\text{-}\overset{O}{\overset{\|}{C}}\text{-}R\text{-}\overset{O}{\overset{\|}{C}}\text{-}\overset{Br}{CH_2} \longrightarrow$$

R is: nil, C_6H_4,

Ar is: m- or p-C_6H_4 or

Method *Poly(p-phenylenethiazole) by procedure 1* [110] A solution of 3.20 g (0.0100 mol) of p-bis(bromoacetyl)benzene [128] and 1.963 g (0.0100 mol) of dithio-1,4-benzenedicarboxamide [35] in 300 ml of DMF

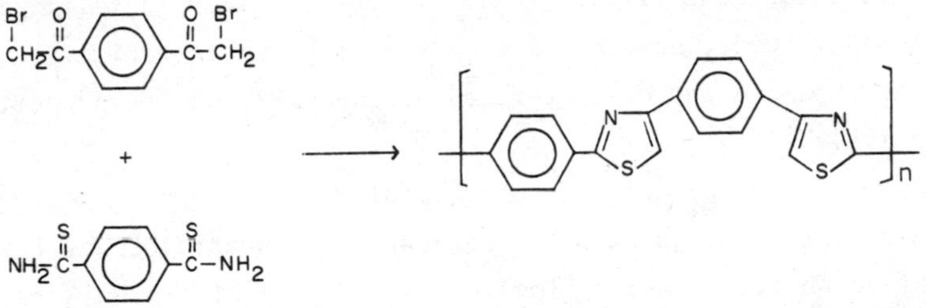

is heated with stirring to reflux for 105 hr. A fine yellow powder begins to separate early in the reaction. Upon cooling of the reaction mixture the powder is collected by filtration, washed with methanol, and dried under vacuum to yield 2.810 g with inherent viscosity of 0.12 dl/g in sulfuric acid.

The second general type (procedure 2) yields the fused benzothiazole by cyclization of an ortho-mercapto aromatic amine with an acid derivative. This can be accomplished in one step in PPA at 200°C. An alternate process involves two steps; the first step is a room temperature condensation to polyamide in solution (DMF, DMAc, or pyridine). The second step is a thermal (~160°C) cyclization of the yellow precursor. The precursor may be formed into a final shape before postcure. The polythioester can also be envisioned as an intermediate but is the less favored one.

Procedure 2

Type A

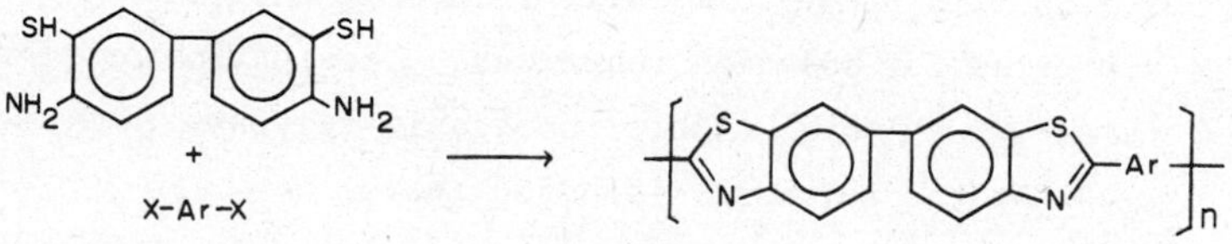

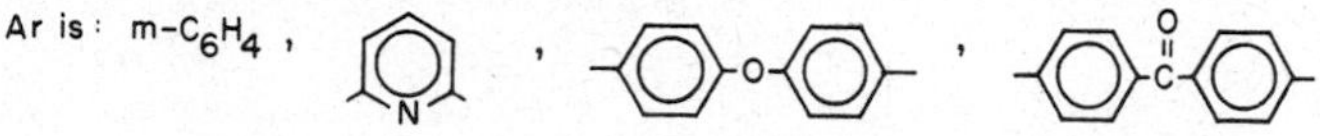

Ar is: m-C_6H_4 ,

X is: CO_2H , CO_2R , $CONH_2$, CN , COCl , $C(=NH)NH_2$

Type B

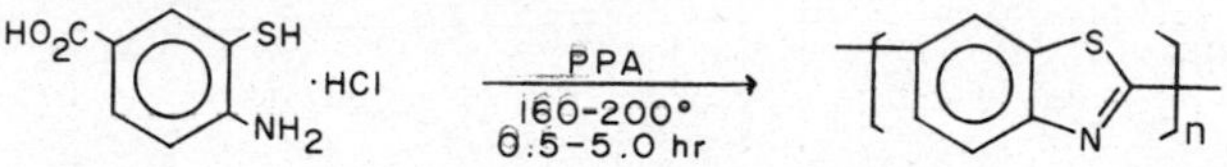

PPA, 160–200°, 0.5–5.0 hr

Method *Synthesis of a polybenzothiazole by procedure 2* [60] The dihydrochloride of 3,3'-dimercaptobenzidine [65] is stirred in PPA under argon with heating until a clear yellow solution results. To this is added an equimolar amount of isophthalic acid, and the mixture is stirred at 200°C for 1 hr. When cooled the mixture is poured into hot water in a high-speed blendor to precipitate the polymer. The solid is washed with an aqueous solution of sodium carbonate and then water to give a yellow polymer with an inherent viscosity of 1.51 dl/g in concentrated sulfuric acid. This product showed no weight loss by TGA at 593°C and a Tg of 910°C.

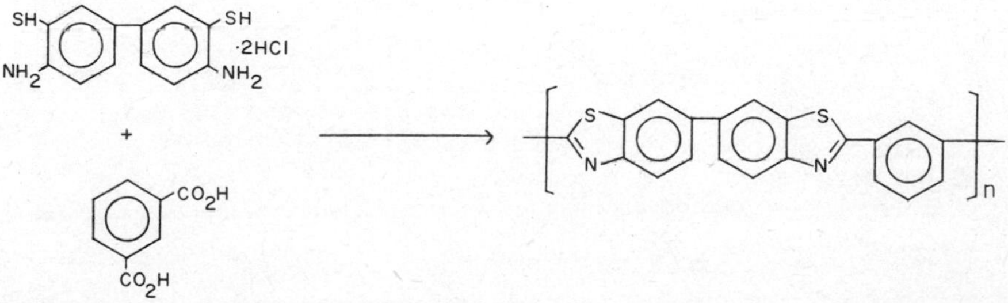

The third approach (procedure 3), also yielding benzothiazole, is one in which the benzothiazole nucleus is formed prior to polymerization, an advantage in that one is not concerned with complete cyclization, but a disadvantage since no easily soluble prepolymer is available. Here a chloro or phenoxy leaving group is displaced by electrophilic substitution. Generally the phenoxy substituent performs better in melt conditions, while the chloro group is preferred for solution reactions. These amino or ether thiazoles are generally much more soluble in organic solvents (NMP, tetrachloroethane, etc.) than the polymers discussed above.

Procedure 3

Type A

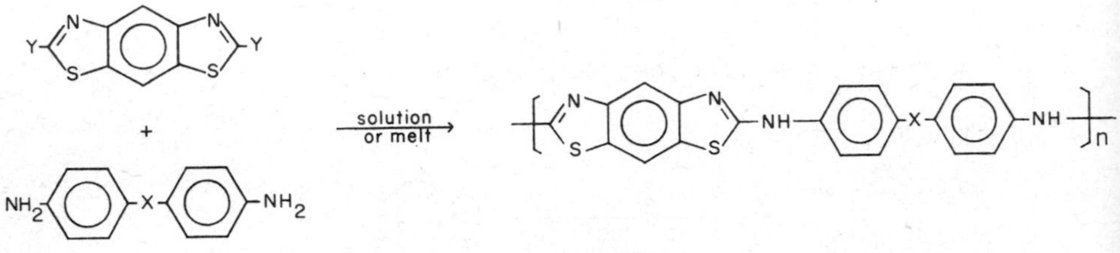

solution or melt

Y is: Cl, C_6H_5O

X is: SO_2, O, CH_2

Type B

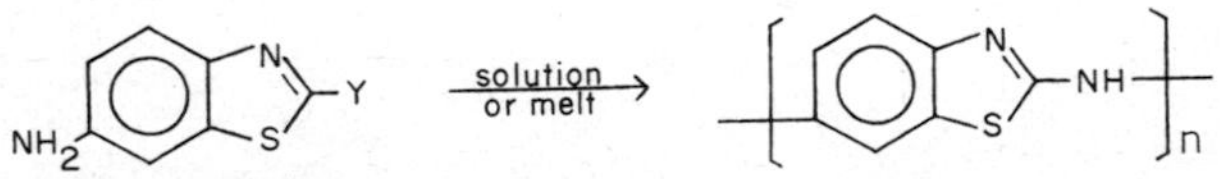

solution or melt

Y is: C_6H_5O or Cl

Type C

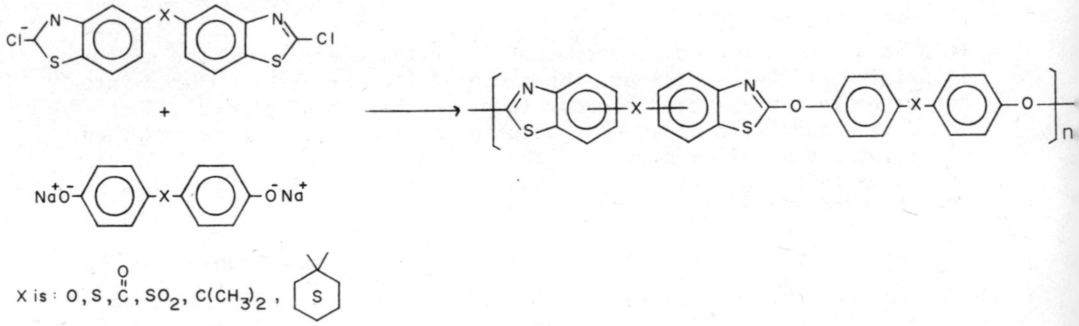

X is: O, S, C=O, SO_2, $C(CH_3)_2$

Method *Poly(biphenylene ether aminobenzothiazole) by procedure 3* [32] *Monomer synthesis* To a hot solution of 10.7 g (0.05 mol) of 2-chloro-6-nitrobenzothiazole in 500 ml of 95% ethanol is added a solution of 7.0 g (0.075 mol) of phenol and 4.2 g (0.075 mol) of KOH in 50 ml of water. The solution is heated to reflux overnight and treated with charcoal. The filtrate is diluted with 200 ml of water, the volume is reduced, and the solution cooled to induce crystallization (yield 12 g of 2-phenoxy-6-nitrobenzothiazole, orange needles, mp 114-115°C).

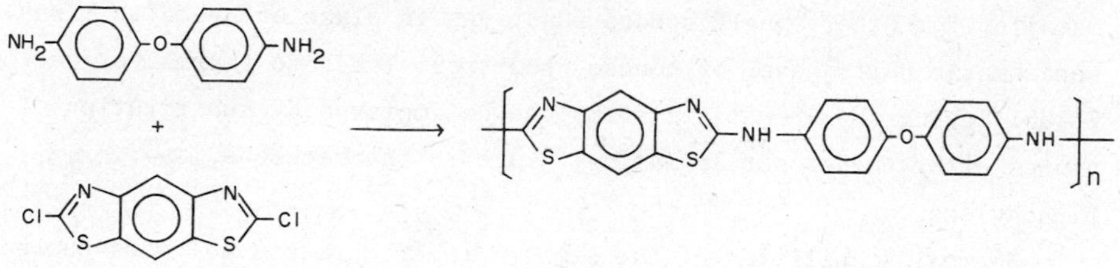

A mixture of 25 g of powdered iron, 7.0 ml of glacial acetic acid, and 150 ml of water is heated at 85°C for 15 min. To this, 12.0 g of the above 2-phenoxy-6-nitrobenzothiazole is added in portions for 30 min, and stirring is continued for 1 hr at 90°C. The mixture is cooled to 0°C and filtered, and the residue is dried and then extracted with hot benzene. The benzene extract is treated with charcoal, washed with dilute aq. potassium hydroxide and then with water, and finally dried over anhydrous sodium sulfate. Hydrogen chloride gas is bubbled through the dry solution to precipitate the amine salt, which is isolated and dried; mp 233°C (dec.). The salt is dissolved in water, the solution filtered, and the free amine precipitated by the addition of dilute aq. potassium hydroxide. The product (7.0 g) is a light tan powder, mp 78-79°C.

Polymerization The above 2-phenoxy-6-aminobenzothiazole (5.0 g) is heated under nitrogen in a polymerization tube at 260-280°C for 1 hr. The tube is cooled and the yellow solid is powdered and heated an additional 30 min at 280°C. The polymer is then dissolved in DMAc (70% soluble), the solution is filtered, and the polymer precipitated into a tenfold excess of methanol. After being isolated, dried, and ground to a powder, the product is extracted (Soxhlet) for 100 hr with methanol and dried again at 150°C for 8 hr to yield 2.8 g of a gree-yellow powder (inherent viscosity 0.29 dl/g in hexamethylphosphoramide).

The solution self-condensation of 2-chloro-6-aminobenzothiazole is not entirely successful in that low yields of low molecular weight polymer are obtained.

In synthesizing and processing polythiazoles with both aromatic and aliphatic linkages, Sheehan demonstrated properties that remain today as nemeses of polymer chemists and engineers [148]. Briefly, these are that thermal stability is inversely related to all other desirable properties of the system (solubility, tractibility, and film- and fiber-forming ability). This is so because to interject tractability to a backbone, one must introduce methylene or other singly bonded functions in place of aromatic rings. These nonaromatic units have, of course, poor resistance to thermal oxidation. Solubility of the aromatic systems can be improved by substitution of pendant groups or by the proper aromatic linkage in the backbone, e.g., 2,2'-biphenylene.

Thermal stabilities of the simple linear thiazole polymers depend largely on the connecting units, the thiazole function itself being resistant to thermooxidation and electrophilic substitution. Aliphatic backbone inclusion gives melting points of 164-250°C, with decomposition beginning below 500°C. Specifically, a 10% weight loss by TGA in nitrogen occurs at 475°C, and the isothermal analysis shows a 4.2% loss at 300°C in 24 hours.

For the aliphatic thiazole backbone, it is possible to spin fibers by melt, wet, or dry techniques and to cast films from solutions of formic acid [149]. The highly crystalline fibers shown an elongation of 20-60%, depending on molecular weight, modulus of 18-25 g/den (at 4% elongation), and Tm of 242°C. The films were flammable, yellow, and transparent with tensile strength of 45 MPa (6500 psi), 4-25% elongation, 1.06 Pa (1.5×10^5 psi) modulus, 3-8% water absorption (a value similar to cellulose acetate, decreasing with higher molecular weights), and a high water vapor transmission rate (up to 42 g m^{-2} day^{-1}).

Aromatic links have the immediately obvious advantages of higher thermal stability and melting point, 520°C (TGA 20% loss in nitrogen at 585°C, isothermal 20% loss in 1 hr at 500°C). It has been shown that up to 625°C the aromatic polybenzothiazoles crosslink with elimination of hydrogen, 1/10 of the sulfur (as H_2S), and 1/28 of C=N links (as HCN), to leave a Schiff base backbone link [31]. However, the inherent disadvantage of aromatic links is the chain stiffness which they contribute. This molecular property translates to the macroproperty of brittle films or fibers, if these forms can indeed be produced. The surface of these materials has been known on light exposure to turn to a pink hue due to photooxidation [20].

The benzo systems show solubility similar to the linear simple thiazoles which are aromatically linked, i.e., soluble only in concentrated H_2SO_4. They have good resistance to acidic or basic hydrolysis (boiling sulfuric acid or 20% aq. NaOH), inherent viscosities up to 0.6 dl/g, and quite good thermal stability if totally aromatic (TGA 10% loss in air at 560°C). Their Tg's are around 365°C. These polymers can be postcured at 400°C to improve their TGA results but with a lowering of the carbon content to 8% below theoretical. That is to say, postcuring probably changes the molecular structure considerably from ideal.

6.12 Poly(1,3,4-oxadiazoles)

The 1,3,4 isomer of the oxadiazole family is far more widely used in polymer backbones than the 1,2,4 or 1,2,5 isomers, which are discussed later. The oxadiazole ring has been known for some time, but not until about 1958 did Huisgen and co-workers [67,135] begin its detailed exploration. Then in 1961 Abshire and Marvel [1] prepared a long series of the polymers from bistetrazoles and diacid chlorides. Finally, in 1964, Frazer et al. [43] developed one of the most popular synthetic routes, through the cyclodehydration of polyhydrazides. This last method offered the advantage of a soluble and processable intermediate polyhydrazide polymer, which can be cast to films, spun to fibers, etc. Once formed, articles are converted by heat treatment to the more thermally stable poly(1,3,4-oxadiazoles).

Five reaction schemes have been successful in producing poly(1,3,4-oxadiazoles). Each of these will be discussed in the following text and the most common or useful example of a synthesis will be given in detail. The first two presented are by far the most widely used: (1) cyclodehydration of a polyhydrazide or (2) polymerization of a diacid (or its derivative) with hydrazine.

The first approach, cyclodehydration of a polyhydrazide, was first used by Frazer et al. [43], although Stolle [160] had many years earlier used the analogous reaction to synthesize 2,5-diaryl-1,3,4-oxadiazoles. The procedure is carried out as a two-step process. In the first step a polyhydrazide is synthesized and isolated. At this point, the polyhydrazide may be processed into a fiber or film or perhaps applied as a hot-melt adhesive. The polyhydrazide is then converted by cyclodehydration to a polyoxadiazole. Conversion of greater than 95% of the hydrazide links is typical.

Cyclization is most often done by heating the polyhydrazide in vacuo or under nitrogen atmosphere, but can also be accomplished by heating in a high-boiling liquid which promotes dehydration, such as concentrated sulfuric or polyphosphoric acid. A temperature of 200-300°C is usually required. Conversion of polyhydrazide fibers to polyoxadiazole has also been done by exposing the fibers to the hot vapor of a refluxing liquid such as diphenylmethane (bp 283°C) [39]. The more aromatic, crystalline, and ordered polyhydrazides require more rigorous conditions in order to achieve cyclodehydration.

$$\left[-NH-NH-\overset{O}{\overset{\|}{C}}-(R)-\overset{O}{\overset{\|}{C}}- \right]_n \xrightarrow{\Delta} \left[-C\overset{N-N}{\underset{O}{\diagdown\diagup}}C-(R)- \right]_n + nH_2O$$

R is: m-or p-C_6H_4, $-C_6H_4-\overset{O}{\overset{\|}{P}}(C_6H_5)-C_6H_4-$, C_5F_{10},

$-C_6H_4-Si(C_6H_5)_2-C_6H_4-$, 1,1'-ferrocenylene, $-C_5F_{10}-$

To shorten the time needed for thermal cyclization, it is tempting to use very high temperatures, perhaps considerably in excess of 300°C. It has been demonstrated, however, that the properties of the product polyoxadiazole are markedly influenced by the cyclization temperature [40]. This influence is probably caused by competition from unwanted side reactions as the temperature and reaction rate increase. Thermal analysis of the precursor polyhydrazide can serve as a rough guide to the correct cyclization temperature. In general, a satisfactory isothermal dehydration in vacuo or under a dynamic nitrogen atmosphere, as carried out in a muffle furnace, may be accomplished at 25-50°C below the temperature at which the transition due to dehydrocyclization is observed in temperature-programmed, thermal analysis.

Conversion to oxadiazole can be followed spectrophotometrically by observing the disappearance of the 1650-cm^{-1} band (hydrazide C=O stretching) and appearance of a 1550-cm^{-1} band (oxadiazole C=N stretching) and a weak band at 960-990 cm^{-1} (oxadiazole).

Many different aromatic polyoxadiazoles have been synthesized by the following procedure or one very similar. The reader is referred to the work by Cotter and Matzner [21] for a tabulation of polymers which have been synthesized by this route. Frazer and Reed have given a very explicit presentation of the following procedure [39].

Method *Poly[1,4-phenylene-diyl-2,5-(1,3,4-oxadiazole)-1,3-phenylene-diyl-2,5-(1,3,4-oxadiazole)]* [98] The polyhydrazide precursor is prepared by the method of Frazer and Wallenberger [41]. The dried

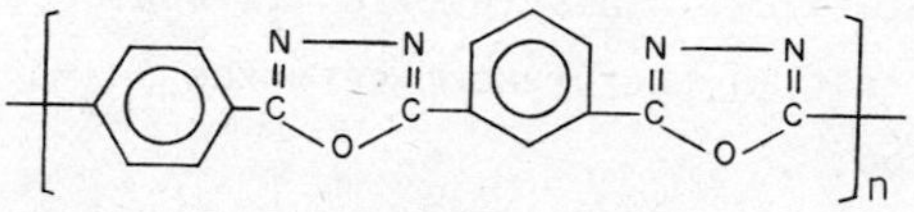

polyhydrazide is ground to a fine powder and heated in a muffle furnace under dry N_2 at 280°C for 24 hr to give virtually quantitative conversion to poly(1,3,4-oxadiazole) soluble in conc. H_2SO_4. The molecular weight of the product will, of course, depend on the molecular weight of the precursor polyhydrazide. A polyoxadiazole with η_{inh} = 1.9 dl/g (DMSO) has given by light-scattering techniques a molecular weight of 173,600 [39].

By the second process, which is also widely used, a poly(1,3,4-oxadiazole) is obtained in a one-step, solution polymerization by reaction of a dicarboxylic acid, or the corresponding nitrile, amide, or ester, with hydrazine or its salt in polyphosphoric (PPA) or fuming sulfuric acid (oleum).

Iwakura et al. [73] were the first to use this synthetic route. Their method is an adaptation of Neugebauer's synthesis in which hydrazine sulfate is reacted with p-aminobenzoic acid in polyphosphoric or fuming sulfuric acid to give bis(p-aminophenyl)-1,3,4-oxadiazole [115].

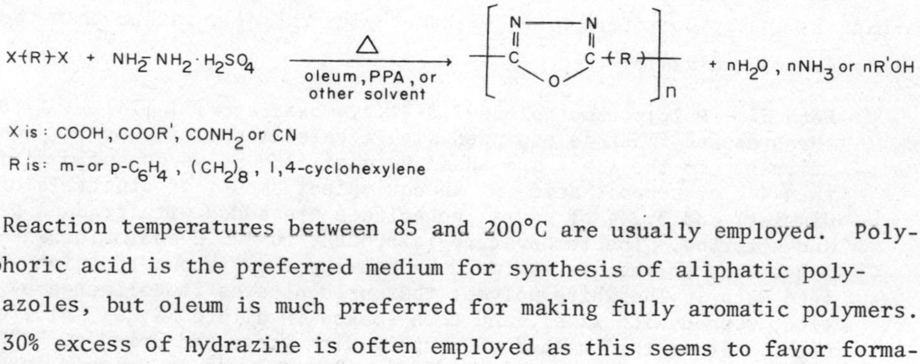

X⁅R⁆X + NH_2–$NH_2 \cdot H_2SO_4$ →(Δ; oleum, PPA, or other solvent) [oxadiazole–⁅R⁆]$_n$ + nH_2O, nNH_3 or nR'OH

X is : COOH, COOR', $CONH_2$ or CN

R is: m-or p-C_6H_4 , $(CH_2)_8$, 1,4-cyclohexylene

Reaction temperatures between 85 and 200°C are usually employed. Polyphosphoric acid is the preferred medium for synthesis of aliphatic polyoxidiazoles, but oleum is much preferred for making fully aromatic polymers. A 20-30% excess of hydrazine is often employed as this seems to favor formation of higher weight polymers.

This procedure provides a very convenient synthesis of rather high molecular weight polymers. For example, a poly(1,3,4-oxadiazole) prepared by reacting equimolar amounts of terephthalic acid and hydrazine sulfate in oleum for 3 hr at 85°C results in a polymer with an inherent viscosity of 3.7 dl/g as measured in concentrated sulfuric acid at 30°C [73]. Tough films and fibers can be formed from this polymer.

When dicarboxylic acids are used in starting materials, it is assumed that the reaction proceeds through an intermediate polyhydrazide which then loses water to form poly(1,3,4-oxadiazole). This reaction is analogous to the method discussed above, except that in the present case the polyhydrazide is generated in situ and not isolated.

It is conceivable that the reaction of nitriles or amides with hydrazine sulfate proceeds thru an intermediate poly(N-acyl hydrazidine):

$$\left[R-\overset{NH}{\overset{\|}{C}}-NH-NH-\overset{O}{\overset{\|}{C}}-R \right]_n$$

It is well known that polyphosphoric acid is an excellent reagent for hydrolysis of nitriles to amides and that hydrazine can add directly to nitriles [16] to form hydrazidines. Also, reactions of amides with hydrazine form amidrazones or the tautomeric hydrazidines.

When carboxylate esters are used in this process, the reaction probably proceeds from ester to acid to polyhydrazide, which then dehydrates to polyoxadiazole. However, esters are less satisfactory as starting material because of the possible formation of N-substituted hydrazide links as well as oxadiazole links in the product. For example, the reaction of dimethyl terephthalate with hydrazine sulfate in polyphosphoric acid gives a product which is mostly poly(terephthaloyl-N-methylhydrazide), rather than the desired poly(1,3,4-oxadiazole).

Method *Poly*[*octamethylene-2,5-(1,3,4-oxadiazole)*] [74] A 100-ml three-necked flask is equipped with a calcium chloride drying tube and stirrer. To the flask are added 90 g of 116% polyphosphoric acid and then 4 g of sebacic acid (or an equivalent amount of dinitrile or diamide) and 3.2 g of hydrazine sulfate are added with gradual heating and stirring. The temperature is brought to 140°C and maintained for 3 hr. The polymer is isolated by pouring the viscous reaction solution into water. The white polymer that precipitates is collected by filtration, washed with water, and then soaked in dilute Na_2CO_3 solution overnight. It is then washed thoroughly with water and dried at 50°C in vacuo to give 3.6 g of polymer [95]. The polymer is soluble in concentrated H_2SO_4 and formic acid (80%) at room temperature and in dimethylsulfoxide, N,N-dimethylacetamide, and N-methylpyrrolidone at elevated temperature (η_{inh} = 0.95 dl/g, 0.2 g/100 ml in conc. H_2SO_4).

The third approach to poly(1,3,4-oxadiazoles) is also due to Iwakura et al. [74,75] and involves self-polymerization of a dihydrazide in polyphosphoric or fuming sulfuric acid. Presumably, this reaction proceeds through an intermediate polyhydrazide formed by elimination of hydrazine.

Again, polyphosphoric acid is preferred for the synthesis of aliphatic and alicyclic products; fuming sulfuric acid is preferred for synthesis aromatic products.

$$NH_2NH-\overset{O}{\overset{\|}{C}}-(R)-\overset{O}{\overset{\|}{C}}-NH-NH_2 \xrightarrow[\text{PPA or oleum}]{\Delta} \left[\text{oxadiazole ring}-(R) \right]_n + nNH_2NH_2 + nH_2O$$

R is aliphatic, aromatic, or carborane

A large number of aliphatic and alicyclic polyoxadiazoles have been synthesized using this method. The size of the aliphatic residue that can be incorporated is limited by the fact that aliphatic dihydrazides become less soluble in polyphosphoric acid as their methylene content increases. Nevertheless, dihydrazides containing 20-carbon aliphatic chains have been successfully polymerized by this method. Other potential problems are possible side reactions that can occur in the preparation of aliphatic polymers, i.e., cyclization of the monomer to form imides or cyclic hydrazides. Of course, all of these limitations are overshadowed by the fact that these polymers are not thermooxidatively stable compared to other systems.

An interesting one-step preparation of the simplest polyoxadiazole using N-methylpyrrolidone and lithium chloride in place of strong acids has been reported by Lehtinen and Sundquist [95].

$$NH_2NH-\overset{O}{\overset{\|}{C}}-\overset{O}{\overset{\|}{C}}-NH-NH_2 \xrightarrow[\text{LiCl, 160–180°}]{\text{N-methyl-pyrrolidinone}} \left[\text{oxadiazole ring} \right]_n$$

(mol. wt. = 6000)

Another variation was used by Korshak et al. [83]. Diacid chlorides and dihydrazides were copolymerized in polyphosphoric acid to form boron-containing polyoxadiazoles with good thermal stability. For example, copolymerization of 1,2-bis(4-hydrazinocarbonylphenyl)carborane with isophthalic acid dichloride in polyphosphoric acid gave the following poly(1,3,4-oxadiazole):

$$\left[C_6H_4-\underset{B_{10}H_{10}}{C-C}-C_6H_4-\text{oxadiazole}-C_6H_4-\text{oxadiazole} \right]_n$$

This polymer was soluble in dimethylacetamide and resisted thermal decomposition to 440°C.

Method *Poly[1,4-cyclohexane-2,5-(1,3,4-oxadiazole)]* [78]

$$NH_2\text{-}NH\text{-}\overset{O}{\overset{\|}{C}}\text{-}\langle S \rangle\text{-}\overset{O}{\overset{\|}{C}}\text{-}NH\text{-}NH_2 \xrightarrow[PPA]{\Delta} \left[\langle S \rangle \text{-} \underset{O}{\overset{N—N}{C \quad C}} \right]_n$$

Preparation of dihydrazide Dimethylhexahydroterephthalate is obtained as a mixture of cis and trans isomers by catalytic hydrogenation of dimethyl terephthalate [151]. The trans isomer of the diester separates as a solid upon allowing the mixture to stand at room temperature.

trans-Hexahydroterephthalic dihydrazide is obtained by reaction of equimolar quantities of the trans-dimethylester with hydrazine hydrate. The dihydrazide is recrystallized from water and does not melt below 350°C.

Polymerization of trans-hexahydroterephthalic dihydrazide To 25 g of 116% polyphosphoric acid in a three-necked flask equipped with a $CaCl_2$ drying tube and stirrer is added 2 g of trans-hexahydroterephthalic dihydrazide. The temperature is increased gradually to 140°C, maintained there for 3 hr, and then increased to 160°C for another 3 hr. The product is isolated by pouring the viscous reaction mixture into water and collecting the precipitated polymer by filtration. The polymer is washed in water and soaked in dilute Na_2CO_3 solution overnight. It is then washed thoroughly with water and dried in vacuo at 50°C to give 1.3 g of product which is soluble only in conc. H_2SO_4 or 80% formic acid. This polymer is capable of forming tough films. (η_{inh} = 0.38 dl/g at 0.2 g/100 ml in 95% H_2SO_4).

The fourth method to be discussed has been little used but in some cases has given a quite satisfactory result. This procedure, which is primarily due to Saga and Shono [133] and to Hergenrother [57], involves cyclodeammonation of poly(N-acylhydrazidines), also referred to as N-acylamidrazones, to form poly(1,3,4-oxadiazoles) by heating in strong acids such as refluxing trifluoroacetic acid or polyphosphoric acid at 200°C.

$$\left[NH\text{-}NH\text{-}\overset{NH}{\overset{\|}{C}}\text{-}(R)\text{-}\overset{NH}{\overset{\|}{C}}\text{-}NH\text{-}NH\text{-}\overset{O}{\overset{\|}{C}}\text{-}(R')\text{-}\overset{O}{\overset{\|}{C}} \right]_n \xrightarrow[CF_3COOH \text{ or } PPA]{\Delta} \left[\underset{O}{\overset{N—N}{C \quad C}}\text{-}(R)\text{-}\underset{O}{\overset{N—N}{C \quad C}}\text{-}(R') \right]_n + 2nNH_3$$

For example, poly[2,2'-(2,6-pyridinediyl)-5,5'-(m-phenylene)-di(1,3,4-oxadiazole)], where R is 2,6-pyridinediyl and R' is m-phenylene, is obtained in near-quantitative yield by refluxing the corresponding hydrazidine in trifluoroacetic acid for 18 hr. N-acylhydrazidines also give 1,2,4-triazoles under proper reaction conditions, i.e., treatment with hot, strong acids. Thus the above poly(N-acylhydrazidine) can yield the following poly(1,2,4-triazole) [57].

$$\left[\text{triazole(N—N, N-H)} \text{-} \text{2,6-pyridinediyl} \text{-} \text{triazole(N—N, N-H)} \text{-} \text{m-phenylene} \right]_n$$

Mixed polytriazole and polyoxadiazole are obtained by heating polyacylhydrazidine under inert atmosphere. Triazole formation seems to be favored, but very rapid heating increases the amount of oxadiazole formed. The precursor poly(N-acylhydrazidine) is usually obtained by reaction of a dihydrazidine with a difunctional acid chloride. Dihydrazidines may be obtained by reaction of dinitriles with hydrazine in alcohol solution [16,56].

Method [57] *Poly[2,2'-(2,6-pyridinediyl)-5,5'-(m-phenylene)di-(1,3,4-oxadiazole)]*

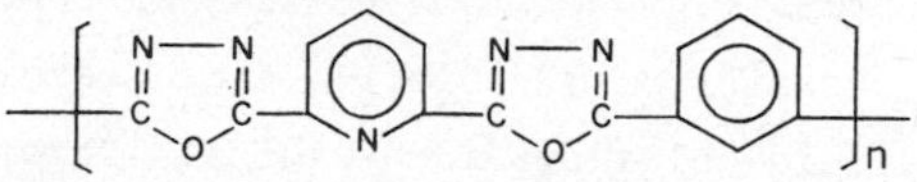

Monomer synthesis *2,6-Pyridinediyldihydrazidine* [56]

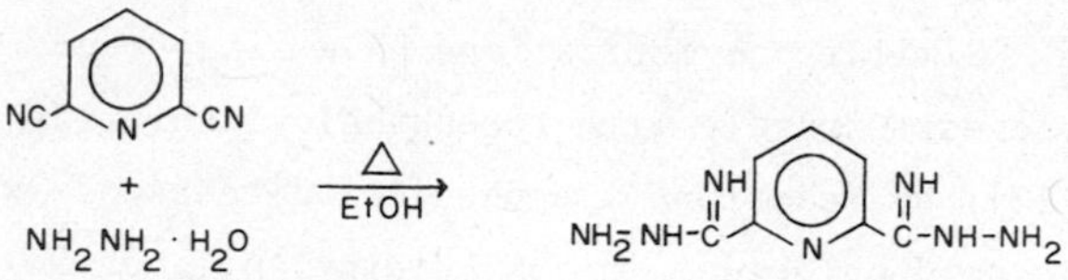

Hydrazine hydrate (95%, 20 ml) is added to a stirred solution of 2.8 g 2,6-dicyanopyridine in 250 ml ethanol at 40°C. The resulting clear yellow solution is stirred at 50-55°C for 2 hr to form a white suspension which is cooled and filtered. The white solid obtained is recrystallized from water (250 ml) to give pale yellow needles (3.0 g, 75% yield) of 2,6-pyridinediyldihydrazidine, mp 230-231°C with decomposition.

Polymerization The following reaction using solvent and reactants of high purity is performed under conditions to exclude moisture. A cold solution of isophthaloyl chloride (4.060 g, 0.020 mol) in 24 ml hexamethylphosphoramide is added under nitrogen over 0.5 hr to a vigorously stirred slurry of 3.862 g (0.020 mol) of 2,6-pyridinediyldihydrazidine, 2.50 g anhydrous lithium chloride, and 3.280 g (0.040 mol) anhydrous sodium acetate in 47 ml hexamethylphosphoramide at 5-7°C. After complete addition, the viscous yellow mixture is stirred for 3 hr at 5°C and then at room temperature for an additional 18 hr. The resulting highly viscous, yellow mixture is poured into water in a laboratory blendor to precipitate a fibrous yellow solid. This solid is washed thoroughly with water and methyl alcohol to yield 6.5 g (95% based on the monohydrate) of solid yellow polymer. An inherent viscosity of 0.71 dl/g is achievable when determined using a 0.5% solution in sulfuric acid at 25°C. The polymer is soluble in dimethylacetamide, dimethylsulfoxide, and hexamethylphosphoramide immediately after synthesis. After drying at 100°C, the polymer is only partially soluble in these solvents. This behavior has been attributed to induced ordering of the polymer upon drying rather than to partial cyclodehydration.

Cyclodehydration A yellow solution of 1.0 g poly(N-acylhydrazidine) in 20 ml trifluoroacetic acid under nitrogen is heated at reflux for

18 hr. The resulting orange solution is poured into water to precipitate a white solid which is isolated by filtration and washed successively with aq. sodium carbonate, water, and methanol. The poly(1,3,4-oxadiazole) is obtained in quantitative yield as a white solid which exhibits an η_{inh} of 0.63 dl/g (0.5% solution in H_2SO_4 at 25°C). The polymer is stable in air up to about 420°C.

The last route to poly(1,3,4-oxadiazoles) to be presented here is mostly of historical interest. Huisgen [66] was apparently the first to synthesize a polyoxadiazole by self-polymerization of a molecule which contained both an acid chloride and tetrazole group. Shortly thereafter, Abshire and Marvel [1] synthesized a series of 12 poly(1,3,4-oxadiazoles) by reaction of bistetrazoles with diacid chlorides in pyridine. Unfortunately, all polymers obtained by this route were of low molecular weight (<6000) as evidenced by inherent viscosities under 0.2 dl/g when measured in concentrated sulfuric acid. Similar polymerizations have since been carried out by reaction of m-phenyleneditetrazole with terephthaloyl chloride in hexamethylphosphoramide [112], by reaction of p-phenyleneditetrazole with 1,1'-ferrocene dicarboxylic acid chloride, in N,N'-diethylaniline [98], and by reaction of 5-aminotetrazole with diacid chlorides in dry pyridine [159]. The latter reaction yielded a polymer with both amide and oxadiazole functions in the backbone. In general, low molecular weight polymers have resulted from the tetrazole route; therefore, this method has not been widely adopted.

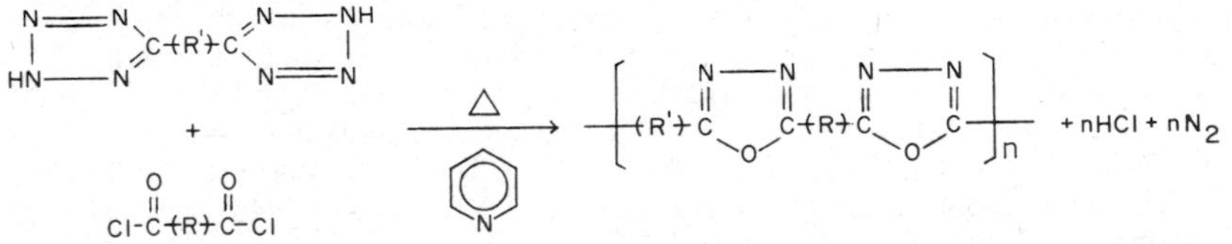

N=N, HN—N, C+R'+C, N—NH, N=N + Cl-C(=O)+R+C(=O)-Cl →(Δ, pyridine) +R'+C(N—N, O)C+R+C(N—N, O)C+$_n$ + nHCl + nN_2

R and R' are: nil, m-or p-C_6H_4, 2,2'-or 4,4'-biphenylene, $(CH_2)_7$, 2,6- or 3,5-pyridylene

Method [1] *Poly*[*1,4-phenylenediyl-2,5-(1,3,4-oxadiazole)-1,3-phenylenediyl-2,5-(1,3,4-oxadiazole)*]

[structure: N—N, C, O, C, N—N, C, O, C, n]

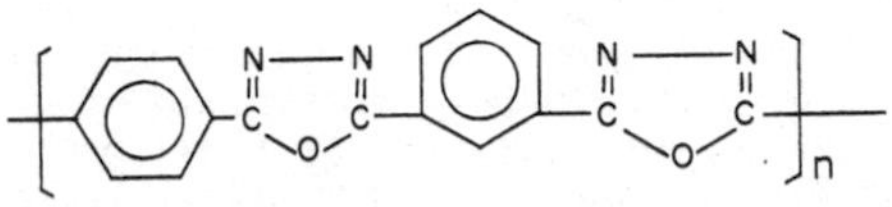

To a 250-ml three-necked flask equipped with a stirrer and reflux condenser are added 100 ml of pyridine, 4.000 g of purified p-phenylenebistetrazole [55], and 3.794 g of isophthaloyl chloride. The mixture is refluxed for 72 hr and then poured into 2 liters of CH_2Cl_2, which precipitates the polymer and dissolves pyridine hydrochloride. The polymer is collected on a filter, washed with additional CH_2Cl_2, and dried to give 3 g of white powder (η_{inh} = 0.148, 0.2% solution in conc. H_2SO_4). The product loses 12% of its weight after 12 hr at 350°C and 26% after 12 hr at 400°C in air.

Several examples of polyoxadiazoles, some of which have been alluded to earlier in this discussion, deserve special mention. In the following few paragraphs, two general types will be covered: those with organometallic or metalloid functions and those with other heteroatom functions, such as amide and imide, in the backbone.

Oxadiazole polymers containing boron, phosphorous, silicon, or iron have been synthesized. Boron, as carborane, was incorporated by Korshak et al. [83] by copolymerization of 1,2- and 1,7-bis(4-hydrazinocarbonyl-phenyl)carborane with various aromatic diacid chlorides in polyphosphoric acid. Several self-extinguishing, phosphorous-containing polyoxadiazoles have been prepared by Konya and Yokoyama [79] and by Yoshida et al. [174].

With the intention of obtaining a soluble polyoxadiazole, Kovacs et al. [88-90] have synthesized a poly(1,3,4-oxadiazole) with silicon, as tetra-phenylsilane, in the backbone. The polymer is readily soluble in dimethyl-acetamide, benzene, and chloroform, whereas the polyhydrazide starting material is insoluble in the latter two solvents. Thermally stable, flexi-ble films exhibiting good adhesion to metals and glass can be cast from solutions of this poly(silane oxadiazole).

Lorkowski et al. [97-98] have synthesized and determined the thermal stability of various aromatic, ferrocenylene-containing poly(1,3,4-oxadia-zoles). They determined that the presence of ferrocenylene groups decreases thermal stability. This phenomenon of lower stability with ferrocenylene units has been confirmed in numerous other polyaromatic heterocycles. On heating in air it is likely that the first degradation process to occur in ferrocenylene polymers is caused by oxidation of iron atoms to the plus three oxidation state. This is accompanied by marked darkening in color.

A large number of poly(oxadiazole amides) have been made. A detailed synthesis of poly[2,5-bis(p-phenylene-1,3,4-oxadiazole)isophthalamide] has been published by Preston and Black in Vol. 3 of *Macromolecular Syntheses* [124]. Patents have also been issued for processes to synthesize polymers containing various combinations of imide, oxadiazole, and imidazopyrrolone rings [27,161]. These polymers are obtained by thermal cyclization of the corresponding polyhydrazides. Polymers containing oxadiazole, amide, and
corresponding polyhydrazides. Polymers containing oxadiazole, amide, and

imide functional groups have also been synthesized by Bruma et al. [12]. A process for synthesizing a novel polymer containing phenylene ether and oxadiazole groups in a backbone crosslinked through amide groups has been patented by Steinmann [152]. This synthesis is based on simultaneous polymerization of p-phenylene-bis(N-aminocarbamate), isophthaloyl chloride, and melamine to form a polymeric precursor which is thermally cyclized to a polyoxadiazole.

The properties and uses of polyoxadiazoles have been extensively investigated, perhaps more so by researchers at Furukawa Electric Company, Ltd. (Japan) than by any other single source, as evidenced by the 25 patents issued to them in the period 1971-1973. The polymers are hydrolytically stable, partially crystalline, and by drawing can be oriented to a higher degree of crystallinity. They are somewhat more stable than polytriazoles but less so than polythiadiazoles.

Most aliphatic and alicyclic polyoxadiazoles are soluble in dimethylsulfoxide and a few other highly polar solvents such as N,N'-dimethylacetamide and N-methylpyrrolidone. This gives them additional processability not found in fully aromatic polyoxadiazoles, which are soluble only in strong acids such as concentrated sulfuric or trifluoroacetic acid. Aliphatic polyoxadiazoles are, of course, also soluble in concentrated sulfuric acid. But this is not the solvent of choice for these polymers.

Aromatic backbone units are used much more widely than aliphatic ones because they are more thermally stable (400-450°C compared to 300-350°C by TGA), and they have significantly higher melting temperatures (above 400°C compared to 100-200°C). Although aliphatic polyoxadiazoles are more soluble than the aromatic species, the latter can be processed by virtue of the tractable polyhydrazide intermediate which is available through some synthetic routes. This technique was the key to the development of these polymers.

The most widely known use of this type of polymer is in films and fibers. Polyoxadiazoles form yellow or brown transparent films with a percent elongation-to-break reported as high as 138%, but more commonly in the 25-50% range. Tensile strengths are near 118 MPa (1200 kg/cm^2), and upon weathering or heat aging this value drops to about 78 MPa (800 kg/cm^2). Immersion of the polyoxadiazole film in N,N-dimethylacetamide followed by drying increases the tensile properties significantly [116].

An interesting film has been produced by compounding a polyoxadiazole with a thermoplastic resin (polyethylene, polyacrylonitrile, or PVC) in sulfuric acid [64]. The solution is cast to a film, gelled in water, and the thermoplastic particles extracted with hot aromatic solvents. The result is a porous polyoxadiazole film. Another approach to a porous, heat resistant film is gelling a film from a sulfuric acid solution by immersing it in water then freezing the water-impregnated gel. The interstitial water is then removed by lyophilization [138].

The polyoxadiazoles can be made into candidates for flame- or fire-resistant or self-extinguishing applications. One approach to decreasing flammability is to include phosphorus in the backbone. This was accomplished by the use of triphenylphosphine oxide dicarboxylic acid halides [174] or hydrazides as monomers [175]. Phosphorus-containing polyoxadiazoles retained good thermal stability (300-450°C) even though aliphatic units were present. Another and more direct method is to simply include a gypsum filler in the polyoxadiazole mixture [137]. Simple aliphatic polyoxadiazoles have been shown to impart a flame resistance to silk fibers when coated onto the fiber from a 10% solution and dried [164].

Moldings have been made from polyoxadiazoles by casting a film from a solution in concentrated sulfuric acid, gelling the film in a dilute (54%) sulfuric acid, and washing the gelled film with water. This film is then pressed in a mold at 80-350°C. Neutralization of the gel prior to molding is reported to increase elongation of the final product [131]. In the gel form, the film can also be laminated with itself [145] or with paper to give an electrical insulator [141].

The water content of fibers and films is important for two reasons. A certain water content is necessary in fibers to impart comfort when in contact with skin. However, the presence of water decreases the thermooxidative stability of polymers [46]. The treatment of polyoxadiazole films by immersion in chlorinated or fluorinated aliphatic carboxylic acid for 1 day followed by drying for 2 hr at 200°C decreases its moisture absorption by a factor of 4 [139]. A sixfold difference (decrease) is realized by a similar treatment of the fiber with xylene [140]. A threefold increase in water retention is induced by treating fibers with a silicone emulsion [144].

Fibers of poly(aryl-1,3,4-oxadiazoles) are spun from sulfuric acid solution, den 2-3. They demonstrate tensile strengths from 1.4 to 5.2 g/den, elongation 5-40%, and Young's modulus of 42-54 g/den, values which, of course, depend on the composition of connecting units and processing variables [70,117]. Blends and copolymers of polyoxadiazoles, primarily with amides, are utilized to improve fiber properties [69,125,132,146].

The following table lists other applications which have been developed for poly(1,3,4-oxadiazoles).

Application	Ref.
Glass laminate composite	17
Metal adhesive	126,127
Dyeable films and fibers	91
Cation exchange resin	91
Weather-resistant, pigmented film	142,143
Graphitized fibers	34

6.13 Poly(1,2,4-oxadiazoles) and Poly(1,2,5-oxadiazoles)

The 1,2,4- and 1,2,5-oxadiazole systems have received very little attention since the mid-1960s, when syntheses of many high-temperature polymers were rampant. The characteristic functional groups of these polymers are as follows:

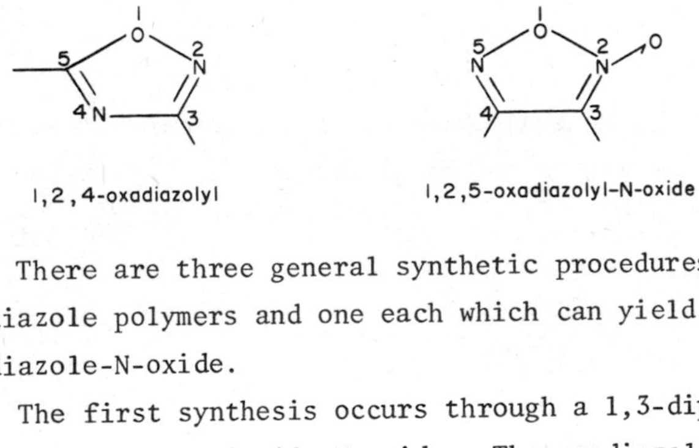

1,2,4-oxadiazolyl

1,2,5-oxadiazolyl-N-oxide

There are three general synthetic procedures which can lead to 1,2,4-oxadiazole polymers and one each which can yield the 1,2,4- or 1,2,5-oxadiazole-N-oxide.

The first synthesis occurs through a 1,3-dipolar cycloaddition of a cyano group to a nitrile N-oxide. The oxadiazole nucleus occurs in a random arrangement in the backbone in all probability when R is phenylene, although only one isomer is shown [119].

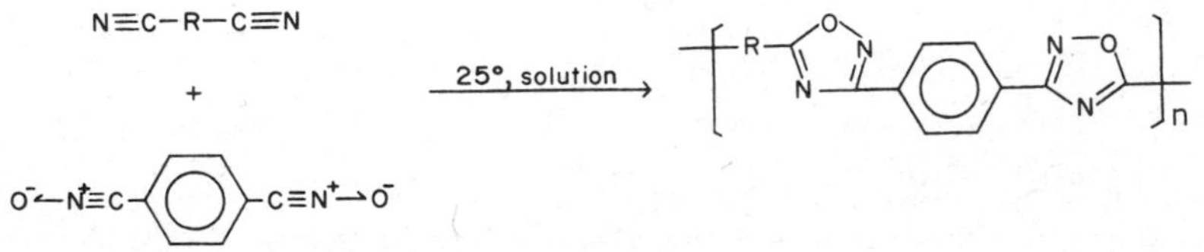

N≡C—R—C≡N + $O^- \leftarrow N^+ \equiv C-C_6H_4-C \equiv N^+ \rightarrow O^-$ —25°, solution→

R is: p-C_6H_4 or polyperfluoroethylene oxide oligomer

An R group which is an oligomeric perfluoroether has also been used [18]. In this case the reaction is run in Freon TF (difluorotrichloroethane) at room temperature. The components of this reaction are 1,3,4-tricyanobenzene, terephthalonitrile oxide, and a cyano-terminated poly(perfluoroethylene oxide) oligomer. The result is a crosslinkable copolymer of 1,2,4-oxadiazole and perfluoroether units connected through phenylene links. The product is a tough crepe gum with properties dependent upon the size of the ether segment. It is readily cured (crosslinked) by conducting another dipolar cycloaddition with the cyano group pendant in the prepolymer gum.

Method [18] *Copolymer of perfluoroethylene oxide and phenylene-1,2,4-oxadiazole*

NC CN CN + O←NC– –CN→O + $NC-CF_2(OCF_2CF_2)_x O(CF_2)_5 O(CF_2CF_2O)_{6-x} CF_2CN$

1. Freon TF, room temp., 4 days
2. CH_2Cl_2, 3 days

N O–N N N–O N O–N N N–O (oligomeric fluoroether) n CN

O←NC– –CN→O

Crosslinked elastomer

Polymerization To a 250-ml round-bottom, one-neck flask equipped with a magnetic stirring bar are added N,N'-terephthalonitrile oxide (TPNO, 2.36 g, 14.8 mmol) and 1,3,5-tricyanobenzene (0.551 g, 3.6 mmol) along with Freon TF solvent (50 ml). A solution of oligomeric fluoroether dinitrile (PCR, Inc.) (12.84 g, 11.35 mmol) in Freon TF solvent (20 mol) is added all at once to the TMNO/trinitrile suspension. The flask is capped with a vacuum adapter and, with stirring, the flask is alternately evacuated (to 300 mm) and flushed with dry nitrogen. The flask is then sealed off, disconnected from the nitrogen line, and allowed to stir at ambient temperature for 4 days. Forty milliliters of methylene chloride is then added to the suspension, and stirring is continued for an additional 3 days. Thirty milliliters of Freon TF is then added, and the suspension is filtered on a Buchner funnel, giving a clear polymer solution. The clear solution is concentrated at ambient temperature in vacuo, and the polymer is precipitated by the addition of methylene chloride (110 ml). After the supernatant liquid is decanted the polymer is dried for 20 min at 120°C in an air oven and for 1.5 hr at 115°C/1.0 mm. The tough, pale yellow gum (10.8 g, 70%) shows an inherent viscosity of 0.22 dl/g in hexafluoroisopropanol at 25°C.

The IR spectrum (film cast from Freon TF) shows bands at 1610, 1550, 1460, 1330, 1250-1110, 990, 960, 900, 880, and 860 cm^{-1}. A portion of the gum is further dried at 180°C/0.1 mm for 6 hr to provide a sample for Tg determination and elemental analysis. The polymer gum shows a Tg of -64°C with a possible Tm at 33°C.

Curing The gum is blended with 3% (w/w) TPNO on a laboratory micromill and then press-cured 20 min at 93°C (200°F). A TGA analysis in air of a cured sample of gum shows no weight loss up to 270°C and a 6% weight loss at 300°C.

A novel process for obtaining poly(1,2,4-oxadiazoles) by 1,3-dipolar cycloaddition has appeared in the literature [2] and has been patented [107]. p-Cyanobenzonitrile-N-oxide undergoes solid-state self-polymerization upon standing at room temperature. Heat or UV light may be used to accelerate the reaction. Interestingly, solution polymerizations yield little or no oxadiazole.

$$NC{-}Ar{-}\overset{\overset{NOH}{\|}}{C}{-}Cl \xrightarrow[25^\circ]{Et_3N} N{\equiv}C{-}Ar{-}C{\equiv}\overset{+}{N}{\rightarrow}\overset{-}{O} \xrightarrow[53\ days]{25^\circ} \left[\text{1,2,4-oxadiazole} {-}Ar{-} \right]_n$$

Ar is: m-or p-C_6H_4, $C_6H_4C_6H_4$, or 1,5-naphthylene

It has been hypothesized [107] that the molecules are positioned in the crystal lattice so that each N-oxide group can react only with a nitrile group and not with another nitrile N-oxide group. This reaction results in formation of the 1,2,4-oxadiazole moiety rather than 1,2,4-oxadiazole-N-oxide, which has been derived from solid-state reaction of two nitrile N-oxide groups [47,122].

This reaction can be monitored by IR spectroscopy most readily by observing the loss of monomer bands, 2300 cm^{-1} (nitrile oxide) and 2260 cm^{-1} (nitrile).

Method [2,107] *Poly(p-phenylene-1,2,4-oxadiazole)* *Monomer synthesis* Triethylamine (0.280 g) in 5 ml methanol is carefully added to a solution of 0.500 g of p-cyanobenzhydroxamic acid chloride in 5 ml methanol. Crystals of p-cyanobenzonitrile-N-oxide form in the resulting solution. These are collected, washed with methanol, and dried (yield 70%).

Polymerization Crystals of p-cyanobenzonitrile-N-oxide are allowed to stand at room temperature for 53 days. A polymer which has a melting point of 360°C results. Heat treating this polymer by gradually warming it in air to 250°C over 1 hr and then at 250°C for 2 additional hr results in an increased melting point and an inherent viscosity of 0.75 dl/g, as measured at a concentration of 0.002 g/ml in conc. H_2SO_4 at 30°C.

The second general method involves the condensation of aromatic diacyl halides with bisamidoximes. This reaction has been well characterized and shown to proceed through the poly-O-acylamidoxime [28]. Performing an initial condensation at low temperatures obviates side reactions.

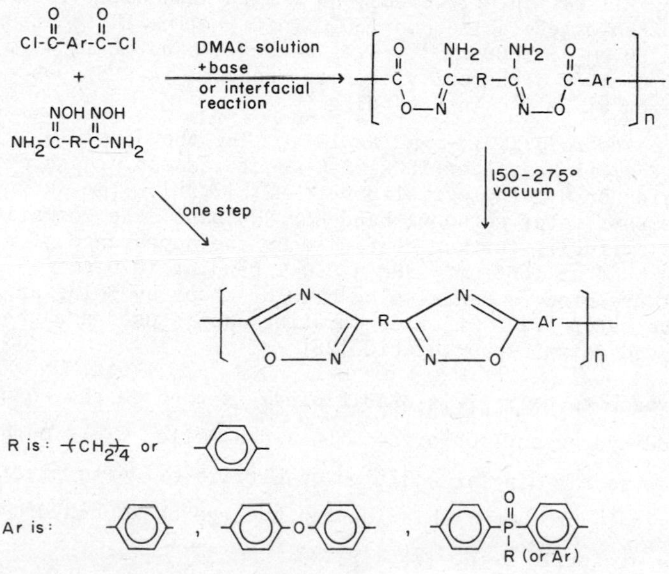

The cyclization mechanism involves nucleophilic attack of the free amino nitrogen to the carbonyl carbon. The resulting unstable intermediate was proposed to be in equilibrium with a structure which could also be envisioned as arising from acyl halide attack on the amide nitrogen:

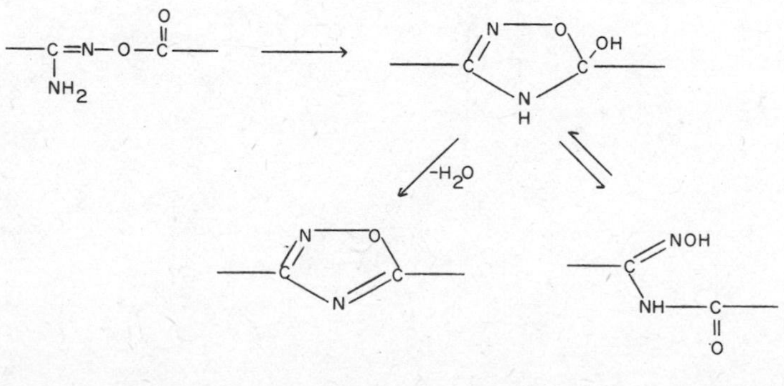

Method [28] *Polymerization of terephthalamidoxime with diphenylether-4,4'-dicarboxylic acid chloride* *Monomer synthesis* A 30.6-g (0.44 mol) portion of hydroxylamine hydrochloride is dissolved in 14 ml of water. To this solution are added 17.6 g (0.42 mol) of sodium hydroxide dissolved in 27 ml water at 0-5°C and 25.6 g (0.2 mol) of terephthalonitrile suspended in 350 ml of ethanol. The mixture is heated at reflux for 24 hr. The solvent is removed and recrystallization from water affords white platelets of terephthalamideoxime, mp 238°C. The yield is 20 g (52%). The IR spectrum (KBr) shows bands at 3250 cm^{-1} (-OH), 3300-3500 cm^{-1} ($-NH_2$), and 1650 cm^{-1} (-C=NO).

Polymerization To a solution of 0.827 g (0.038 mol) of terephthalamidoxime in 15 ml of purified N,N'-dimethylacetamide is added 1.082 g (0.038 mol) of diphenylether-4,4'-dicarboxylic acid chloride with stirring. The addition takes place over 5 min while cooling the reaction in Dry Ice acetone. 1.21 g (0.01 mol) portion of dimethylaniline is added with ice bath cooling. The mixture is then stirred at room temperature for 2 hr and at 50-60°C for 2 hr. The intermediate polymeric precipitate is collected by filtration, filtered, and washed with water to afford 1.47 g (77%) of colorless solid, mp 230°C (dec.). The IR spectrum shows NH_2 bands at 3350 cm^{-1}, 3500 cm^{-1}; ester carbonyl band at 1725 cm^{-1}; and C=N band at 1620 cm^{-1}. Inherent viscosity measured in 0.5% concentration in H_2SO_4 solution at 25.0°C in 0.04 dl/g.

Cyclodehydration to poly(1,2,4-oxadiazole) The above poly(O-acylamidoxime) (0.2 g) is heated at 150-180°C under reduced pressure in dimethylsulfoxide or N-methylpyrrolidone (NMP) [7,28]. The IR spectrum (KBr) shows neither ester carbonyl band nor NH_2 band; the formation of 1,2,4-oxadiazole ring is further confirmed by the appearance of a new C=N absorption band at 1608 cm^{-1} and a C-O-C band at 1070 cm^{-1}.

This postcure process can also be brought about by using an azeotrope to remove water [5]. Other dehydrating agents used are acetic anhydride [7] and polyphosphoric acid [28].

The third approach to poly(1,2,4-oxadiazoles) is through the condensation of terephthalodihydroxamoyl chloride with a dinitrile. This method may be quite similar to the dipolar addition of nitrile oxide to nitrile, since the terephthalodihydroxamoyl chloride can undergo basic dehydrochlorination to the dinitrile oxide [75,119].

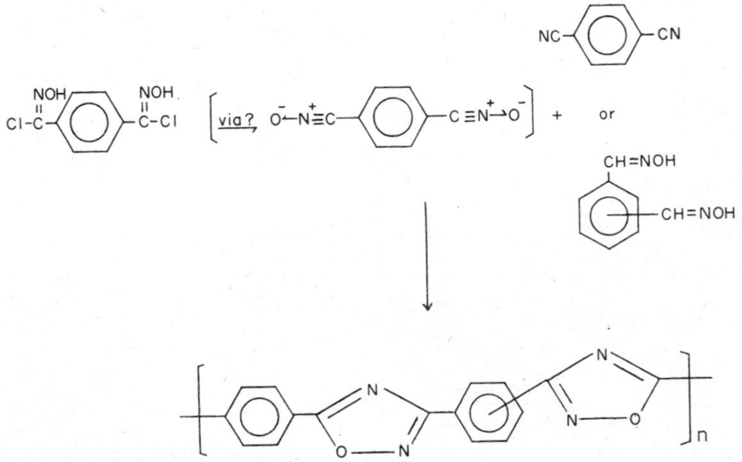

It is not clear whether the dinitrile oxide is an intermediate or whether the dihydroxamoyl chloride reacts directly with terephthalonitrile or a phthalaldehyde dioxime. The polymers were found to be more crystalline when para catenation was present in the chain compared to meta.

Little work has appeared which relates to the processing and use of these materials. Most recently a perfluoroether-1,2,4-oxadiazole polymer has been synthesized as a tough, crepe gum which can be crosslinked for use as an oil-resistant, low-temperature, flexible, thermooxidatively stable elastomer [18].

In terms of solubility and tractability these polymers are quite similar to poly(1,3,4-oxadiazoles); however, they possess considerably less thermal stability and lower Tg. For example, poly(diphenylether-1,2,4-oxadiazole) begins to decompose at 310°C in air (incipient weight loss by TGA) [76], whereas the analogous 1,3,4 polymer decomposes above 400°C. This lack of extreme thermooxidative stability in 1,2,4 polymers is no doubt the reason these polymers have been studied much less than the 1,3,4-oxadiazoles.

Dinitrile-N-oxides have been shown to self-condense also by 1,3-dipolar cycloaddition to 1,2,4- or 1,2,5-oxadiazole-N-oxide polymers, also known as polyfuroxans [47,49,68,72,107,119]. This reaction can be conducted in solution, with or without Lewis acid catalyst, to result in the 1,2,5 system or in the solid-state at room temperature or with mild heating to yield the 1,2,4 isomer. The mechanism for the formation of the 1,2,4 polymer has been elucidated by Fujimoto [47].

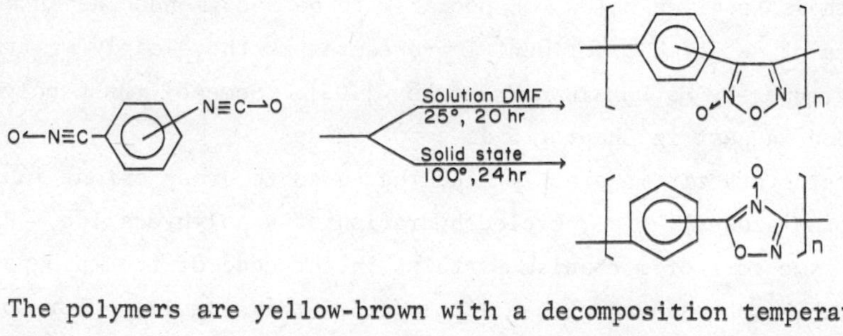

The polymers are yellow-brown with a decomposition temperature of 217-223°C and inherent viscosities up to 0.18 dl/g in sulfuric acid. Yields of polymers are 68-89% from solution but quantitative from solid-state polymerizations. The low decomposition temperature is not unexpected in view of the fact that the simplest example, poly(1,2,5-oxadiazole-N-oxide), explodes when heated [49]; i.e., the thermal stability is less than satisfactory:

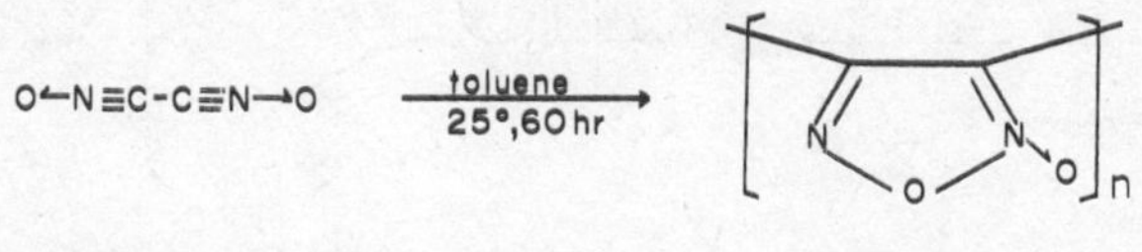

Method [107] *Poly(m-phenylene-1,2,5-oxadiazole-N-oxide)* Isophthalonitrile-N-oxide (0.705 g) is placed in a 20-ml flask with 0.5 ml of HMPA (hexamethylphosphoramide) and 4.5 ml of dioxane, and the mixture is heated at 70°C for 7 hr. It is then poured into 100 ml of acetone or methanol to precipitate the polymer, which is washed with ether and dried under vacuum to yield 0.571 g (81%) of polymer; inherent viscosity 0.12 dl/g in H_2SO_4.

Other reaction conditions are (1) dimethylacetamide (DMAc) solvent, 15°C, 44 hr with p-toluene sulfonic acid catalyst and (2) dimethylformamide (DMF), room temperature, 20 hr. The polymers are soluble in DMF.

Poly(p-phenylene-1,2,4-oxadiazole-N-oxide) [107,119] The monomer is either heated in a sealed glass tube at 100°C for 24 hr or stored at room temperature for 5 months. The yellow-brown polymer is dissolved in DMF, precipitated into methanol, collected, and dried.

6.14 Polythiadiazoles

Two isomers of the thiadiazole ring have been incorporated into the backbone of polymers. These have either a 1,2,5 or 1,3,4 structure as follows:

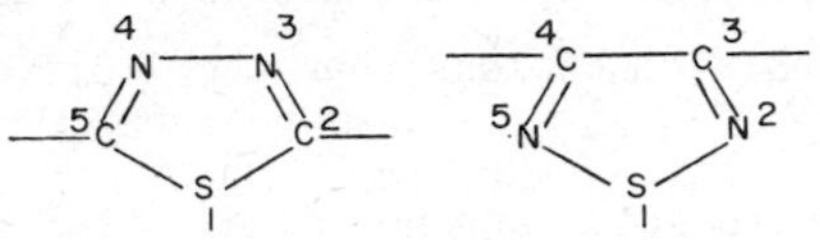

Thiadiazoles have received much less attention than oxadiazoles, not because their properties are in any way inferior to their oxo relatives (indeed they are not), but because precursors for the thio polymers are less accessible.

Numerous synthetic approaches, some with several variations, have been developed for polythiadiazoles. The general procedure and an example of each of these methods is given below. Especially to be recommended as an additional source of detailed experimental procedures is the clearly written and thorough paper by MacDonald and Sharkey [103]. Some of these polymers were presented in part in Chapter 5 as copolyimides.

The first synthesis is exactly like that used to great extent in making poly(1,3,4-oxadiazoles), i.e., cyclodehydration of a polyhydrazide. Presumably the same sort of mechanism pertains in the case of the thio polymers. Frazer used this approach to obtain a poly(1,3,4-thiadiazole) prior to 1965 [44]. Thus, reacting dimethyltetrathioisophthalate with isophthalic acid dithiohydrazide in hexamethylphosphoramide (HMPA) yields the polydithiohydrazide. This can be converted to the corresponding poly(1,3,4-thiadiazole) by heating at 285°C under vacuum for 18 hr.

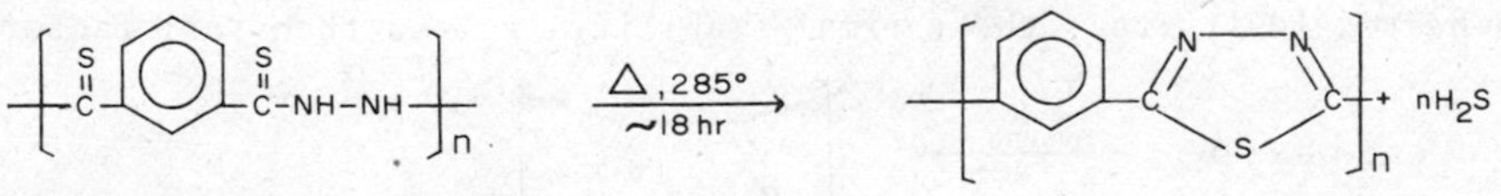

$$\left[\overset{S}{\overset{\|}{C}}-C_6H_4-\overset{S}{\overset{\|}{C}}-NH-NH \right]_n \xrightarrow[\sim 18\,hr]{\Delta,\,285^\circ} \left[C_6H_4-\text{(1,3,4-thiadiazole-2,5-diyl)} \right]_n + nH_2S$$

The second general method involves condensation of 2,5-difunctional thiadiazoles with other difunctional molecules and is the approach most often used to make poly(1,3,4-thiadiazoles). However, the kinds of functional groups used in these condensations have varied widely, and to date no particular synthetic route seems to have achieved dominance.

Hirsch [62] has condensed diaminothiadiazole with pyrazinetetracarboxylic dianhydride to produce a copoly(imide thiadiazole) devoid of hydrogen, thereby eliminating any weak position on which thermooxidative degradation can take place.

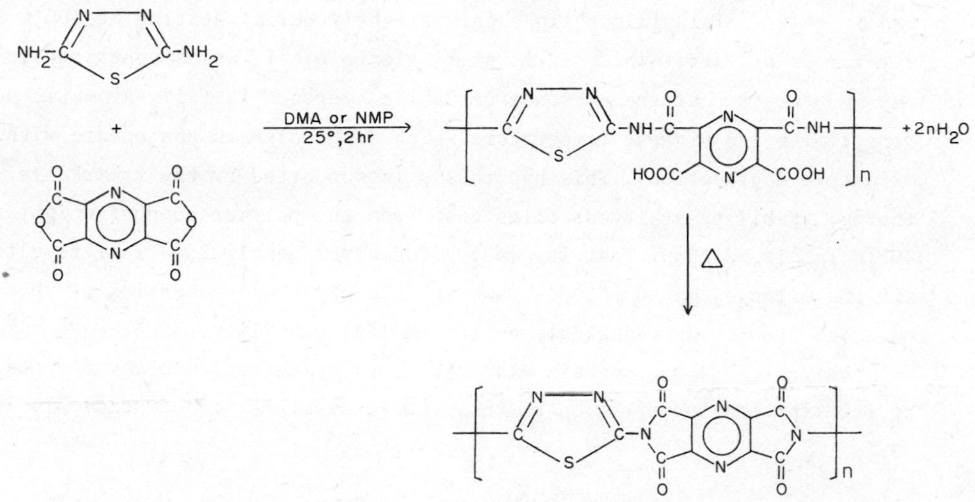

DMA or NMP
25°, 2 hr
$+ 2nH_2O$
△

Method *Poly[pyrazineimide-2,5-(1,3,4-thiadiazole)]* [62] *Preparation of pyrazinetetracarboxylic acid dianhydride* Pyrazinetetracarboxylic acid is synthesized according to the method of Mager and Berends [104] and purified by crystallization from constant-boiling hydrochloric acid. The tetraacid (16 g) and acetic anhydride (400 ml) are stirred together under nitrogen while the temperature is brought slowly to 80-85°C. A higher temperature must not be used. At 80°C the reaction commences with concomitant dissolution of the acid. After all acid is dissolved, the temperature is maintained at 80°C for an additional 0.5 hr. Then the reaction mixture is concentrated to a slush by evacuation with a water aspirator. The slush is further dried at 80°C under strong facuum overnight. The product is collected by sublimation from a loosely covered dish at 180°C. A first fraction is discarded and then highly pure, very pale green crystals of pyrazinetetracarboxylic acid dianhydride are collected. The crystals decompose at about 180°C.

Polymerization The above monomeric dianhydride (0.4624 g) is weighed into a thoroughly dry, 1-oz screw-cap bottle having a polyethylene-lined cap. Dimethylacetamide (10 ml, dried over Linde 5A molecular sieves) is added from a syringe and the bottle swirled until solution is complete. The 0.2440 g of 2,5-diamino-1,3,4-thiadiazole [6] is added, and the mixture is shaken vigorously by hand. The reaction sets in rapidly, and the color of the solution darkens. Mixing of the reaction solution is continued until gelation appears imminent. At room temperature this requires approximately 1.5 hr. (Continuing the reaction will increase the molecular weight of the product; however, once the reaction gels, it cannot be redissolved.) Films of the amic acid intermediate may be cast directly from the dimethylacetamide solution. Thermal curing of films in air to form the polythiazole is done using the following schedule: room temperature to 250°C at 4°/min; 250-300°C at 1°/min; 300°C for 1 hr; 300-350°C at 1°/min; 350° for 1-1/4 hr; 350-375°C at 1°/min; finally, 375°C for 2 hr. This slow heating avoids decarboxylation, which causes CO_2 bubbles to be trapped in the film.

In the form of a nonoriented film, the above polymer is clear and deep red in color. This film shrinks only slightly during heating at 400°C in air for 25 hr, after which it is still flexible. It was synthesized to test the hypothesis that the presence of labile hydrogen in fully aromatic polymers limits the highest temperature which such polymers can endure without oxidative degradation. This hypothesis is supported by the remarkable thermal stability achieved; films cast from the polymer support weight to 592°C in air, without charring [62]. TGA gives nearly identical results in air and nitrogen up to 600°C. However, the physical properties of this material are not as remarkable as its thermal stability.

Polyamides which contain also the 1,3,4-thiadiazole moiety are possible by the expected condensation of thiadiazole diamines with diacid chloride [93]:

NMP, 0-25°
~4 hr

This polymer decomposes above 340°C without melting.

Frazer and Memeger obtained patents for various aromatic and aliphatic 1,3,4-thiadiazole hydrazide copolymers [44,45]:

HMPA
~17 hr

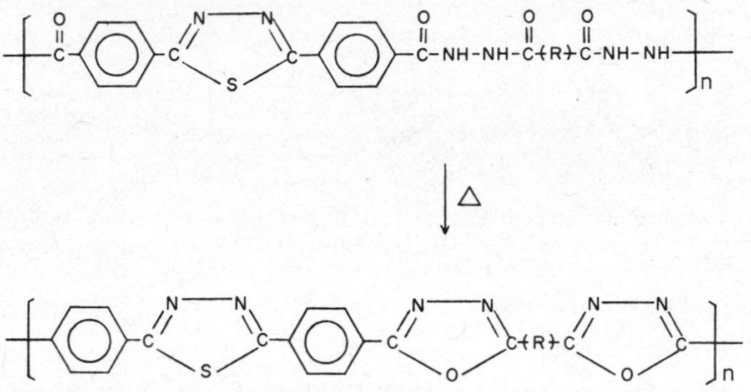

$\xrightarrow{\Delta}$

The copolyhydrazide intermediate can be spun from N-methylpyrrolidone solution. The resulting fibers have a tensile strength of 1.04 g/den at 71.6% elongation, and initial modulus of 38 g/den.

Kossmehl et al. reported a Knoevaenagel-like synthesis in 1974. The highly conjugated polymers which result are claimed not to melt or soften below 400°C [86]:

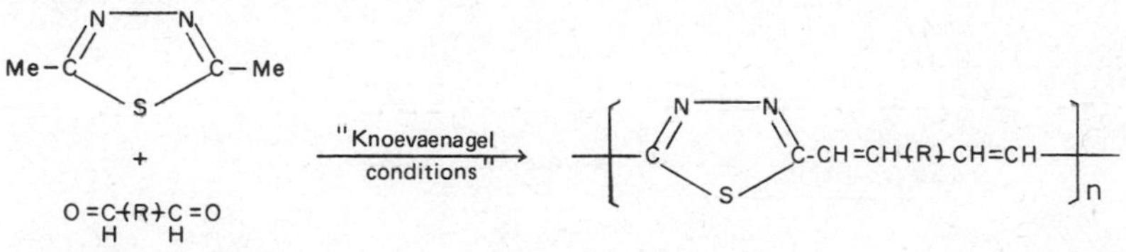

"Knoevaenagel conditions"

An interesting extension of the Williamson ether synthesis gives a poly(1,3,4-thiadiazole) by refluxing bis(chloromethylphenyl)ether and disodium 1,3,4-thiadiazolidine-2,5-dithiolate in dioxane [176]:

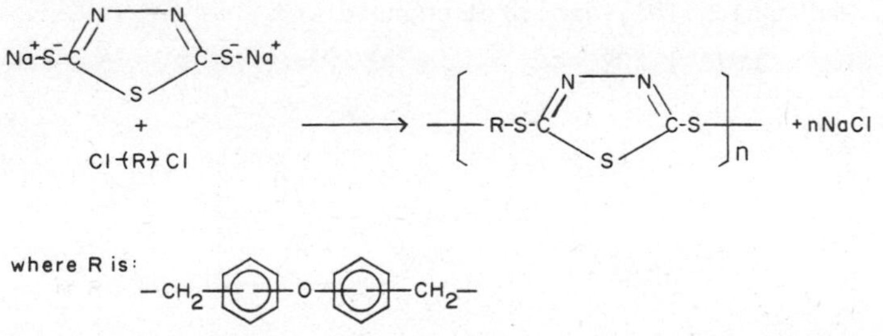

+ nNaCl

where R is: $-CH_2-C_6H_4-O-C_6H_4-CH_2-$

The above polymers all encompass the 1,3,4-thiadiazole moiety. Also, the 1,2,5 isomer has been incorporated into polymers largely by virtue of the availability of 1,2,5-thiadiazole-3,4-dicarboxylic acid, the synthesis of which was developed by researchers at E. I. duPont de Nemours and Company about 1970 [45,101,102,130]:

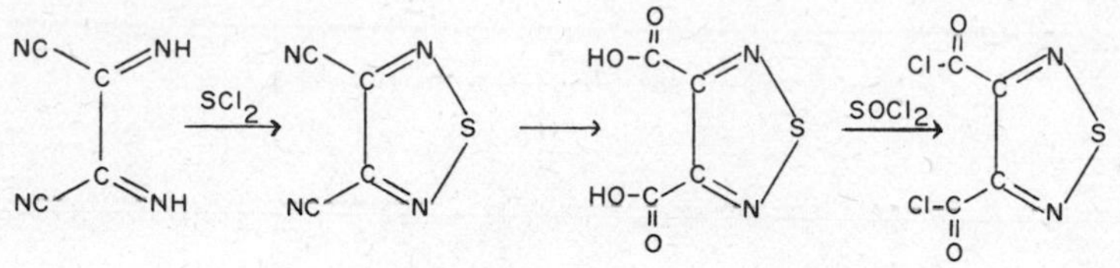

The 3,4-diacid decarboxylates above 160°C to the mono acid and above 200°C to 1,2,5-thiadiazole. Therefore the diacid is not useful for conventional melt polymerizations and requires relatively gentle treatment in other types of polymerization.

Russo et al. reacted the diacid with N,N'-diphenyl-p-phenylenediamine by heating them together for 30 hr at 150°C under nitrogen in dichlorobenzene. This was reported to give an 82% yield of transparent polyamide with moderately high molecular weight [129].

The reaction below illustrates formation of a 1,2,5-thiadiazole polyester by titanium isopropoxide catalyzed transesterification between the dimethyl ester of the above described acid and ethylene glycol [101].

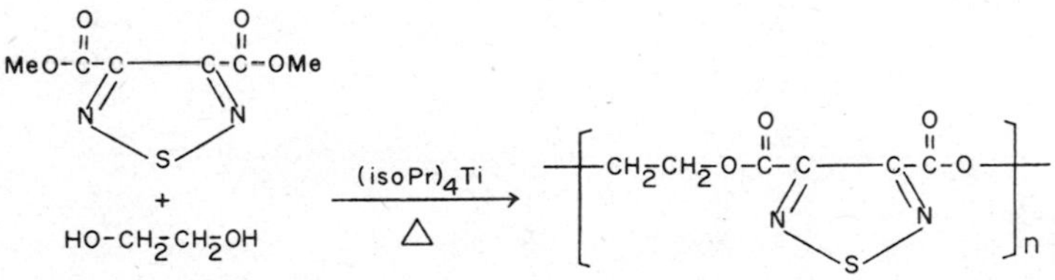

Polycondensation between the 3,4-diacid chloride of 1,2,5-thiadiazole and either a diamine or diol has been used to make a large number of 1,2,5-thiadiazole-containing polyamides and polyesters. The acid chloride route is mainly due to MacDonald [101] and to MacDonald and Sharkey [103] but has been used by others as well [130].

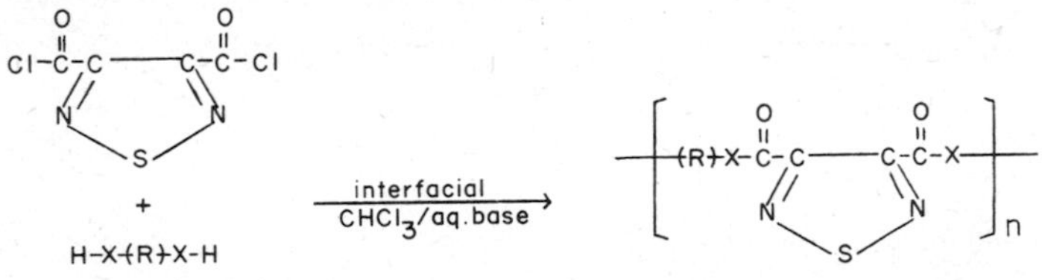

where X is: NH or O

For example, nonflammable thiadiazole polyesters are obtained by interfacial polymerization of 2,2-bis(3,5-dichloro-4-hydroxyphenyl)propane with 1,2,5-thiadiazole-3,4-dicarbonyl chloride, the resulting structure being [103] the following:

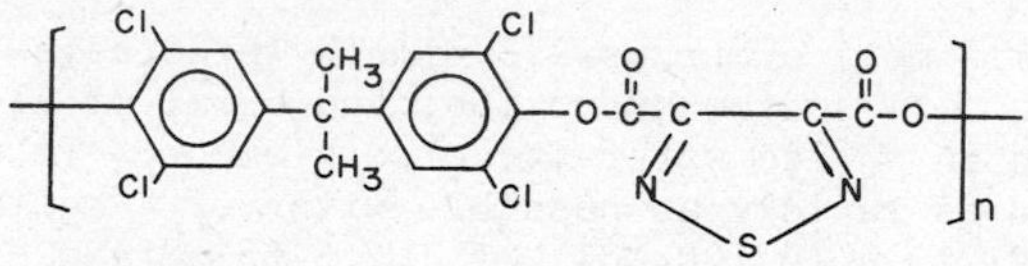

Clear, strong films of this poly(ester thiadiazole) are obtained from chloroform solution. These films require a 50% oxygen atmosphere to sustain combustion (i.e., oxygen index = 50). The glass transition temperature of this polymer is about 200°C, and the softening temperature is about 250°C. The analog of the polymer with trifluoromethyl instead of methyl groups behaves similarly, but the glass transition and softening temperatures are both lower by about 35°C.

In preparation of polyamides from the diacid chloride, primary diamines give lower molecular weight products than do secondary amines. This is caused by chain termination through imide formation. Thus:

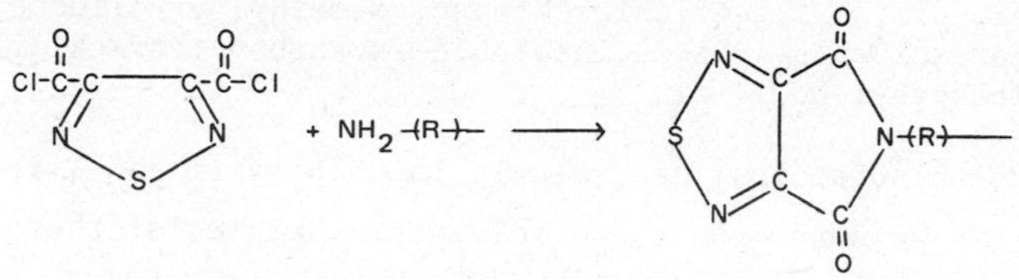

Adding the acid chloride to an excess of primary amine suppresses imide formation. Nevertheless, secondary amines, which cannot chain-terminate, invariably give higher molecular weight polyamides.

The following synthesis is illustrative of polyamide formation from thiadiazole diacid chloride. Note that the procedure calls for addition of benzoyl chloride. This serves as a chain length regulator, without which the molecular weight attained is so high as to render the polymer insoluble in any convenient solvent. The same procedure with minor modifications can be used to make many different polyamides.

Method *Interfacial polycondensation of 3,4-dicarbonyl chloride-1,2,5-thiadiazole, with trans-2,5-dimethylpiperazine* [103]

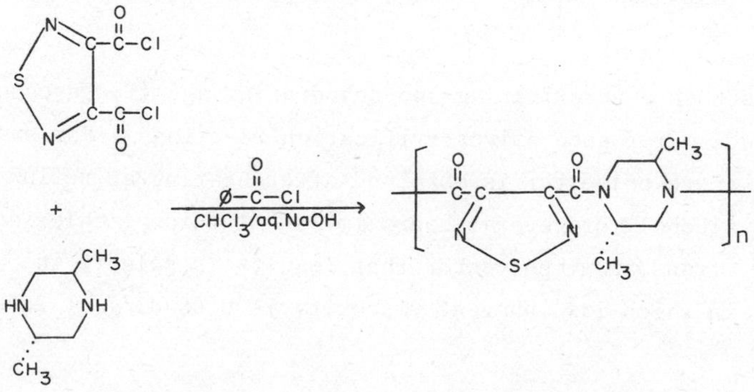

$\emptyset$–C(O)–Cl

$CHCl_3$/aq.NaOH

A mixture of 12.92 g (0.113 mol) of trans-2,5-dimethylpiperazine, 340 ml water, 8.00 g (0.2 mol) sodium hydroxide, and 60 ml of alcohol-free chloroform is placed in an ice water jacketed blendor. To this mixture at 5.5°C is added as rapidly as possible with very vigorous agitation a solution of 21.10 g (0.100 mol) of 1,2,5-thiadiazole-3,4-dicarbonyl chloride and 1.4 ml (0.012 mol) of benzoyl chloride in 70 ml of anhydrous, alcohol-free chloroform. After the addition, the temperature rises rapidly to 25°C and in 5 min the reaction mixture is a thick emulsion. Acetone (500 ml) is added to coagulate the product, and the reaction is stirred 1 min longer. The coagulated polymer is allowed to settle, and the supernatent liquid is decanted. The polymer which remains in the blendor is stirred with an additional 500 ml of acetone and collected by filtration. While on the filter, the product is rinsed with 250 ml of acetone and then washed seven times in the blendor with successive 500-ml portions of water. The polymer is again collected by filtration and washed with 250 ml water. The water is removed by blending the polymer with 500 ml of acetone, collecting the solid on a filter, and washing it with 250 ml acetone. The snow-white polymer so obtained is dried overnight in a vacuum oven at 80°C. The yield of polyamide is 17 g (67%). The product is soluble in chloroform, 1,1,2-trichloroethane, N-methylpyrrolidone, and other similar solvents. It is insoluble in carbon tetrachloride and in tetrachloroethylene.

These poly(amide thiadiazoles) are soluble in chloroform and 1,1,2-trichloroethane and can be spun from these solvents. Samples of these fibers, which have been oriented by drawing, exhibit tensile strengths up to 4.3 g/den and moduli up to 41 g/den. Further, these fibers show almost complete retention of room temperature physical properties at 90°C and 100% relative humidity. Work recovery values are very high, being greater than the corresponding values for either silk or wool. These good mechanical properties are somewhat unexpected since there are no N-H functions to provide hydrogen bonds.

The final synthetic method leading to poly(1,2,5-thiadiazoles) that will be discussed is the self-polymerization of 3-carboxy-4-hydroxy-1,2,5-thiadiazole, in thionychloride solution.

$$\mathrm{HO{-}C{=}N{-}S{-}N{=}C{-}\overset{O}{\overset{\|}{C}}{-}OH} \xrightarrow[\Delta,\ \sim 4\,hr]{SOCl_2} \left[\mathrm{{-}C{=}N{-}S{-}N{=}C{-}\overset{O}{\overset{\|}{C}}{-}O{-}}\right]_n$$

Thionyl chloride is both a chlorinating and dehydrating agent; consequently, it is a suitable medium for some polyesterification reactions. For example, a 51% yield of the above polyester is obtained after heating at reflux 17.6 g of the acid alcohol for several hours in 260 ml thionyl chloride [102]. The glassy, transparent polyester that results is soluble in dimethylacetamide, in which its inherent viscosity is 0.65 dl/g at 25°C.

The thiadiazoles exhibit the same desirable chemical and thermal stability as the 1,3,4- and 1,2,5-oxadiazoles but are somewhat more tractable. Thio analogs of other backbone systems also have exhibited increased solubility and suffered similarly from precursor availability (e.g., benzoxazole/benzothiazole and imidine/thioimidine).

Polyamides, polyesters, polyethers, polyurethanes, and polyhydrazides, all containing 1,2,5-thiadiazole rings, are claimed to be useful as packaging materials, fibers, and electrical insulators [101,102]. Poly(1,2,5-thiadiazole amides) that are press-moldable and extrudable have been patented [130].

6.15 Polytriazoles

Patents and a paper describing aminotriazole polymers [4] and partially hydrogenated triazole polymers [3,169] appeared in 1950 and 1963. In 1961, however, Abshire and Marvel described the first wholly aromatic backbone containing the triazole moeity [1]. Their approach was through the loss of nitrogen between a bistetrazole and a bisimidoyl chloride.

pyridine, reflux, 5 days

Ar is: m-or p-C_6H_5

If an aryldiacylhalide is reacted with the bistetrazole, an oxadiazole polymer results. The polytriazoles from this reaction were of fairly low molecular weight as indicated by the inherent viscosity, 0.24 dl/g, 0.2% in H_2SO_4. They were, however, white powders and soluble also in formic acid. The moderate yields and low molecular weight can be attributed in part to the sensitivity of the N,N'-diphenylphthalamide chloride to moisture.

Thermal analysis showed softening points of 240-250°C for the all-meta catenation but 290-310°C for the alternating meta- and para-phenylene-type backbone. TGA data indicated that these polytriazoles were less heat-stable than the corresponding oxadiazoles; i.e., they relinquish 10% of their weight below ~400°C in either air or nitrogen.

The second method of obtaining fully unsaturated triazole nuclei in the backbone was published in 1966 by Holsten and Lilyquist [63]. This method consisted of the cyclization of a polyhydrazide with aniline. A detrimental side reaction which occurs is the backbone cleavage by aniline so that molecular weights of prepolymer are lowered in the cyclization. This degradation can be minimized by providing the proper balance between aniline and PPA or by lowering the reaction temperature. The latter approach therefore results in higher molecular weights commensurate, unfortunately, with incomplete conversion.

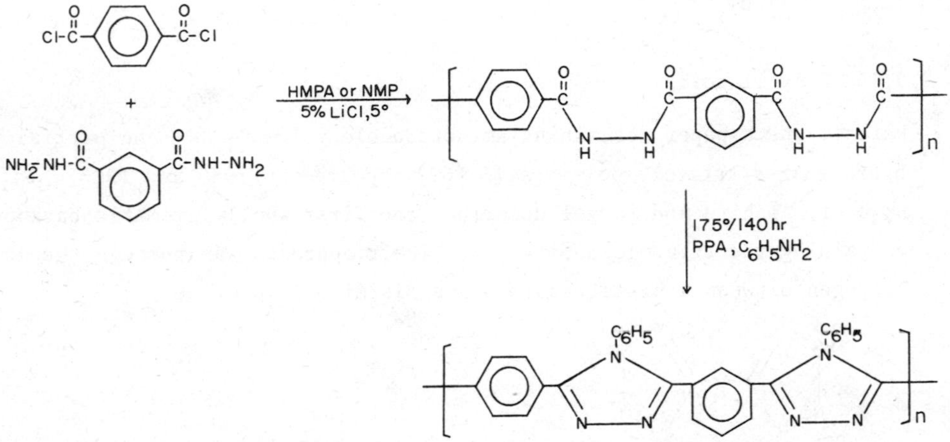

There are two interesting structural features of this polymer: (1) It has no N-H bonds or aliphatic groups which would weaken it toward thermooxidation and (2) it has a regular, alternating meta and para catenation on the backbone phenylene units which should serve to increase tractability.

The polytriazoles were prepared in different molecular weights depending on composition of the medium and temperature of the reaction. Lower temperatures and higher aniline/PPA ratio decreased the reaction rate. The relationship among number average molecular weight, intrinsic, and inherent viscosities is as follows:

$\overline{M}_n$	Intrinsic viscosity $[\eta]$ (dl/g)	Inherent viscosity η_{inh} (dl/g)
13,670	1.79	1.31
↓	↓	↓
26,750	2.56	1.84

Formic acid was a good solvent for the polytriazoles, and from it films could be cast (10% solution) and fibers could be wet-spun (20% solution). Drawn films and fibers show zero strengths at 490 and 465°C, respectively. The fiber exhibits a fairly low strength (2.52 g/den tenacity) owing to the lack of hydrogen bonding.

The TGA of the polymer shows a stability limit of 512°C in nitrogen (by extrapolation of the break in the TGA curve).

> *Method* [63] *Polyhydrazide* [42] Scrupulously dry and pure conditions are necessary. A solution is prepared under nitrogen from 97.0 g (0.5 mol) of isophthalohydrazide and 1250 ml of HMPA at 50°C. It is then cooled to 5°C, and 101.5 g (0.5 mol) of terephthaloyl chloride is added over 2 hr with stirring. After the solution is stirred for 2 more hr at 25°C, it is poured into water to precipitate the polymer. The solid is washed thoroughly with water in a blendor and dried; inherent viscosity 2.5-3.2 dl/g (5% in DMSO). The solid poly(m,p-phenylenehydrazide) is ground to 30 mesh to aid solubility.
>
> *Cyclization* PPA (3000 g) is heated to 150°C under nitrogen, and 751 g (8.1 mol) of aniline is added slowly with stirring to keep the temperature below 190°C. Then with the temperature of the solution at 175°C, 97.2 g of the above poly(phenylenehydrazide) is added, and stirring and heating (175°C) is continued for 140 hr. The end of the reaction is indicated by the dissolution of the polyhydrazide into a clear, viscous, light orange solution. The hot mixture is poured into an excess of stirred water, and then sodium hydroxide pellets are added carefully to precipitate and agglomerate the polymer. The solid is isolated, washed well with 5% aq. sodium hydroxide and water, extracted (Soxhlet) with ethanol, and finally dried under vacuum at 80-120°C; $\overline{M}_n$ = 28,000.

The third method to be discovered for the preparation of triazoles first appeared in 1966 by Saga and Shono [133] and was extended by Korshak et al. in 1968 [81] and Hergenrother in 1970 [57], the last reported investigation of this system. The route used by these three groups was to condense a bisamidrazone (bishydrazidine) at low temperatures with a diacyl chloride. The resulting poly(N-acylamidrazone) is then cyclized either to the traizole or the oxadiazole, depending on reaction conditions. The polytriazole contains some (<4%) oxadiazole and amidrazone moieties, however.

The intermediate poly(N-acylamidrazone) is soluble in DMF, DMAc, NMP, DMSO, H_2SO_4, and HMPA, and has a moderate molecular weight, inherent viscosity 0.35-0.80 dl/g at 0.5% in H_2SO_4. Chain cleavage will occur on standing in H_2SO_4. It forms clear, yellow, tough, flexible films which embrittle on being heated to 300°C. Owing to the crystallinity development, the precursor polymer also becomes insoluble when it is heated to 100-130°C.

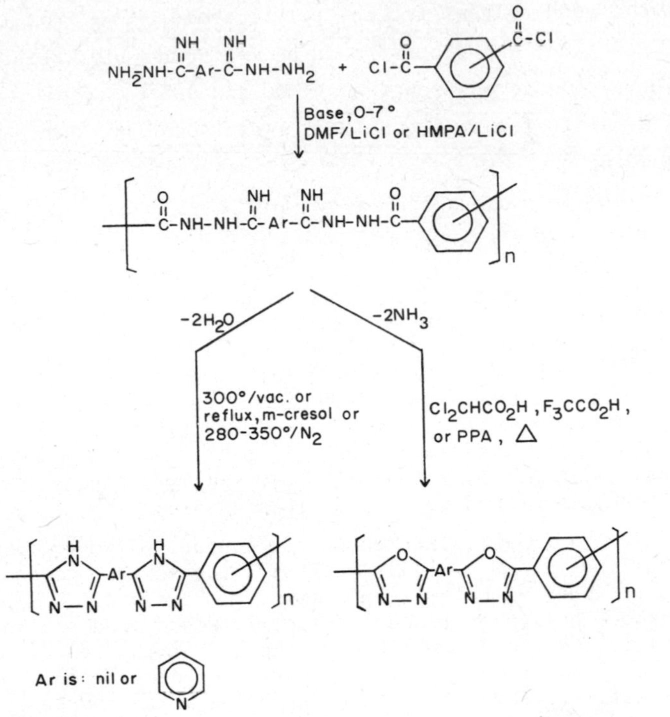

Base, 0-7°
DMF/LiCl or HMPA/LiCl

$-2H_2O$

$-2NH_3$

300°/vac. or
reflux, m-cresol or
280-350°/N_2

Cl_2CHCO_2H, F_3CCO_2H,
or PPA, Δ

Ar is: nil or

The final, cyclized polytriazole is soluble only in sulfuric acid, inherent viscosity 0.65 dl/g (0.5% concentration). Thermal analysis by TGA in nitrogen shows a 10% weight loss at 465°C in either air or nitrogen. Of course, in air this loss goes to 100% at 650°C but only to 50% at 900°C in nitrogen.

Method [57] *Poly(N-acylamidrazone)* A cold (0-7°C) slurry is prepared with 3.862 g (0.020 mol) of 2,6-pyridinediyldihydrazine, 3.280 g (0.040 mol) of anhydrous sodium acetate, 2.50 g of anhydrous lithium chloride, and 47 ml of HMPA. To this vigorously stirred slurry is added a cold solution of 4.060 g (0.020 mol) of isophthaloyl chloride in 24 ml of HMPA over 0.5 hr under nitrogen. The mixture is stirred at 5°C for 3 hr and at 25°C for 18 hr to result in a viscous, yellow solution. The reaction mixture is poured into vigorously stirred water to precipitate a fibrous, yellow solid of poly[N,N'-(2,6-pyridinediimidyl)-N'',N'''-isophthaloyldihydrazine]. The polymer is isolated, washed well with water and methanol, and dried below 100°C; yield 6.5 g (95.5%); inherent viscosity 0.71 dl/g at a concentration of 0.5% in H_2SO_4.

Still another process is known for the synthesis of polytriazoles, this one, however, yielding the 1,2,3-triazole rather than the 1,2,4 as discussed above. The process used here is the dipolar cycloaddition of 4-azido-1-butyne in either bulk or solution [77].

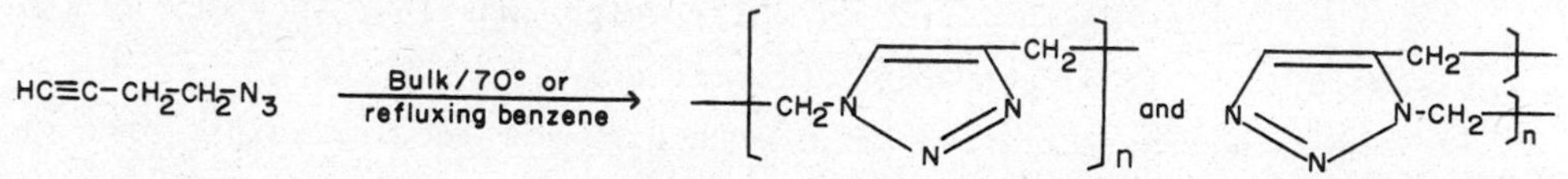

$$HC\equiv C-CH_2-CH_2-N_3 \xrightarrow[\text{refluxing benzene}]{\text{Bulk/70° or}}$$ [poly(1,4- and 1,5-disubstituted 1,2,3-triazole)]$_n$

The polymer is an opaque white glass from bulk polymerization and a yellow powder from solution. They are soluble in DMSO, DMF, and H_2SO_4, although less so if prepared in solution. The molecular weights are low, however, as evidenced by the inherent viscosity (0.08 dl/g).

Thermal analysis shows a Tg of 63-71°C and a decomposition temperature of 320°C.

Polytriazoles are also available from the 1,3-dipolar addition of dinitriles with bisnitrilimines [158]. Terephthaloylphenylhydrazide chloride in the presence of a base generates in situ the double dipolarophile bisnitrilimine with immediately adds to perfluoroglutaronitrile. The intermediate can also be generated from tetrazoles [156].

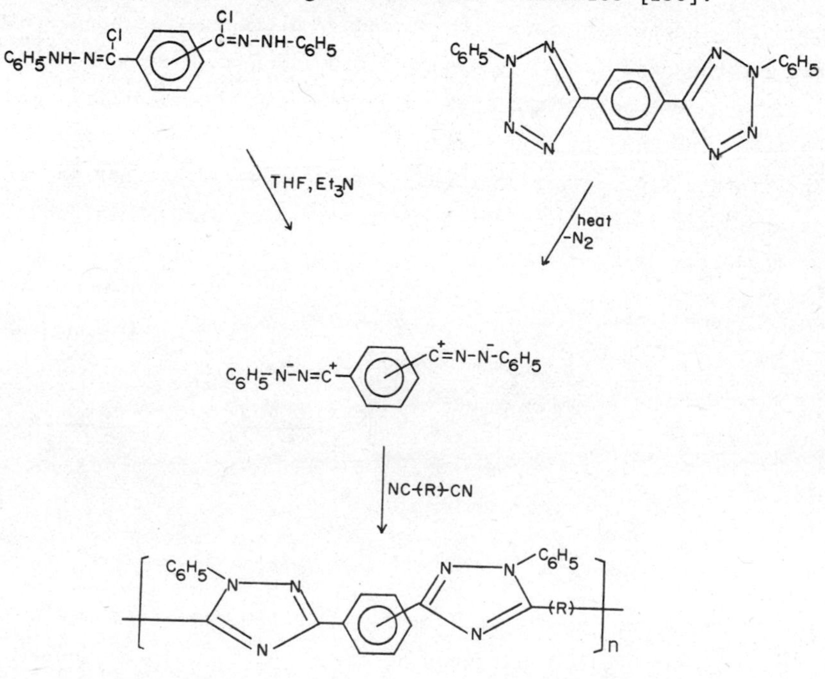

$C_6H_5NH-N=C(Cl)-C_6H_4-C(Cl)=N-NH-C_6H_5$ (THF, Et_3N) or bis(2-phenyltetrazolyl)benzene (heat, $-N_2$) →

$C_6H_5-\bar{N}-N=\overset{+}{C}-C_6H_4-\overset{+}{C}=N-\bar{N}-C_6H_5$

↓ NC-(R)-CN

[polytriazole]$_n$

R is: $(CF_2)_3$, $p-C_6H_4$, $p-C_6F_4$, or p,p'-biphenylene

The acid hydrazide chloride gave a dark brown polymer which does not melt below 300°C and which shows a break in the TGA curve at 350°C in either air or nitrogen. Its molecular weight was low, shown by an inherent viscosity of 0.13 dl/g (0.25% in formic acid).

However, the tetrazole afforded a higher molecular weight product, intrinsic viscosity up to 0.4 dl/g, which was soluble in chlorobenzene and 1,2,4-trichlorobenzene but not in acid. More important, the thermal stabilities of these products were at 450-480°C (TGA break) in air or nitrogen.

The superiority of the tetrazole route to triazoles compared to the acid hydrazide chloride route is attributed to the higher reaction temperatures required for tetrazole decomposition (195°C). At this temperature the nitrilimine dipole possesses a much greater reactivity than at 65°C.

The method used for this polymerization is identical to that described for the synthesis of polypyrazoles in Section 6.3.

6.16 Polytriazolines

A nonaromatic functional group with secondary nitrogens has been synthesized in a polymer backbone [58]. As expected from such an thermooxidatively reactive group, the polytriazolines do not possess the stability required to make them important in this field.

The synthesis of polytriazolines is analogous to that for polytriazoles in that a dihydrazidine (diamidrazone) is condensed with a dialdehyde or diacid, respectively:

$NH_2NH-C(=NH)-C_5H_3N-C(=NH)-NH-NH_2$ + OHC-Ar-CHO $\xrightarrow[\text{solution, 25-100°}]{\text{DMAc, DMSO or m-cresol,}}$ [polytriazoline]$_n$

Ar is: m- or p-C_6H_5 or $-C_6H_4-O-C_6H_4-$

The products are soluble in formic and trifluoroacetic acids from which tough, transparent, yellow films can be cast. Films can also be molded from the powder at 275°C and 1.4 MPa (200 psi). Inherent viscosities are about 0.60 dl/g in DMSO (0.5%), indicating a high molecular weight.

One of the reasons for synthesizing triazoline polymers was to oxidize them to triazoles. This result was, however, not realized.

TGA demonstrated that polytriazolines begin weight loss at 280°C in air and 400°C in nitrogen, not unexpectly poor results. Weight losses of 10% are reached at 310°C in air and 420°C in nitrogen.

Method [58] To a solution of 1.932 g (0.010 mol) of pyridinediyldihydrazidine [56] in 100 ml of m-cresol at 10°C is added 1.341 (0.010 mol) of isophthalaldehyde, and the mixture is stirred at 25°C for 30 min. The hot solution is poured into excess methanol to yield 29 g of poly(m-phenylenepyridinediyltriazoline) (100%).

6.17 Polytetrazoles

Low molecular weight (<5000) polymers which contain the tetrazole moiety have been synthesized but showed unimpressive properties compared to other heterocycles [30]. The method comprises the condensation of an aliphatic dihalide with the disodium salt of a bistetrazole. The reaction is followed conveniently by the appearance of tetrazole bands in the IR region at 971, 1087, and 1585 cm^{-1}. The polymers are soluble in only H_2SO_4 except where Ar is CH_2, when they are soluble also in DMF. Inherent viscosities in H_2SO_4 (at 0.5% concentration) range only from 0.08 to 0.17. They are white to yellow powders and discolor on being heated to 250°C.

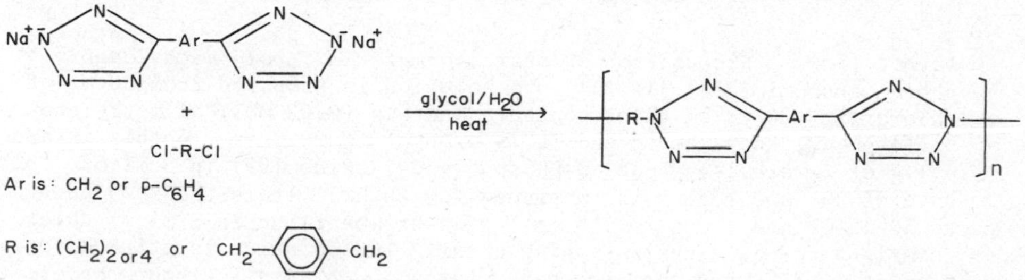

Ar is: CH_2 or p-C_6H_4

R is: $(CH_2)_{2 or 4}$ or CH_2–C_6H_4–CH_2

TGA studies show poor thermal stabilities as would be expected because of the aliphatic carbons present and the lack of conjugation in the backbone. Rather precipitous weight losses occur from 220 to 245°C in nitrogen.

Method [30] To a solution of 2.20 g (0.0085 mol) of the disodium salt of 5,5'-p-phenylenebistetrazole [36] in 25 ml of ethylene glycol and 25 ml of water is added 1.60 g (0.0085 mol) of 1,2-dibromoethane. The solution is heated to reflux for 48 hr, whereupon 0.18 g (0.001 mol) more of dibromoethane is added and the reaction is continued for an additional 12 hr. After the mixture is cooled, the precipitate is isolated, washed well with hot water and then methanol, and dried. It is then dissolved in concentrated H_2SO_4, reprecipitated into water, washed with water and dried in vacuo to yield 1.4 g (68%).

Polyaminotetrazoles were prepared by the same research group [30] through the addition of hydrazoic acid with polycarbodiimides at room temperature. The precursor polymer was arrived at by the self-condensation of diisocyanates with a phospholene oxide [14,15].

These polymers were less soluble than the simple polytetrazoles (only in H_2SO_4) but molecular weights and inherent viscosities were higher

(0.12-0.78 dl/g at 0.5% in H_2SO_4). Owing to hydrogen bonding, the tan to brown powders were high melting (>300°C). However, the same NH group which contributes a high melting point leads to poor resistance to thermal degradation. TGA investigations show incipient weight loss at 190°C with 10% losses from 260 to 280°C in nitrogen.

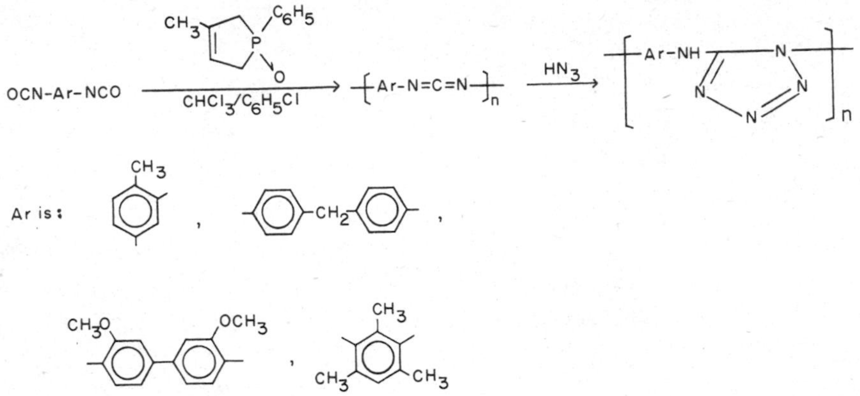

Method [30] *Preparation of starting material, poly*[*methylenebis(4-phenylcarbodiimide)*] [14,15] A solution is prepared from 80 ml of chloroform, 20 ml of chlorobenzene, and 5 g (0.02 mol) of methylenebis(4-phenylisocyanate), and it is brought to reflux. To this is added 0.1 g of 3-methyl-1-phenyl-3-phospholene-1-oxide [99] in 5 ml of chloroform, and reflux is continued for 20 hr. It is then cooled to 30-35°C, and 1.0 g (0.058 mol) of p-bromoaniline in 5 ml of chloroform is added as a chain-capping agent. The solution is stirred for 2 hr at 30-35°C and then is stored for as much as a few hours before precipitation occurs.• This solution is used in the addition described below.

To isolate the polycarbodiimide, the above reaction solution is poured into stirred acetone. The white polymeric precipitate is isolated, washed with acetone, and dried in a vacuum to yield 2.69 g (87%), mp >300°C.

Polyaminotetrazole synthesis A 25-ml portion of the above polycarbodiimide solution is added to a solution of 4.6 g (0.11 mol) of hydrazoic acid at 25°C, and stirring is continued for 48 hr. A precipitate forms which is isolated, washed with chloroform, and dried. It is then purified by dissolution in H_2SO_4 and reprecipitation into water. It is collected, washed well with water, and dried in vacuo at 100°C to yield 1.1 g (88.7%) of white powder, mp >300°C.

6.18 Recent Developments

6.18-1 Polybenzoxazoles

A fluorocarbon ether bibenzoxazole backbone is known which shows a 10% weight loss by TGA at 500°C in air [186,187]. Tgs are from -31 to +106°C and Tm's from 53 to 187°C. Of course, the Tg and thermal oxidative stabilities decreas with increasing proportion of fluorocarbon ether over the benzoxazole nuclei. They are generally soluble in HFIP (hexafluoro-i-propanol) and have inherent viscosities up to 9.94 dl/g. These polymers appear to be tough, rubbery, and fibrous.

6.18-2 Polybenzimidazoles

Research on benzimidazoles has continued with some new methods and structures being discovered. Low molecular weight species were possible from a two-step process: (1) the high-temperature (190°C) condensation with aluminum chloride of aromatic diamines and dinitriles, followed by (2) a cyclization of the amidine intermediate [181]. The product exhibits a quite rigid backbone but is soluble in DMF (inherent viscosity of 0.25 dl/g).

Bisorthoesters have also yielded PBIs when reacted with tetraamines in DMSO at 100°C [184]. Very high yields and good viscosities (0.8 dl/g) are notable. This publication also presents a comprehensive background discussion of PBI synthesis, valuable for the tyro.

Pyridine-2,6-dicarboxylic acid was incorporated into a PBI backbone [179]. However, solubilities were poor, and full cyclization was not achieved. Nonetheless, stabilities were in excess of 500°C.

Rod-like PBI is possible by the self-condensation of 2-(p-carboxyphenyl)-5,6-diaminobenzimidazole derivatives (salts or esters) in PPA at 180-250°C [189]. The same polymer is produced from the reaction of terephthalic acid with a new bistosylamide of tetraaminobenzene derivative, 1,3-diamino-4,6-(p-toluenesulfamido)benzene. The former method gave intrinsic viscosities up to 1.3 dl/g, while the latter gave 5.0 dl/g, a sixfold increase over previously reported molecular weights for this polymer. The yellow product dissolved in methane sulfonic acid (MSA) to give a blue opalescence.

Several modifications of PBIs were discovered. Polymers with cyano or carboxylic acid end groups were possible by the previously discussed reaction between diamines and dicyano compounds [182]. Sulfonic acid pendant groups resulted from using potassium sulfonate salts of dicarboxylic acids in the condensation with amines [192]. Phosphorus-containing PBIs were synthesized from bis(p-carboxyphenoxymethyl)methylphosphine oxide or its chlorinated analog and tetraamines in PPA [191]. Thermal stabilities in nitrogen of these phosphorus-PBIs were only moderate, 360-390°C for the chlorinated version and 430-500°C for the nonchlorinated polymer. However, both showed excellent flame resistance.

Some rather straightforward structure-property relationship studies have appeared recently. One of these compares the polymers from iso- and terephthalic acid [196], and the other looks at aromatic aliphatic systems [197].

6.18-3 Polyoxadiazoles

A review of polyoxadiazoles and analogous structures has appeared [183]. This review covers polymers with 1,3,4- and 1,2,4-oxadiazole, thiadiazole, thiazole, and 1,2,4- and 1,2,5-oxadiazole-N-oxide functions. Two syntheses have been reported, one via the thermal cyclization of a polyhydrazide from a bishydrazide and N,N'-bisisomaleimide [188]:

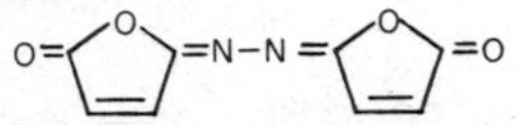

O= =N–N= =O

The other occurs by a common condensation of terephthalic acid with hydrazine sulfate in sulfur trioxide at 100°C [190]. Interestingly, a 4-hr 100°C post-cure increases the inherent viscosity of this polymer from 1.3 to 14.5 dl/g.

As was to be expected, liquid crystal formation was found in concentrated solutions in sulfuric acid of poly(1,4-phenylene-1,3,4-oxadiazole) [185].

Structure-property studies of polyoxadiazoles with ethenyl functions in the backbone revealed that (1) replacing phenylene groups with vinylene increases the solubility of the polymer and (2) the wholly vinylene oxadiazole is quite similar in stability (300°C vs 310-350°C, respectively) to the vinylene-phenylene oxadiazole copolymer [193-195]. The latter datum means, of course, that thermal instability rests entirely in the vinylene link.

High modulus fibers from copoly(p-phenyleneoxidiazole-N-methyl-hydrazide) were comparable to glass, steel, and polyester in performance as tire cores [177,178]. The copolymer overcame the intractability of the wholly aromatic homopolymer and the poor coagulation of the hydrazide homopolymer. The white to pale yellow product gave the following data:

Yarn tenacities	16-21 gpd (grams per denier)
Moduli	350-450 gpd
Elongation	5.5-6.5%
dpf (denier per filament)	0.5-1.0
Decomposition temperature (by DTA)	460°C

6.18-4 Miscellaneous

Polyspiro(indoline isoxazoline) polymers were synthesized as follows [199]:

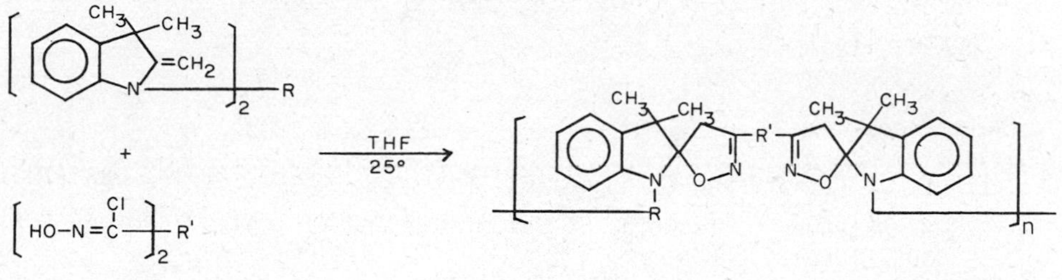

R is: $-C_4H_8-$ or $-CH_2-C_6H_4-CH_2-$

R' is: nil, m- or p-C_6H_4

No thermal data are given; however, the product was said to photorearrange to the following:

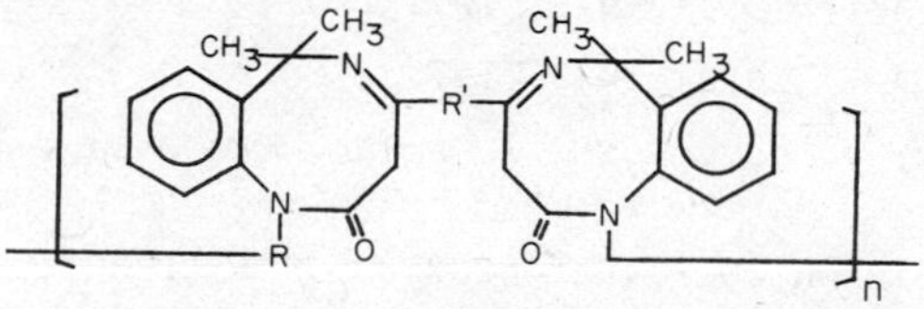

Poly(amide pyrazolones) result from the reaction of bishydrazines, with 1,4-bis(4-ethoxymethylene-5-oxazolon-2-yl)benzene [198]:

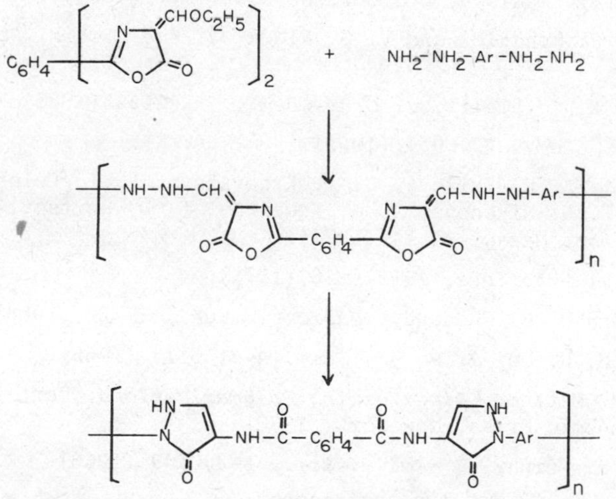

The yellow transparent, tough, film products are obtained in quantitative yields under relatively mild conditions. Their stabilities are only fair, however: 280°C in air, 320°C in nitrogen. A new polyimidazolone was also reported [184].

References

1. C. J. Abshire and C. S. Marvel, *Makromol. Chem., 44/46,* 388 (1961).
2. M. Akiyama, Y. Iwakura, S. Shiraishi, and Y. Imai, *J. Polym. Sci., B4,* 305 (1966).
3. BASF, Belgium Patent 628,618 (1963).
4. H. Bates, J. W. Fisher, and E. W. Wheatly, U.S. Patents 2,512,599-601 and 2,512,624-630 (1950).
5. W. Batzill and W. Funke, *Makromol. Chem., 99,* 1 (1966).
6. H. Beyer, *Chem. Ber., 82,* 143 (1949).
7. D. C. Blomstrom, U.S. Patent 3,044,994 (1962).
8. W. Bracke, *J. Polym. Sci., A-1, 10*(4), 975 (1972).
9. W. Bracke, *J. Polym. Sci., A-1, 10*(4), 983 (1972).
10. L. W. Breed and J. C. Wiley, *J. Polym. Sci.: Polym. Chem. Ed., 14,* 183 (1976).
11. K. C. Brinker and I. V. Robinson, U.S. Patent 2,895,948 (1959).
12. M. Bruma, L. Stoicescu-Crivet, and I. Zugravescu, *Rev. Roum. Chim., 16*(10), 1591 (1971).
13. C. D. Burton and N. L. Madison, U.S. Patent 3,560,438 (1971).
14. T. W. Campbell, J. J. Monagle, and V. S. Flodi, *J. Amer. Chem. Soc., 84,* 3673 (1962).
15. T. W. Campbell and K. C. Smeltz, *J. Org. Chem., 28,* 2069 (1963).
16. F. H. Case, *J. Org. Chem., 30,* 931 (1965).
17. A. Y. Chernikhov, L. A. Rodivilova, E. I. Kraevskaya, L. I. Golubenkova, B. M. Kovarskaya, S. N. Nikonova, V. N. Tsuetkov, L. D. Pertsov, and G. V. Bogachev, *Plast. Maaay, 4,* 24 (1973).
18. R. E. Cochoy, *Polym. Preprints, 16*(2), 18 (1975).
19. A. B. Conciatori and E. C. Chenevey, *Macromol. Syn., 3,* 24 (1968).
20. D. T. Congone and H. H. Un, *J. Polym. Sci., A-3,* 3117 (1965).
21. R. J. Cotter and M. Matzner, *Ring-Forming Polymerizations,* Part B-1, Chaps. II-III, Academic Press, New York, 1972.
22. B. Culbertson and R. Murphy, *J. Polym. Sci., B-4,* 249 (1966).
23. G. F. D'Alelio, U.S. Patent 3,518,233 (1970).
24. G. F. D'Alelio, D. M. Feigl, W. A. Fessler, Y. Giza, and A. Chang, *J. Macromol. Sci. Chem., A3*(5), 927 (1969).
25. G. F. D'Alelio, W. A. Fessler, Y. Giza, D. M. Feigl, A. Chang, and S. Saha, *J. Macromol. Sci. Chem., A5*(2), 383 (1971).

26. G. F. D'Alelio, W. A. Fessler, Y. Giza, D. M. Feigl, A. Chang, and S. Saha, *J. Macromol. Sci. Chem.*, *A5*(6), 1097 (1971).

27. N. Dogoshi, S. Toyama, M. Kurihara, K. Ikeda, and N. Yoda, Japan Patent, 69, 20, 113 (1969).

28. N. Dokoshi, Y. Bamba, M. Kurihara, and N. Yoda, *Makromol. Chem.*, *108*, 170 (1967).

29. duPont, British Patent 811,758 (1959); and K. C. Brinker, D. D. Kameron, and I. M. Robinson, U.S. Patent 2,904,537 (1959).

30. E. Dyer and P. E. Christie, *J. Polym. Sci.*, *A-1*, *6*, 729 (1968).

31. G. L. Ehlers, K. R. Fish, and W. R. Powell, *J. Polym. Sci.*, *C*, *Polym. Symposia*, No. 43, 55 (1973).

32. R. C. Evers, *J. Polym. Sci.*, *A-1*, *8*, 563 (1970).

33. R. C. Evers, *Polym. Preprints*, *15*(1), 685 (1974).

34. H. M. Ezekiel, *Amer. Chem. Soc. Div. Org. Coat. Plast. Chem.*, *31*(1), 415 (1971).

35. A. E. S. Fairfull, J. L. Lowe, and D. A. Peak, *J. Chem. Soc.*, 742 (1952).

36. W. G. Finnegan, R. A. Henry, and R. Lofquist, *J. Amer. Chem. Soc.*, *80*, 3911 (1958).

37. R. T. Foster and C. S. Marvel, *J. Polym. Sci.*, *A*, *3*, 417 (1965).

38. A. H. Frazer, *High Temperature Resistant Polymers*, Interscience, New York, 1968.

39. A. H. Frazer and T. A. Reed, *J. Polym. Sci.*, *C*, *19*, 89 (1967).

40. A. H. Frazer and I. M. Sarasohn, *J. Polym. Sci.*, *A-1*, *4*, 1649 (1966).

41. A. H. Frazer and F. T. Wallenberger, *J. Polym. Sci.*, *A*, *2*, 1137 (1964).

42. A. H. Frazer and F. T. Wallenberger, *J. Polym. Sci.*, *A-2*, 1147 (1964).

43. A. H. Frazer and F. T. Wallenberger, *J. Polym. Sci.*, *A*, *2*, 1157 (1964).

44. A. H. Frazer, French Patent 1,473,595 (1967).

45. A. H. Frazer and W. Memeger, Jr., U.S. Patent 3,476,719 (1967).

46. P. N. Fribkova, T. N. Balykova, L. A. Glivka, P. M. Valetskii, S. V. Vinogradova, and V. V. Korshak, *Vysokomol. Soedin.*, *Series A*, *15*(7), 1506 (1973).

47. S. Fujimoto, *J. Polym. Sci.*, *B*, *5*, 301 (1967).

48. J. K. Gillham and C. K. Schoff, *J. Appl. Polym. Sci.*, *20*, 1875 (1976).

49. C. J. Grundmann, V. Mini, J. M. Dean, and H. D. Frommeld, *Justus Liebigs Ann. Chem.*, *687*, 191 (1965).

50. A. S. Hay, *J. Org. Cham.*, *25*, 635 (1960).

51. A. S. Hay, *J. Polym. Sci.*, *A-1*, *7*, 1625 (1969).

52. F. L. Hedberg and C. S. Marvel, *J. Polym. Sci.: Polym. Chem. Ed.*, *12*, 1823 (1974).

53. H. Hehr and H. Erlenmeyer, *Helv. Chim. Acta*, *27*, 489 (1944).

54. T. E. Helminiak, F. E. Arnold, and C. L. Benner, *Polym. Preprints*, *16*(2), 659 (1975).

55. R. M. Herbst and K. R. Wilson, *J. Org. Chem.*, *22*, 1142 (1957).

56. P. M. Hergenrother, *J. Polym. Sci.*, *A-1*, *7*, 945 (1969).

57. P. M. Hergenrother, *Macromolecules*, *3*(1), 10 (1970).

58. P. M. Hergenrother and L. A. Carlson, *J. Polym. Sci.*, *A-1*, *8*, 1003 (1970).

59. P. M. Hergenrother, W. Wrasidlo, and H. H. Levine, *Polym. Preprints*, *5*(1), 153 (1964).

60. P. M. Hergenrother, W. Wrasidlo, and H. H. Levine, *J. Polym. Sci.*, *A*, *3*, 1665 (1965).

61. J. Higgins and C. S. Marvel, *J. Polym. Sci.*, *A-1*, *8*(1), 171 (1970).

62. S. S. Hirsch, *J. Polym. Sci.*, *A-1*, *7*, 15 (1969).

63. J. R. Holsten and M. R. Lilyquist, *Polym. Sci.*, *A*, *3*, 3905 (1965).

64. K. Hosoda, H. Sekiguchi, and K. Sadamitsu, Japan Patent 73 00,178 (1973).

65. *Houben-Weyl's Methoden der Organoischen Chemie* (E. Miiller, O. Bayer, H. Meerwein, and R. Ziegler, Eds.), George Thieme, Stuttgart, 1955, Vol. IX, p. 39.

66. R. Huisgen, *Angew. Chem.*, *72*, 359 (1960).

67. J. Saurer, R. Huisgen, and H. J. Sturm, *Tetrahedron*, *11*, 241 (1960).

68. Y. Imai, M. Akiyama, K. Uno, and Y. Iwakura, *Makromol. Chem.*, *95*, 275 (1966).

69. T. Imai, S. Hara, and M. Uchida, Japan Patent 70 37,791 (1970).

70. Y. Imai, I. Taoka, K. Uno, and Y. Iwakura, *Makromol. Chem.*, *83*, 167 (1965).

71. Y. Imai, K. Uno, and Y. Iwakura, *Makromol. Chem.*, *83*, 179 (1965).

72. Y. Iwakura, M. Akiyama, and K. Nagakubo, *Bull. Chem. Soc., Japan*, *37*, 767 (1964).

73. Y. Iwakura, K. Uno, and S. Hara, *J. Polym. Sci.*, *A*, *3*, 45 (1965).

74. Y. Iwakura, K. Uno, and S. Hara, *Makromol. Chem.*, *94*, 103 (1966).

75. Y. Iwakura, K. Uno, and S. Hara, *Makromol. Chem.*, *95*, 248 (1966).

76. Y. Iwakura, K. Uno, and Y. Imai, *Makromol. Chem.*, *77*, 33 (1964).

77. K. E. Johnson, J. A. Lovinger, C. O. Parker, and M. G. Baldwin, *J. Polym. Sci.*, *B-4*, 977 (1966).

78. J. I. Jones, *Macromol. Sci.*, *C2*, 303 (1968).

79. S. Konya and M. Yokoyama, *Kogakuin Daigaku Kendyu Kokoku*, *31*, 162 (1972).

80. V. V. Korshak, *Heat Resistance Polymers*, Izdatel'stvo "Nauka," Moscow, 1969; Israli translation, Keter Press, Jerusalem, 1971, pp. 244-248.

81. V. V. Korshak, Ye. S. Krongauz, and A. L. Rusanov, Izv. Akad. Nauk, SSSR, *Ser. Khim.*, No. 11, 2663 (1968).

82. V. V. Korshak, A. L. Rusanov, D. S. Tugushi, and G. M. Cherkasova, *Macromolecules*, *5*(6), 807 (1972).

83. V. V. Korshak, S. V. Vinogradova, L. A. Glivka, P. M. Valetskii, and V. I. Stanko, *Vysokomol. Soedin*, Series A, *15*(7), 1495 (1973).

84. V. V. Korshak, T. M. Frunze, and A. A. Izyneev, Izv. Akad. Nauk, SSSR., *Ser. Khim.*, (11), 2104 (1964).

85. V. V. Korshak, I. F. Manucharova, A. A. Izyneev, and T. M. Frunze, *Vysokomol. Soedin, 8*, 777 (1966).

86. G. Kossmehl, I. Dornacher, and G. Manecke, *Makromol. Chem.*, *175*(5), 1359 (1974).

87. D. Kourtides and J. P. Parker, *Polym. Eng. Sci.*, *15*(6), 415 (1975).

88. H. N. Kovacs, A. D. Delman, and B. B. Simms, *J. Polym. Sci.*, *A-1, 6*, 2103 (1968).

89. H. N. Kovacs, A. D. Delman, and B. B. Simms, U.S. Patent 3,567,698 (1971).

90. H. N. Kovacs, A. D. Delman, and B. B. Simms, *J. Polym. Sci.*, *A-1, 8*(4), 869 (1970).

91. T. V. Kravchenko, M. N. Bogdanov, G. I. Kudryavtsev, and N. P. Okromchedlidze, *Vysokomol. Soedin*, Series B, *15*(9), 667 (1973).

92. T. Kubota and R. Nakanishi, *Polym. Lett.*, *2*, 655 (1964).

93. H. E. Kuenzel, F. Bentz, G. Lorenz, and G. Nischk, British Patent 1,238,511 (1967).

94. J. R. Leal, *Modern Plast.*, August 1975, p. 60

95. A. Lehtinen and J. Sundquist, *Makromol. Chem.*, *168*, 33 (1973).

96. H. H. Levine, *Encycl. Poly. Sci. Technol.*, *11*, 188 (1969).

97. H. J. Lorkowski and R. Pannier, *J. Prak. Chem.*, *311*, 958 (1969).

98. H. J. Lorkowski, R. Pannier, and A. Wende, *J. Prak. Chem.*, *35*, 149 (1967).

99. W. B. McCormick, U.S. Patent 2,663,737 (1955).

100. M. Maienthal, M. Hellmann, C. P. Haber, L. A. Hymo, S. Carpenter, and A. J. Carr, *J. Amer. Chem. Soc.*, *76*, 6392 (1954).

101. R. N. MacDonald, U.S. Patent 3,664,986 (1969).

102. R. N. MacDonald, U.S. Patent 3,855,183 (1974).

103. R. N. MacDonald and W. H. Sharkey, *J. Polym. Sci.: Polym. Chem. Ed.*, *11*, 2519 (1973).

104. H. I. X. Mager and W. Berends, *Rec. Trav. Chim.*, *76*, 28 (1957).

105. B. S. Marks, L. E. Shoff, and G. W. Watsey, "Study and Production of Polybenzimidazole Billets, Laminates and Cylinders," Tech. Report NASA CR-1723 (May 1971).

106. C. S. Marvel, *SPE J.*, *20*, 220 (1964).

107. Mitsubishi Petrochemical Co., British Patent 1,073,325 (1967).

108. W. W. Moyer, C. Cole, and T. Anyos, *J. Polym. Sci.*, *A, 3*, 2107 (1965).

109. J. E. Mulvaney, J. J. Bloomfield, and C. S. Marvel, *J. Polym. Sci.*, *62*, 59 (1962).

110. J. E. Mulvaney and C. S. Marvel, *J. Org. Chem.*, *26*, 95 (1961).

111. J. E. Mulvaney and C. S. Marvel, *J. Polym. Sci.*, *50*, 541 (1961).

112. M. V. Murhina, G. A. Makorushina, and I. Y. Pastorskii, *Vysokomol. Soedin*, Series B9, 284 (1967).

113. T. Nakajima and C. S. Marvel, *J. Polym. Sci.*, *A-1*, *7*, 1295 (1969).

114. T. V. Narayan and C. S. Marvel, *J. Polym. Sci.*, *A-1*, *5*, 1113 (1967).

115. W. Neugebauer, Japan Patent 256,946 (1959).

116. G. Oda, K. Hirasa, H. Sekiguchi, and K. Sadamitsu, Japan Patent 73 66,655 (1973).

117. J. Oda, K. Hirasa, H. Sekiguchi, and K. Sadamitsu, Japan Patent 73 88,139 (1973).

118. A. Oku, M. Okano, and R. Oda, *Makromol. Chem.*, *78*, 186 (1964).

119. C. G. Overberger and S. Fujimoto, *J. Polym. Sci.*, *B-3*, 735 (1965).

120. D. I. Packham, J. D. Davies, and H. M. Paisley, *Polymer*, *10*(12), 923 (1969).

121. T. L. Patton, *Polym. Preprints*, *12*(1), 162 (1971).

122. T. L. Patton, U.S. Patents 3,547,897 (1970); 3,591,562 (1971); 3,635,905 (1972); 3,661,859 (1972); and 3,684,773 (1972).

123. L. Plummer and C. S. Marvel, *J. Polym. Sci.*, *A-2*, 2559 (1964).

124. J. Preston and W. B. Black, *Macromol. Syn.*, *3*, 93 (1968).

125. J. Preston, W. B. Black, and W. L. Hofferbert, Jr., *Makromol Sci. Chem.*, *7*(1), 45 (1973).

126. G. Pruckmayr, U.S. Patent 3,410,834 (1968).

127. G. Pruckmayr, U.S. Patent 3,376,267 (1968).

128. P. Ruggli and E. Gassenmeier, *Helv. Chim. Acta*, *22*, 496 (1939).

129. M. Russo, V. Zamboni, and E. Monza, *Ger. Offen.* 2,260,130 (1971).

130. M. Russo, V. Guidotti, and L. Mortillaro, *Ger. Offen.* 2,136,931 (1972).

131. K. Sadamitsu, K. Hirasa, and H. Sekiguchi, Japan Patent 73 66,654 (1973).

132. K. Sadamitsu, K. Hirasa, H. Sekiguchi, and Y. Katsuyama, Japan Patent 73 71,455 (1973).

133. M. Saga and T. Shono, *J. Polym. Sci.*, *B*, *4*, 869 (1966).

134. R. Sandler and W. Karo, *Polymer Synthesis*, Vol. 1, Academic Press, New York, 1974, p. 258

135. J. Sauer, R. Huisgen, and H. J. Sturm, *Tetrahedron*, *11*, 241 (1960).

136. C. K. Schoff and J. K. Gillham, *J. Appl. Polym. Sci.*, *19*, 2731 (1975).

137. H. Sekiguchi, K. Hosoda, N. Shiina, and M. Kondo, Japan Patent 72 25,231 (1972).

138. H. Sekiguchi and K. Sadamitsu, Japan Patent 73 08,741 (1972).

139. H. Sekiguchi, K. Sadamitsu, and H. Fukushima, Japan Patent 72 50,780 (1972).

140. H. Sekiguchi, K. Sadamitsu, and H. Fukushima, Japan Patent 72 50,778 (1972).

141. H. Sekiguchi, K. Sadamitsu, and K. Hirasa, Japan Patent 73 69,882 (1973).

142. H. Sekiguchi, K. Sadamitsu, and K. Hirasa, Japan Patent 73 71,456 (1973).

143. H. Sekiguchi, K. Sadamitsu, and K. Hirasa, Japan Patent 73 89,280 (1973).

144. H. Sekiguchi, K. Sadamitsu, and A. Kitamura, Japan Patent 73 00,540 (1973).

145. H. Sekiguchi, K. Sadamitsu, and N. Kitamura, Japan Patent 73 37,453 (1973).

146. H. Sekiguchi, K. Sadamitsu, and T. Nakayama, Japan Patent 73 44,178 (1973).

147. S. W. Shalaby, R. L. Lapinski, and E. A. Turi, *J. Polym. Sci.: Polym. Chem. Ed.*, *12*(2), 2891 (1974).

148. W. C. Sheehan, T. B. Cole, and L. G. Picklesimer, *J. Polym. Sci.*, *A*, *3*, 1443 (1965).

149. W. C. Sheehan, *Polym. Eng. Sci.*, *5*(4), 263 (1965).

150. T. Shono, M. Hachihama, and K. Shinra, *J. Polym. Sci.*, *B*, *5*, 1001 (1967).

151. W. R. Sorenson and T. W. Campbell, *Preparative Methods in Polymer Chemistry*, Interscience, New York, 1961.

152. H. W. Steinmann, U.S. Patent 3,629,198 (1971).

153. J. K. Stille, F. W. Harris, and M. A. Bedford, *J. Heterocyc. Chem.*, *3*, 155 (1966).

154. J. K. Stille and M. A. Bedford, *J. Polym. Sci.*, *A-1*, *5*, 2331 (1968).

155. J. K. Stille and M. A. Bedford, *J. Polym. Sci.*, *B-4*, 329 (1966).

156. J. K. Stille and L. D. Gotter, *J. Polym. Sci.*, *A-1*, *7*, 2492 (1969).

157. J. K. Stille and F. W. Harris, *J. Polym. Sci.*, *B-4*, 333 (1966).

158. J. K. Stille and F. W. Harris, *J. Polym. Sci.*, *A-1*, *6*, 2317 (1968).

159. L. Stoicescu-Crivat and M. Bruma, *Rev. Roum. Chim.*, *12*(10), 1245 (1967).

160. R. Stolle, *J. Prakt. Chem.*, *68*, 30 (1903).

161. S. Toyama, N. Dokoshi, K. Ikeda, M. Kurihara, N. Yoda, R. Nakanishi, and M. Watanabe, Japan Patent 69 23,107 (1969).

162. F. D. Trischler and H. H. Levine, *J. Appl. Polym. Sci.*, *13*, 101 (1969).

163. Y. Tsu, H. H. Levine, and M. Levy, *J. Polym. Sci.: Polym. Chem. Ed.*, *12*, 1515 (1974).

164. T. Unishi, *Kogyo. Kagaku. Zasshi*, *69*(12), 2343 (1966).

165. I. K. Varma and Veena, *J. Polym. Sci.: Polym. Chem. Ed.*, *14*, 973 (1976).

166. M. Veda, K. Kino, T. Hirond, and Y. Imai, *J. Polym. Sci.: Polym. Chem. Ed., 14,* 931 (1976).

167. H. Vogel and C. S. Marvel, *J. Polym. Sci., 50,* 511 (1961).

168. H. Vogel and C. S. Marvel, *J. Polym. Sci., A-1,* 1531 (1963).

169. H. Weidinger and J. Kranz, *Chem. Ber., 96,* 1064 (1963).

170. D. M. White, *Polym. Preprints, 12*(1), 155 (1971).

171. Whittaker Corp. product bulletin, "Imidite Foam Compounds," San Diego, 1967.

172. V. G. Yashunskii and L. E. Kholodov, *Zh. Obshch. Khim., 32,* 3661 (1962) [*Chem. Abstr., 58,* 13939 (1963)].

173. N. Yoda, *J. Polym. Sci., A-1,* 1323 (1963).

174. M. Yokoyama, Japan Patent 74 02,198 (1974).

175. T. Yoshida, S. Konya, and M. Yokoyama, *Kogyo. Kagaku. Zasshi, 74*(9), 1945 (1971).

176. A. Zochniak, *Plaste Kaut, 19*(9), 655 (1972).

177. H. C. Bach, F. Dobinson, K. R. Lea, and J. H. Saunders, *J. Appl. Polym. Sci., 23,* 2125 (1979).

178. H. C. Bach, F. Dobinson, K. R. Lea, and J. H. Saunders, *J. Appl. Polym. Sci., 23,* 2189 (1979).

179. A. Banihashemi and M. Eghbali, *J. Polym. Sci.: Polym. Chem. Ed., 14* (11), 2659 (1976).

180. G. Blinne, C. Cordes, and H. Dorfel, *Macromol. Chem., 177*(6), 1687 (1976).

181. R. A. Brand, M. Bruma, R. Kellman, and C. S. Marvel, *J. Polym. Sci.: Polym. Chem. Ed., 16,* 2275 (1978).

182. R. A. Brand, R. J. Swedo, and C. S. Marvel, *J. Polym. Sci.: Polym. Chem. Ed., 17,* 1145 (1979).

183. P. E. Cassidy and N. C. Fawcett, *J. Macromol. Sci. Rev. Macromol. Chem., C17*(2), 209-266 (1979).

184. C. D. Dudgeon and O. Vogl, *J. Polym. Sci.: Polym. Chem. Ed., 16,* 1831 (1978).

185. S. G. Efimova, N. P. Okromchedlidze, A. V. Volokhina, and M. M. Iovleva, *Vysokomol. Soedin,* Series B, *19*(1), 67 (1977); *Chem. Abstr., 86*(14), 107153 (1977).

186. R. C. Evers, *J. Polym. Sci.: Polym. Chem. Ed., 16,* 2817 (1978).

187. R. C. Evers, *J. Polym. Sci.: Polym. Chem. Ed., 16,* 2833 (1978).

188. Y. L. Fan, *Macromolecules, 10*(2), 469 (1977).

189. R. F. Kovar, *J. Polym. Sci.: Polym. Chem. Ed., 14,* 2807-2817 (1976).

190. F. M. Silver, *J. Polym. Sci.: Polym. Chem. Ed., 17,* 1247 (1979).

191. H. Sivriev and G. Borissov, *Europ. Polym. J., 13,* 25 (1977).

192. K. Uno, K. Niume, Y. Iwata, F. Toda, and Y. Iwakura, *J. Polym. Sci.: Polym. Chem. Ed., 15,* 1309 (1977).

193. I. K. Varma and C. K. Geetha, *J. Appl. Polym. Sci.*, *22*, 411 (1978).

194. I. K. Varma, C. K. Geetha, and Veena, *J. Appl. Polym. Sci.*, *19*, 2869 (1975).

195. I. K. Varma, C. K. Geetha, and Veena, *J. Appl. Polym. Sci.*, *20*, 1813 (1976).

196. I. K. Varma and Veena, *J. Macrol. Sci. Chem.*, *A11*(4), 845 (1977).

197. I. K. Varma and Veena, *J. Polym. Sci.: Polym. Chem. Ed.*, *17*, 163 (1979).

198. M. Veda, M. Funayama, and Y. Imai, *J. Polym. Sci.: Polym. Chem. Ed.*, *15*, 1629 (1977).

199. S. Watarai, H. Katsuyama, A. Vmehara, and H. Sato, *J. Polym. Sci.: Polym. Chem. Ed.*, *16*, 2039 (1978).

Chapter 7

Heterocyclic Polymers: Six-Member Rings

Polymers with six-member rings in the backbone are not nearly as well known as those with five-member rings. With the exception of polyquinoxalines, the most well developed polymer in this category, all publications have appeared after 1969 and mostly in the mid 1970s. These materials are summarized in Table 7.1.

7.1 Polyquinolines

In an effort to produce a rigid yet soluble and thermally stable polymer, a bisketomethylene was condensed with bis-o-benzoyl aromatic amines to yield polyquinolines [67]. The reaction proceeds in m-cresol or PPA at elevated temperature to give high molecular weights (50,000-100,000) and intrinsic viscosities (0.6-4.0 dl/g).

$R-CH_2-CO-Ar-CO-CH_2-R$ + C_6H_5CO, NH_2 ... NH_2, COC_6H_5

$R-CH_2-CO-Ar-CO-CH_2-R$ + NH_2 ... NH_2

Type A

Type B

R is: H or C_6H_5

Ar is: $p-C_6H_5$,

The polymers are soluble in m-cresol and chlorinated hydrocarbons ($C_2H_2Cl_4$ and $CHCl_3$) and can be cast to tough, clear, flexible films or wet-spun to fibers. Tg values are 266-76°C and Tm's are 448-83°C for these polymers when R is H and Ar is diphenylether, with type B being the higher value in each instance. This 200° differential between Tg and Tm allows the polymer to be processed, yet results in a material with a high use temperature. The effect of extra phenyl substituents is to raise the Tg by 40-45° due to increased hindrance to rotation.

TGA shows decomposition at 550°C in air or nitrogen, and isothermal investigations reveal no weight loss at 300°C for 100 hr.

7.2 Polyanthrazolines and Polyisoanthrazolines

Polyanthrazolines were first synthesized in 1969 by Bracke, who applied the Friedlander quinoline synthesis to the condensation of bis-o-aminoaldehydes and bisketomethylenes [10]. Mild conditions (25-80°C) in various solvents (pyridine, DMF, HMPA, and HMPA + 10% LiCl) gave polymers of good yield (90%) but low molecular weight due to precipitation of the product.

OHC CHO
H_2N NH_2
+
$CH_3-\overset{O}{\overset{\|}{C}}-Ar-\overset{O}{\overset{\|}{C}}-CH_3$
solvent
25-80°
KOH/CH_3OH
N N Ar n
Ar is: $p-C_6H_4$, N, or O

The products were yellow and soluble in only H_2SO_4 (inherent viscosity 0.12-0.26 dl/g). Thermal stabilities by TGA in nitrogen are excellent, showing only a 5% loss at 650°C; however, a decomposition of 10% is seen at 490°C in air. In H_2SO_4 solution the polymers fluoresce strongly in a yellow to orange color depending on the Ar structure.

Following this, Higgins and Janovic discovered a new route which resulted in carboxylated polyanthrazolines [38]. However, the same problem of precipitation from the reaction solution was encountered.

Table 7.1 Polymers with Six-Membered Rings

Name	Structure	Comments
Polyquinolines	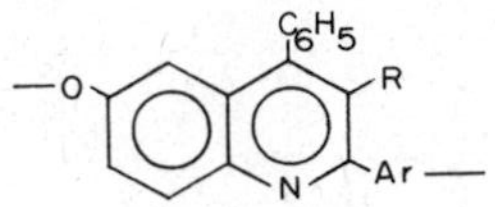	High molecular weight. Soluble. Tough, flexible films and fibers. Tg 266-276°C. Tm 448-483°C. Stable to 550°C.
Polyanthrazolines (and polyisoanthrazolines)	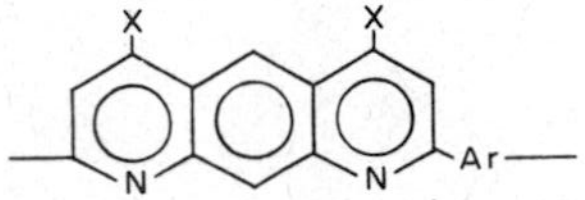	Limited solubility. Yellow. Stable to 500-600°C. Fluoresce in solution. Low molecular weight.
Polypyrazines	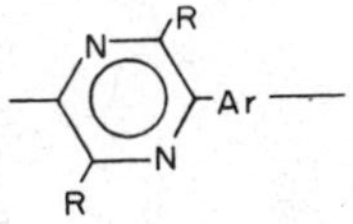	Limited solubility (aided by H_2O_2). Stable to 450-550°C in air and 580-620°C in N_2.
Polyquinoxalines	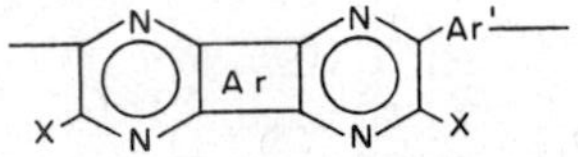	Soluble. High molecular weights. Tg 133-370°C. Stable to 550°C. Used for composites and adhesives.
Polyquinoxalones	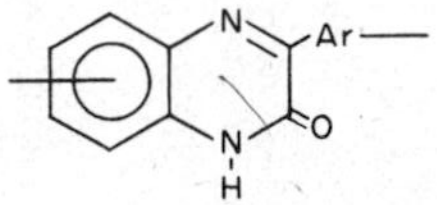	Stable to 400°C. Soluble in aq. base and H_2SO_4.
Polyquinazolones	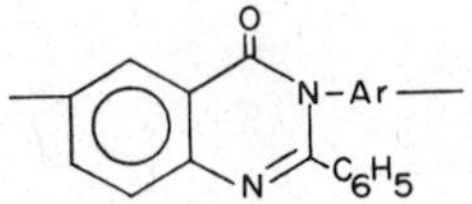	Stable to 425°C/air. Soluble. Forms fibers and films.

Polyquinazolinediones	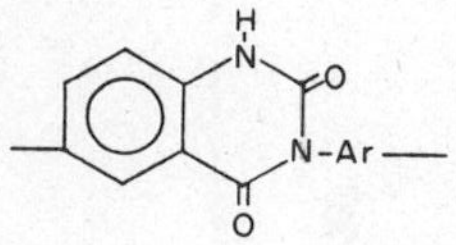	Soluble. Tough, flexible films. Stable to 575-600°C.
Poly-s-triazines	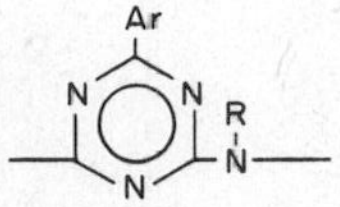	Yellow-brown brittle resins. Soluble. Soften at 50-260°C. Stable to 330°C in N_2.
Poly-as-triazines	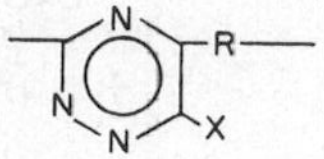	Soluble. Clear, yellow, tough, flexible films. Stable to 380-400°C in air and 530-540°C in N_2. Tg 205-270°C.
Polybenzoxazinones	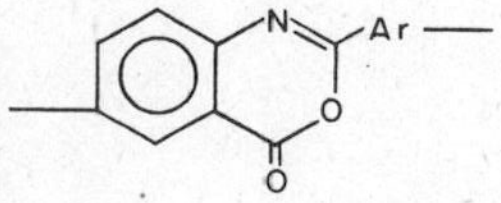	Soluble intermediate. Stable to acid and alkali. Stable to 615°C in N_2 and 580°C in air.
Polybenzoxazinediones	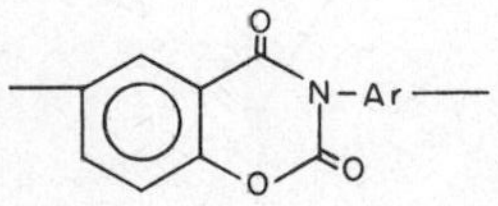	Soluble. Film formers. Soften at 390°C.
Polybenzothiadiazinedioxide	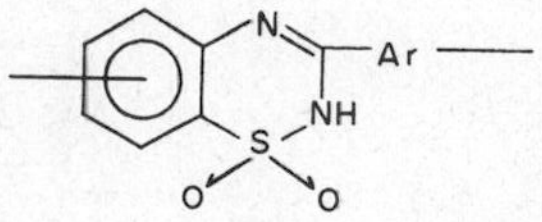	Soluble. Tough films. Stable to 475-485°C. Self-extinguishing.
Polytetrazines	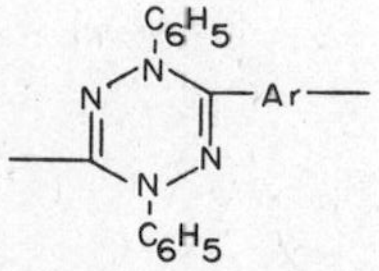	High yield.

H_2O/Dioxane
15-20 hr

$CH_3-\overset{O}{\overset{\|}{C}}-Ar-\overset{O}{\overset{\|}{C}}-CH_3$

Ar is: p-C_6H_4, or $(CH_2)_2$

The red polymers were obtained in good yields (66-93%) and had inherent viscosities (0.24-0.38 dl/g) and solubility (in DMF) superior to the earlier products. Brittle films could be cast, but decomposition began at 350-375°C in air.

Finally, the problem of solubility was solved by using phenyl substituents, a phenomenon which has been shown to be successful with other types of backbones (polyphenylenes, polyquinoxalines, etc.). Phenylated versions of either anthrazoline or isoanthrazoline polymers can be synthesized depending on whether one uses the meta- or para-benzoyl diamines, respectively. The former demonstrated a slightly better solubility [43].

$CH_3-\overset{O}{\overset{\|}{C}}-Ar-\overset{O}{\overset{\|}{C}}-CH_3$

solvent
acid cat.

Ar is: m-or p-C_6H_4 or where X is: nil, O, S, or SO_2

The proper combination of monomers, solvent, and acid catalyst allowed the synthesis of polymers with inherent viscosities commonly of 0.4-0.9 dl/g, with one example of 1.24 dl/g. Solubilities of these products are best in the order of H_2SO_4, m-cresol, and NMP; so this property was improved slightly.

Solvent/catalyst combinations used were the following: m-cresol/HCl, H_2SO_4, H_3PO_4, PPA, or sulfolane/H_3PO_4, one of the best being cresol/PPA.

The use of 1,4-diphenylacetylbenzene as the bisketomethylene results in a polyanthrazoline with four phenyl substituents per anthrazoline nucleus:

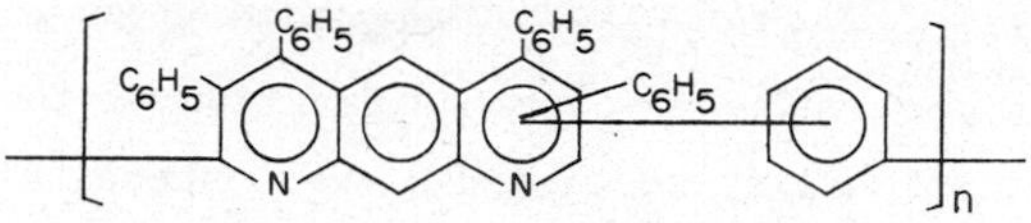

C_6H_5 C_6H_5 C_6H_5 C_6H_5 N N n

TGA shows breaks of 500-560°C in air and 530-600°C in nitrogen. The curves are as expected for aromatic heterocycles showing complete loss of weight in air above 600°C and a residue of about 80% in nitrogen at 800°C.

Method [43] *For the diphenyletherdiphenylisoanthrazoline polymer* An 8% solution of monomers is prepared in a 5:2 m-cresol/PPA solvent by mixing 0.6328 g (2.000 mmol) of 2,5-dibenzoyl-1,4-phenylenediamine [43] and 0.5086 g (2.000 mmol) of 4,4'-diacetyldiphenyl ether [49,60] in 4.0 g of PPA and then adding 10 ml of m-cresol. The mixture is heated and stirred under N_2 for 3 hr at 130°C and then for 10 hr at 170°C. The dark red solution is then cooled, diluted with methanol, and poured into 500 ml of 1N NaOH solution to precipitate the polymer. The solid is collected, washed well with hot water and hot methanol, and then dried at 120°C; inherent viscosity 0.56 dl/g in H_2SO_4.

7.3 Polypyrazines

In 1969-1972 Higgins and co-workers synthesized a series of polymers which contained the pyrazine nucleus [8,39,40]. This was accomplished quantitatively by thermal condensations of bis-α-haloketones in DMAc-NH_3 solutions. One of the unique features of this reaction is that the stoichiometric balance of monomers is of no concern. It is also interesting that a completely aromatic ring and backbone can be formed readily and in one step. Furthermore, air exclusion from the reaction is unnecessary; indeed air is a needed reactant, an unusual circumstance in this field.

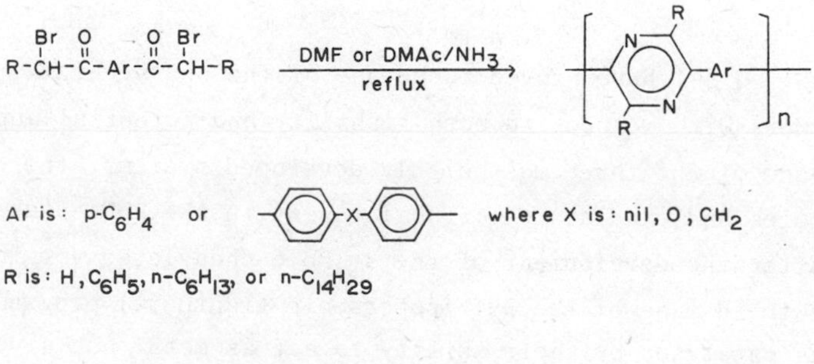

Br O O Br
R-CH-C-Ar-C-CH-R —DMF or DMAc/NH_3, reflux→ [pyrazine ring with R, N, Ar, N, R]n

Ar is: p-C_6H_4 or -C₆H₄-X-C₆H₄- where X is: nil, O, CH_2

R is: H, C_6H_5, n-C_6H_{13}, or n-$C_{14}H_{29}$

The historical development of these materials follows that for many backbones. That is, the unsubstituted backbone (R is H) was synthesized first and found to have limited solubility [40]. The polymers were, however, soluble in formic, sulfuric and phosphoric acids and DMAc when 6% of a 30% solution of H_2O_2, an interesting new solubilizing agent, was present. This phenomenon was attributed to the formation of polar N-oxide functions, the presence of which could be detected by IR spectroscopy (1220-1270 cm^{-1}).

Next, phenyl substituents were added through modification of the monomer, and this improved the solubility so that H_2O_2 was unnecessary [39]. Finally, aliphatic pendant groups were added. Of course this addition lowered the thermal stability of the polymer somewhat [8]. The C_{14} pendant function yielded a tacky but quite soluble product (in ethanol, CCl_4, pyridine, etc.).

The polymers show no melting below 500°C except the phenylated version which softens at 270-300°C and, of course, the aliphatic-substituted backbones. Most polymers show 10% loss by TGA at 580-620°C in nitrogen, while the diphenylether backbone reaches this amount of degradation at ~480°C. In air the materials decompose to this point (10% weight loss) at 450-550°C, the phenylated system being one of the best.

Method *Monomer preparation:* *α,α'-dibenzoyl-α,α'-dibromo-p-xylene* To 150 ml of acetic acid was added dropwise 11.6 g (0.037 mol) of α,α'-dibenzoyl-p-xylene [72] and 12.5 (0.078 mol) of bromine. A spontaneous reflux occurs, after which the product precipitates. The solid is isolated, washed with water, dried, and recrystallized from carbon tetrachloride/benzene to yield 11 g (62%), mp 179-80°C.

Polymerization [*poly 2,5-(1,4-phenylene)-3,6-diphenylpyrazine*] [39] To 150 ml of pure DMAc, which is saturated with NH_3, 2 g of α,α'-dibenzoyl-α,α'-dibromo-p-xylene is added; and the mixture is stirred at 25°C for 1 hr, then at 50-60°C for 1 hr, and finally at reflux for 20 hr in the presence of air. The mixture is cooled and poured into 500 ml of water to precipitate the polymer. The solid is dried in vacuo at 200°C to give a quantitative yield; inherent viscosity 0.18 dl/g in formic acid.

7.4 Polyquinoxalines

Polyquinoxalines ("PQ's") have proved to be one of the better high-temperature polymers with respect to both stability and potential application. They are one of the three most highly developed systems, the others being benzimidazoles and oxadiazoles. The interest in the PQ's peaked in the late 1960s after the development of the soluble phenylated version (PPQ) and before the demise of the SST (supersonic transport) program. This interest was catalyzed by their ability to act as metal adhesives and by the Boeing Company's desire to have a high-temperature adhesive for the titanium wing panels of the SST. The anticipated weight saved by using an adhesive over rivets was so great that a premium price was no obstacle [said to be $990 per kg ($450 a pound) at that time]. There is still great interest in PQ's in spite of the expensive monomers required.

The totally aromatic backbones are derived from the condensation of a tetraamine with a bisglyoxal and were first synthesized in 1964 concurrently by de Gaudemaris and Sillion in France [18], and by Stille and Williamson [64,65] and then extended by the latter group with Arnold [62]. In 1967 the phenylated PQ (PPQ) was prepared by Hergenrother and Levine [21,37]. This material showed significant improvements in solubility, thermal stability and processability. The amines are, of course, very sensitive to air oxidation; and the bisglyoxal is obtained by the selemium dioxide oxidation of an aryl diacetyl compound. Therefore, one can see the difficulties imposed alone by monomer synthesis and purification. Either melt or hot solution polymerizations can be conducted, but the latter is preferred (in m-cresol, dioxane, HMPA, DMF, NMP, and N,N-dimethylaniline) because the products are more soluble. For the PPQ's a tractable, uncyclized precursor is not possible or necessary as it is for other types of backbones. However, the properties of the polymer can be improved by extending the molecular weight of the final polymer by means of a postcure at 375°C. Excellent review articles which include synthesis, properties, and applications are available for the PQ's and PPQ's [23,24,29,61].

NH_2 Ar NH_2 / NH_2 NH_2 + X-C(=O)-C(=O)-Ar'-C(=O)-C(=O)-X

solution heat

Ar is : or

Ar' is : m- or p-C_6H_4 , or where Y is : nil, O, S, SO_2, CO

X is: H or —A where A is : H, halogen, alkyl, OH, CN, alkoxy, aryloxy, ester

It is obvious that two or three isomeric forms of the repeating unit shown are possible depending on the symmetry of the diamine structure. The predominance of one isomer can be reasoned on the basis of the nucleophilicities of the amino groups and reactivities of the aldehydes [23,62,66]. For example the amino group located para to an electron donor will be more nucleophilic (reactive) than one in the meta position. The para amine will then react faster with the more reactive aldehyde (the terminal one in a simple arylene bisglyoxal). The existence of these isomers serves to make the polymer amorphous and therefore aids in its solubility.

A copious amount of property information is available for the quinoxaline polymers; and numerous structure-property studies have been done. Table 7.2 gives a list of publications not referenced elsewhere in this section and the subjects with which they deal.

Some of the representative data are quoted here. Molecular weights can be an impressive--250,000-300,000. Tg's range from 133 to 370°C, above which point the polymers exhibit thermoplasticity. Polymer softening temperatures (PST) are reported to be from 265 to 350°C. Solubilities are good for PQ in m-cresol and H_2SO_4 and for PPQ in chloroform, tetrachloroethane, and other phenolic solvents. Inherent viscosities run as high as 1.7 dl/g in m-cresol. Tough, flexible, yellow films can be cast from a m-cresol solution.

Thermooxidative stabilities by TGA are about 500°C in air or nitrogen, with the p-phenylene-containing backbone being superior. Isothermal analyses at 316°C (600°F) show the following percent weight losses with time: 2% at 25 hr, 7-12% at 600 hr, and 9-31% at 1100 hr, depending on structure [29].

Table 7.2 References for Polyquinoxalines

Topic	Ref.
Synthesis and characterization[a]	30,32,34,70,71
Characterization (molecular)	20
Polymerization kinetics	20
Thermal behavior	3,4
Mechanical properties	6,30

[a]Characterization of bulk properties, e.g., Tg, solubility decomposition temperature, solution viscosity, film-forming ability, etc.

The applications of the PQ systems include glass [23], boron [36], and graphite fiber laminates [23,59]; and adhesives on stainless steel [23,36] and titanium [29]. Seven-ply glass laminates displayed flexural strengths in the vicinity of 69 GPa (100,000 psi) at 25°C and even 120-210 MPa (17,000-30,000 psi) after 10 min and tested at 288°C. Moduli under the same conditions were 25-27 GPa (3.6-3.9 x 10^6 psi) and 15-22 GPa (2.2-3.2 x 10^6 psi), respectively [29].

Because of the amount and quality of mechanical data on the diether PQ (the polymer structure given above where Ar is 3,3',4,4'-diphenylether, Ar' is 4,4'-diphenylether, and X is H) and because a large-scale (1200 g) preparation has been developed, the synthesis for that polymer is given below.

Method [29] In 2 liters of m-cresol under nitrogen, a slurry is prepared with 460.5 g (2.0 mol) of 3,3',4,4'-tetraaminodiphenylether [17]. To the above vigorously stirred slurry is added a slurry of 636.5 g (2.0 mol) of 4,4'-oxybis(phenyleneglyoxal hydrate) [18] in 1.5 liters of m-cresol, over 5 min. Then an additional 888 ml of m-cresol is added, and the mixture is stirred for 1 hr while the temperature is raised to 64°C, where it is maintained for an additional hour. This polymeric solution is stored for use as is.

To isolate the solid polymer, the above solution is poured into methanol to cause precipitation of a beige, fibrous solid which is washed with methanol and dried at 170°C for 1 hr under vacuum; inherent viscosity 0.66 dl/g (0.5% in H_2SO_4); PMT 265-270°C.

The polymer is postcured by immerging it into a preheated bath at 350°C. The temperature is increased to 400°C within 1 hr and held there for an additional hour. The solid melts, foams, and resolidifies to give an inherent viscosity of 1.59 dl/g (0.5% in H_2SO_4).

A more recent interest in polyquinoxalines is to develop a backbone which will crosslink by an addition or oxidative mechanism. Such a polymer would solve two problems: the high-temperature plasticity of PQ's and the volatile byproducts from the normal postcure process. This goal has been accomplished by heating poly(amide quinoxalines) [6] or PQ's which contain p-tolyl pendant groups [70]. It was also brought about in a more controlled fashion by designing polymers with acetylene terminal groups [50]. Yet a third approach was to incorporate as a tridentate a trisglyoxal in the polymerization or to add it to a diamine terminated prepolymer in solution [57]. Although mechanical properties (particularly thermoplasticity) of the crosslinked systems were improved, thermal stabilities were not significantly changed.

Another method whereby properties of a polymer can be improved is to synthesize a copolymer. This approach has been taken with PQ's whereby groups such as imide [2,4,5], amide [12,13], and phenyl-as-triazine [30] are incorporated. The amide function can also be used to provide a crosslinking site. This is done by using a trifunctional acid derivative to synthesize an amide bisglyoxal monomer with a pendant reactive group:

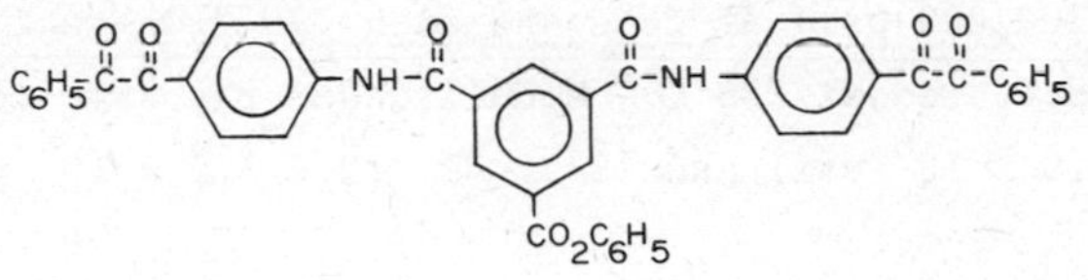

This compound is first polymerized to the PPQ and then crosslinked with diamine. Again, as with the crosslinking discussed above, the use of cofunctions in the backbone with quinoxalines does not improve the thermal stability but does change bulk and mechanical properties from the simple PQ or PPQ.

7.5 Polyquinoxalones

After the polyquinoxalines became known, Higgins and Menon synthesized polyquinoxalones in a similar fashion in an attempt to obtain a water-soluble system [41].

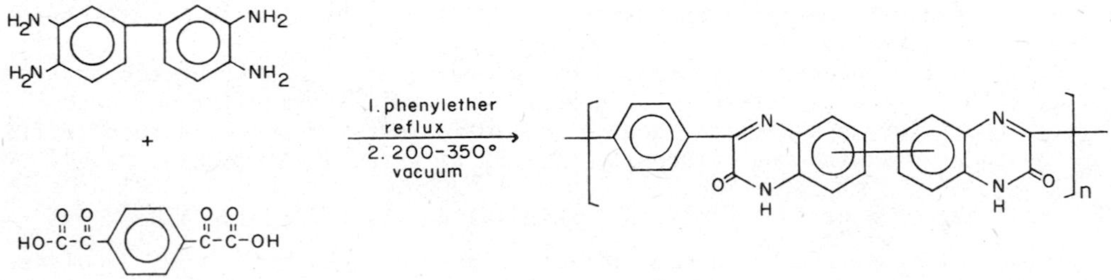

Several difficulties were encountered, however, which detracted from this approach: (1) yield and purity of the p-phenylenediglyoxylic acid (by oxidation of the acetic acid with SeO_2 in dioxane) were poor and (2) the acid monomer decarbonylates during polymerization (at 170°C), leading to low yields and molecular weights.

The polymer displayed an inherent viscosity of 0.45 dl/g in sulfuric acid and thermal stability above 400°C. Due to the formation of the aromatic sodium salt, it was also partially soluble in aqueous base:

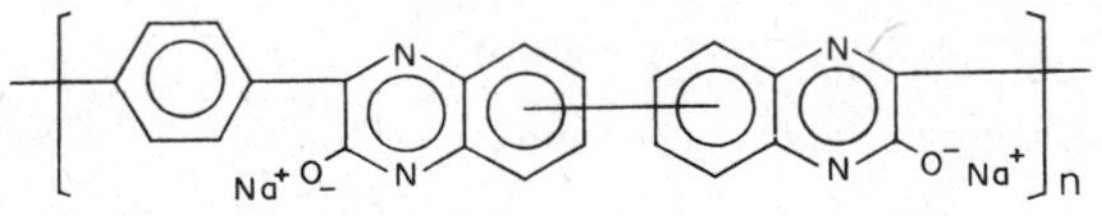

7.6 Polyquinazolones

An isomeric form of polyquinoxalones, the polyquinazolones, has been patented, [7]. The structural differences are the inverted positions of the keto group and one cyclic nitrogen. The keto appears adjacent to the ring, and the nitrogen is directly incorporated into the backbone bonding. This was accomplished (in one step) by heating an aromatic diamine with a bisamic acid.

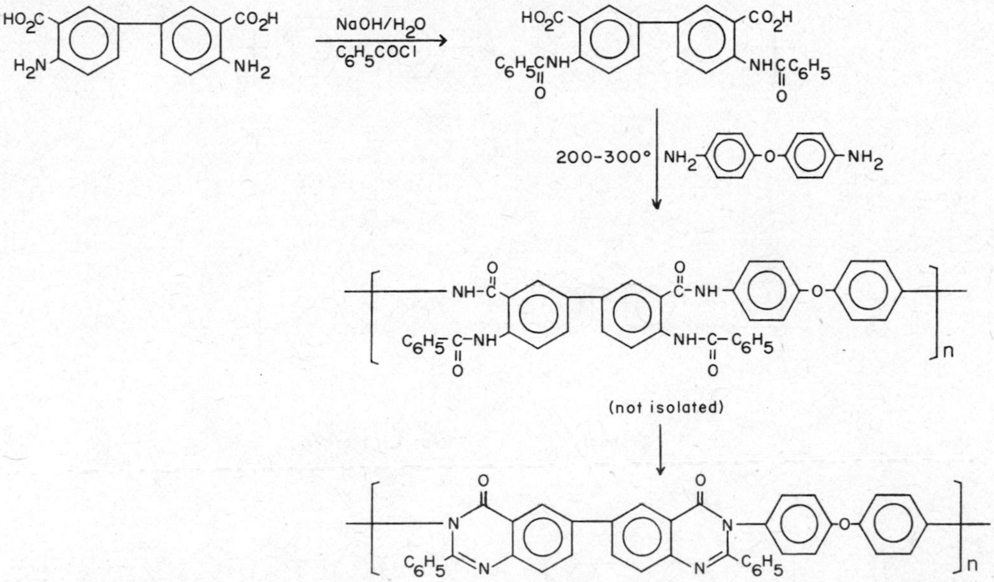

NaOH/H_2O, C_6H_5COCl; 200-300°; (not isolated)

The polymers are stable to 525°C in nitrogen and 425°C in air and are said to be useful as fibers, films, and molded articles. They are soluble in H_2SO_4, DMSO, DMF, and formic acid.

> *Method* [7] *Monomer preparation* Into a solution of 15.9 g of 3,3'-dicarboxybenzidine and 14.1 g of benzoyl chloride in 600 ml of water is added slowly 12.0 g of sodium hydroxide. After the reaction is stirred for 15 min, acetic acid is added to precipitate the dibenzamide of dicarboxybenzidine.
>
> *Polymerization* A mixture of 2.4 g of the dibenzamide and 1.0 g of 4,4'-diaminodiphenylether is heated under nitrogen for 7 hr at 200-300°C to yield the polymer.

A year after the above patented polymerization, a route to polyquinazolones was discovered via a backbone reaction [51]. Thus polybenzoxazinones (see Section 7.9) reacted with primary amines to suffer first ring opening to a poly(amic amide) which is tractable. Then, this intermediate can take one of two cyclization paths upon being heated. If no catalyst is present, it deaminates back to the polybenzoxazinone starting material. If, however, it is heated in the presence of LiCl or $ZnCl_2$, it dehydrates to polyquinazolone. This later reaction is said to be due to the formation of a complex of ortho amide groups with the inorganic salt. The molecular weight of the polymer is retained through this reaction unless excess amine reagent is present.

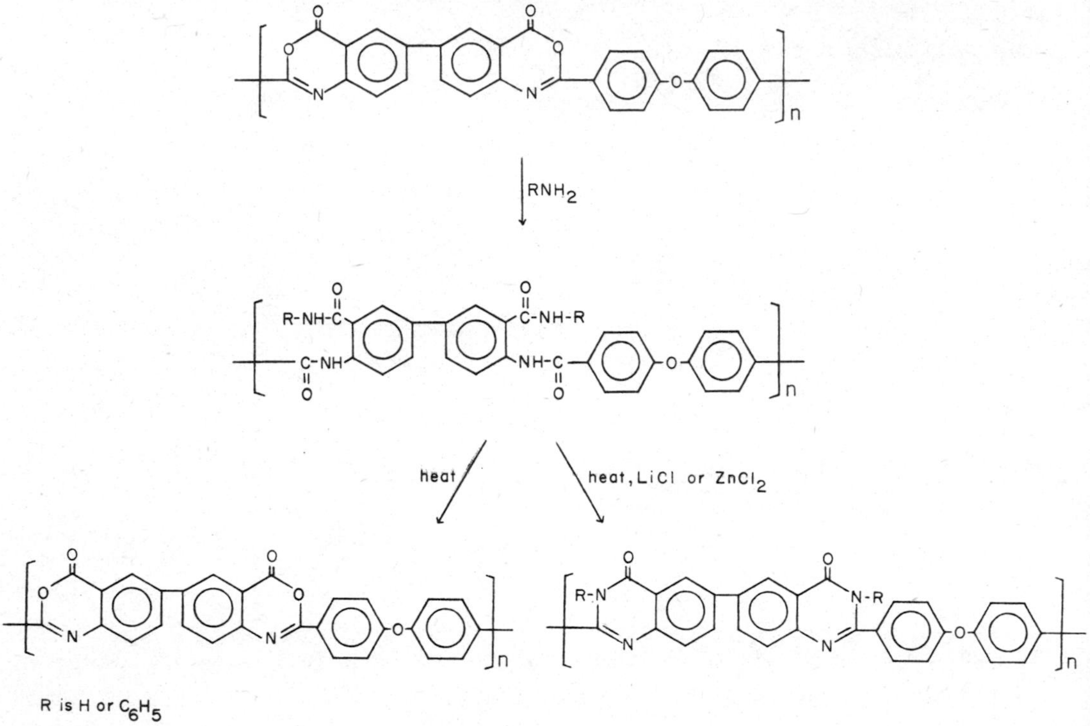

RNH_2

heat

heat, LiCl or $ZnCl_2$

R is H or C_6H_5

The intermediate poly(amic amide) is soluble in DMAc, NMP, DMF, DMSO, and H_2SO_4; whereas, the final polyquinazolone is soluble only in concentrated sulfuric and fuming nitric acids.

Thermal stabilities for the polyquinazolones are fair to moderate as thermally stable polymers go. The phenyl substituent has little effect on this property. They show initial weight loss near 300°C and 10% loss at 440°C in air and 530°C in nitrogen.

Method [51] *Ring opening* A solution of 20 mg of polybenzoxazinone in 50 ml of aniline is heated to reflux and a polymer precipitated. The solid is filtered off, washed, and dried; and the filtrate is poured into chlorobenzene to precipitate the aniline-soluble fraction of the poly(amic amide).

Cyclization A 5% solution of the above poly(amic amide) in NMP is made, and 0.5% zinc chloride is added (i.e., the zinc chloride/polymer ratio is 1:10). A film is cast and heated at 250°C under vacuum for 1.5 hr and then at 300°C under vacuum for 2 hr. Finally, zinc chloride is removed from the film by treating it with a 10% HCl solution in water. For the nonphenylated polymer, lithium chloride is a satisfactory catalyst; and it can be removed from the final film by a water wash.

7.7 Polyquinazolinediones and Polyquinazolinetetraones

Another heterocyclic system similar to the quinoxalones and quinazolines is that of the higher oxidized quinazolinediones. This new class of polymers was reported in 1966 by Yoda et al. as being derived from a diisocyanate and an aromatic bisaminocarboxylic acid [54,68,78]. The process involved two steps with a tractable poly(urea acid) [or poly(ureylene acid) or poly(ureido acid)] intermediate.

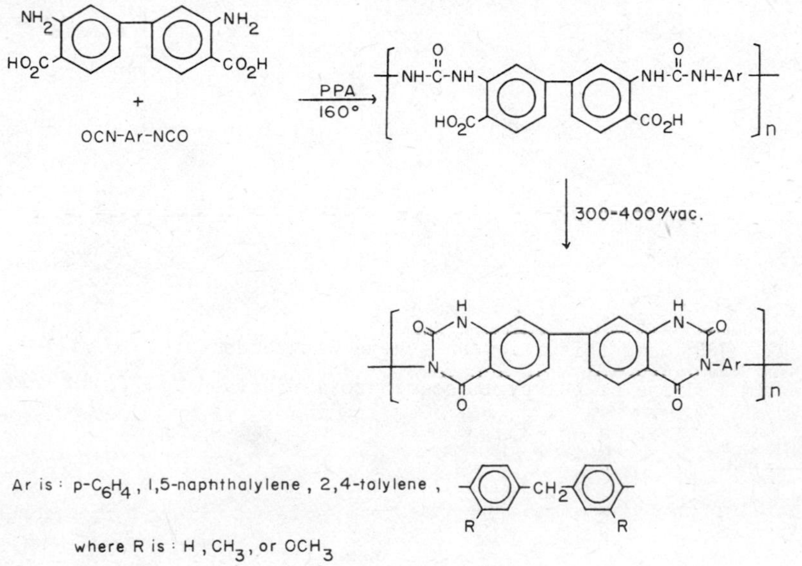

Ar is: p-C_6H_4, 1,5-naphthalylene, 2,4-tolylene,

where R is: H, CH_3, or OCH_3

The intermediate is obtained in essentially quantitative yield and is soluble in H_2SO_4, NMP, DMAc, or DMSO containing 5-10% of LiCl, the latter solution from which tough, flexible films can be cast and post-cured. Inherent viscosities range from 0.6 to 1.8 dl/g in H_2SO_4.

The final cyclized polymer was insoluble and showed susceptibility to hydrolysis in boiling 35% H_2SO_4 and 78% KOH. Its thermal stability by TGA was excellent, showing similar curves in either air or nitrogen (10% weight loss at 575-600°C) [54].

Method [54] A solution is prepared from 80 g of PPA and 1.6 g of p-phenylene diisocyanate and heated with stirring under nitrogen to 130°C. A 2.72 g portion of 4,4'-diamino-3,3'-biphenyldicarboxylic acid is added, and stirring and heating is continued at 140-150°C for 5 hr. The viscous mixture is poured into a large excess of water to precipitate the polymer which is isolated, washed with water and methanol, and then dried. The solid is tan- to brown-colored and is obtained in 90% yield (2.7 g) with inherent viscosity of 1.29 dl/g in H_2SO_4.

The second method of incorporating this functional group into polymers was first indicated by the synthesis of 3-phenylquinazoline-2,4-dione from aniline and N-mesyloxyphthalimide [46].

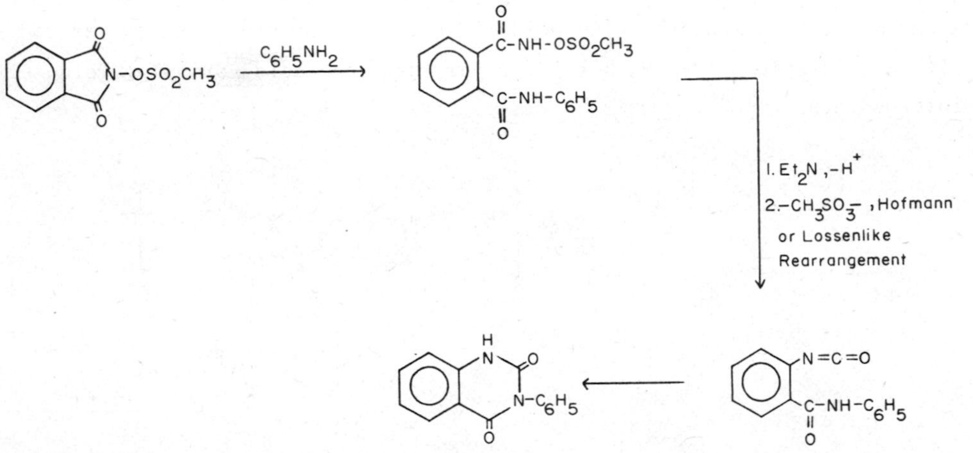
N-OSO2CH3
C6H5NH2
C-NH-OSO2CH3
C-NH-C6H5
1. Et3N, -H+
2. -CH3SO3-, Hofmann or Lossenlike Rearrangement
N=C=O
N-C6H5

Imai et al. then used this reaction scheme with bidentates to yield a fused polyquinazolinedione or polypyrimidequinazolinetetraone [47].

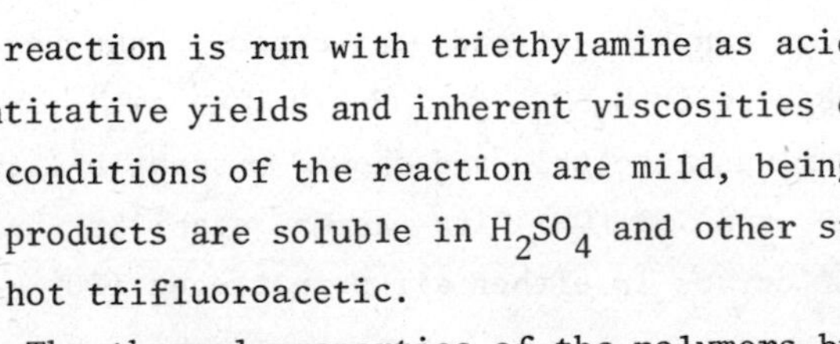
N-Ar
n
Ar is :
where X is -CH2- or -O-

The reaction is run with triethylamine as acid acceptor in NMP or HMPA to quantitative yields and inherent viscosities of 0.17-0.27 dl/g in H_2SO_4. The conditions of the reaction are mild, being room temperature for 7 days. The products are soluble in H_2SO_4 and other strong acids, methanesulfonic, and hot trifluoroacetic.

The thermal properties of the polymers by this method were not as good as those from the earlier synthesis discussed above. TGA shows 10% weight loss levels at 450°C in air and 470-480°C in nitrogen, about 100°C below earlier materials. Furthermore, the solubilities appear to be lower.

Method [47] To a solution of 0.500 g (2.5 mmol) of 4,4'-diaminodiphenylether and 0.51 g (5 mmol) of triethylamine in 10 ml of NMP is added 1.011 g (2.5 mmol) of N,N'-bis(mesyloxy)pyromellitimide [46,47] with stirring at 25°C. After 7 days the mixture is poured into 500 ml of water, and the solid is collected, washed with boiling acetone, and dried at 120°C to yield 1.02 g; inherent viscosity 0.25 dl/g in H_2SO_4; IR: 1715 and 1650 cm^{-1} (C=O).

Yet a third scheme has been discovered for the synthesis of quinazolinedione polymers. In this case an amide copolymer is produced by ammonolysis

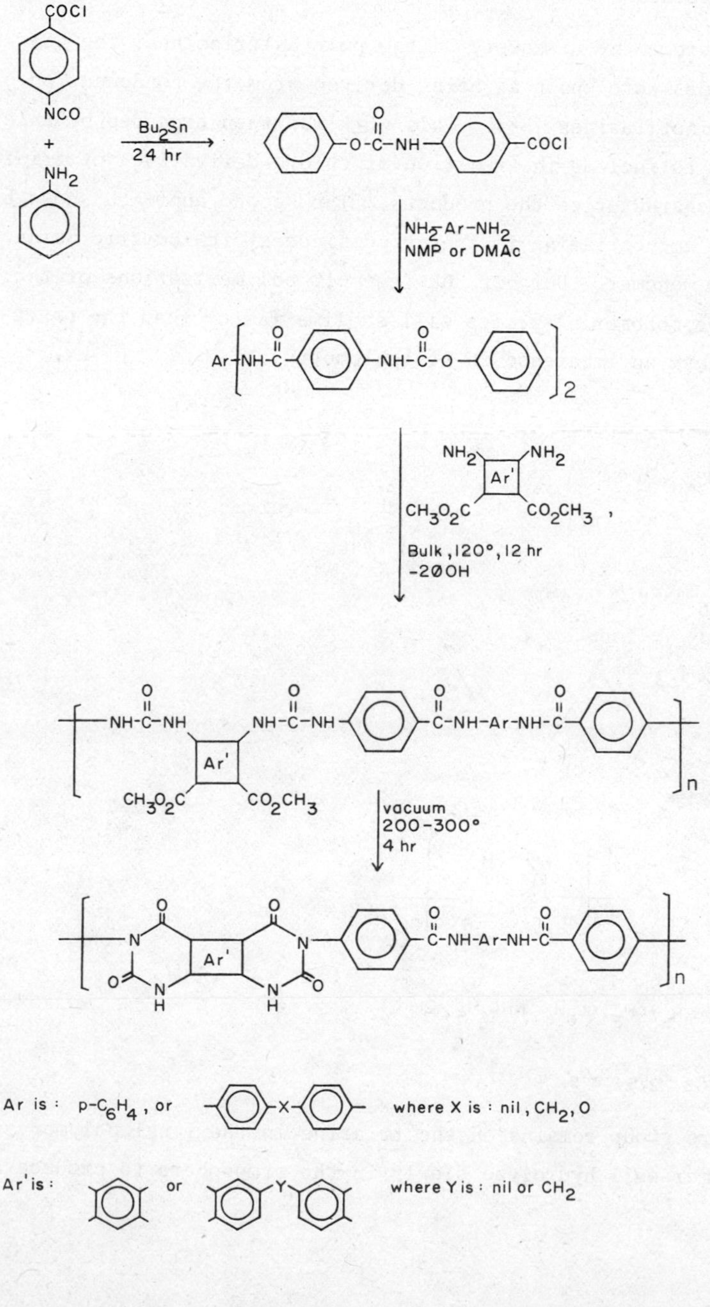

COCl

NCO

\+

NH_2

Bu_2Sn

24 hr

O-C(=O)-NH- -COCl

NH_2-Ar-NH_2

NMP or DMAc

Ar-[NH-C(=O)- -NH-C(=O)-O-]$_2$

NH_2 NH_2

Ar'

CH_3O_2C CO_2CH_3 ,

Bulk, 120°, 12 hr

-2ØOH

-[NH-C(=O)-NH Ar' NH-C(=O)-NH- -C(=O)-NH-Ar-NH-C(=O)-]$_n$

CH_3O_2C CO_2CH_3

vacuum

200-300°

4 hr

-[N Ar' N- -C(=O)-NH-Ar-NH-C(=O)-]$_n$

Ar is: p-C_6H_4, or -⟨⟩-X-⟨⟩- where X is: nil, CH_2, O

Ar' is: or where Y is: nil or CH_2

of a urethane link followed by cyclodehydration of the resulting urea with an adjacent ester group [48]. Chemically this method is similar to that of Yoda et al. [54,78].

The polymers were soluble in DMAc, HMPA, and H_2SO_4. However, thermo-oxidative stabilities were only above 360°C by TGA, which is considerably below the polymers produced by the other two methods.

7.8 Poly-s-triazines

Over 10 years before the discovery of the poly-as-triazines, the poly-s-triazinyleneimides were known as being derived from the condensation of phenoxy- and aminotriazines [58]. This work was then expanded by Ehlers and Ray [15,16] to include the reaction of chloro derivatives of triazine and to further characterize the products. The second approach shown below from the chloro aminotriazine is preferred since stoichiometric balance is inherent in the monomer. Further, during melt polymerizations of the first type, one of the monomeric species will sublime faster than the other and, in so doing, cause an imbalance in stoichiometry.

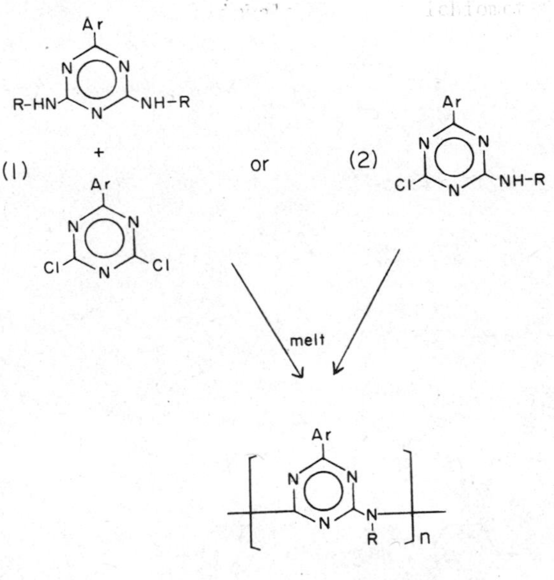

Ar is: H, C_6H_5, CH_3, OC_6H_5, NHC_6H_5, Cl

R is: H, CH_3, C_2H_5, C_6H_5

If a chloro group remains on the triazine in the final polymer structure, the polymer will hydrolyze slowly in the atmosphere to produce a brittle, insoluble powder. Attempts to take advantage of this reactivity in a crosslinking reaction were unsuccessful.

The poly-s-triazinyleneimides are yellow to brown brittle resins. Toughness increases somewhat for the highest molecular weight (16,000) products. They are soluble in phenolic solvents and show low to moderate inherent viscosities (0.07-0.35 dl/g, 0.5% in m-cresol).

In general those polytriazineimides with -NH- functions (R is H) showed the highest heat softening temperatures and some were insoluble. This can readily be attributed to hydrogen bonding. The validity of this reasoning was proven when an insoluble polymer (R is H, Ar is C_6H_5) was reacted with 2,4-diphenyl-6-chloro-s-triazine and the product (R is 2,4-diphenyl-s-triazine-6-yl) was soluble.

Mechanical properties, as indicated by softening curves which are measured by penetration vs temperature on the powder [14] show that softening begins (initial break in the curve) at 50-180°C for most systems but at 150-260°C for the NH varieties. Chloro groups also tend to increase softening temperatures.

Thermal stabilities were measured for these materials by TGA, but only in nitrogen. Generalities were difficult to make from the data. The wholly methyl-substituted polymer was the poorest in thermostability (10% loss at 240-280°C), but the phenylated material was not much better (330°C). A mixture of the two substituents interestingly brought the value to 420°C. The most stable system was where Ar was phenyl and R was diphenyltriazinyl.

Method [15] *Poly(6-phenyl-N-ethyl-2,4-s-triazinyleneimide*

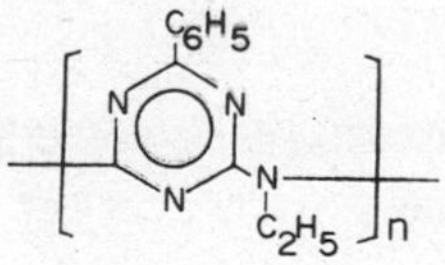

> A mixture of 0.81 g of 2-phenyl-4,5-bis(ethylamino)-s-triazine and 0.75 g of 2-phenyl-4,6-dichloro-s-triazine is heated to 200°C for 43 hr under nitrogen. When no HCl can be detected in the effluent gas, the reaction is cooled and the light yellow brittle resin is ground and extracted for 2 hr with ether; inherent viscosity 0.07 dl/g (0.5% in phenol).

The use of the s-triazine function as a curing or crosslinking species has been known since the early part of the work on thermally stable polymers. In 1960 Brown [11] used this function in the homo- or copolymerization of perfluoroadipodiamidine and/or perfluoroglutarimidine:

$NH_2\overset{NH}{\overset{\|}{C}}{-}(CF_2)_4{-}\overset{NH}{\overset{\|}{C}}{-}NH_2$ and/or [ring: F_2, F_2, F_2, NH, =NH, =NH]

melt
$(-NH_3)$

[triazine ring: N, N, N]

The resulting amber, crosslinked resin was stable to above 350°C. A similar system was reported later from the polymerization of perfluorosebaconitrile and perfluorosebacamidine to give poly(perfluorooctamethylene triazine), with cyano pendant groups [19]. This polymer was then vulcanized with tetraphenyltin to give nonbrittle resins.

In 1966 Anderson and Holovka [1] produced a series of crosslinked resins by reacting various aryl dinitriles with chlorosulfonic acid. Thermal stabilities by TGA were 225-340°C in air (10% weight loss).

Most recently, p-toluenesulfonic acid was shown to be an effective catalyst for trimerization of aromatic dinitriles to produce crosslinked resins [42]. This study also indicated that this process was suitable for obtaining crosslinked polyimides for use in graphite fiber composites.

7.9 Poly-as-triazines

Hergenrother has published extensively since 1969 on the synthesis of poly-as-triazines from the reaction of bisamidrazones with bisglyoxalylarenes [22,25,28,31,33,35,73]. The procedures are quite simple in that they involve stirring the reactants in m-cresol and precipitating the polymer from methanol. The reactions were run at concentrations below 10% to prevent gelation, which is due to branching.

One of the most interesting properties of these polymers is their solubility. They dissolve in up to 30% concentration in m-cresol, chloroform, and sym-tetrachloroethane but not even up to 3% in DMAc, DMSO, or HMPA, hardly the usual behavior for polymeric heterocyles. From the solutions, clear, lemon-yellow, tough, flexible films can be cast. Inherent viscosities in m-cresol (0.5%) range from 0.4 to 2.02 dl/g. Ester function

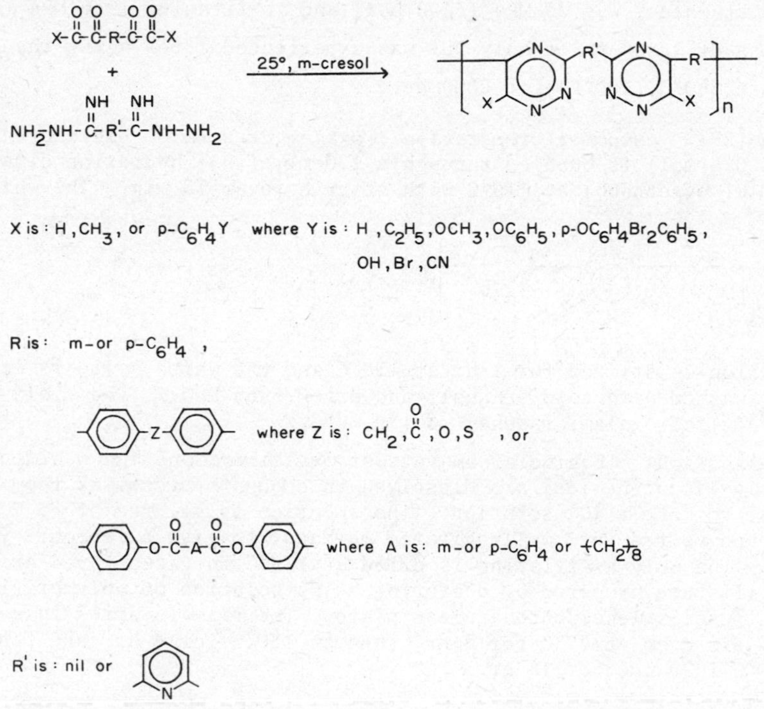

$$X{-}\overset{O}{\overset{\|}{C}}{-}\overset{O}{\overset{\|}{C}}{-}R{-}\overset{O}{\overset{\|}{C}}{-}\overset{O}{\overset{\|}{C}}{-}X \;+\; NH_2NH{-}\overset{NH}{\overset{\|}{C}}{-}R'{-}\overset{NH}{\overset{\|}{C}}{-}NH{-}NH_2 \xrightarrow{25°,\ m\text{-cresol}}$$

X is : H, CH_3, or p-C_6H_4Y where Y is : H, C_2H_5, OCH_3, OC_6H_5, p-$OC_6H_4Br_2C_6H_5$, OH, Br, CN

R is: m-or p-C_6H_4,

where Z is: CH_2, C=O, O, S, or

where A is: m-or p-C_6H_4 or $(CH_2)_8$

R' is: nil or

lower the solution viscosity. It is not surprising that this effect is more pronounced when the ester is in the backbone rather than in a pendant position.

Thermoanalytic data are quite good, showing TGA breaks in air and nitrogen at 380-400 and 530-540°C, respectively. The first of these decomposition points is due to the N-N bond cleavage which releases benzonitrile (when a phenyl pendant group is present) [35]; i.e., the initial degradation does not destroy backbone integrity. When aliphatic groups are present, decomposition begins as expected at 290°C [31]. Isothermal aging demonstrates excellent retention of weight even after 1000 hr at 260°C. At 290°C significant decomposition does occur with time [35]. Glass transition temperatures cover the range of 205-270°C except when aliphatic ester functions are present in the backbone or as pendant groups, whereupon the Tg is lowered to 178°C [31].

Polytriazines have been tested in applications as adhesives, films, and composite matrices [27]. Films had a tensile strength of 217 MPa (18,400 psi), tensile modulus of 2.9 GPa (400,000 psi), and an elongation-to-break of 5%. For adhesives use, a pre-preg, glass tape was prepared and cured between titanium or stainless lap shear panels at 274°C (525°F). The bond strength to stainless was 23 MPa (3400 psi) and to titanium 19.6 MPa (2850 psi). A loss of strength of only 20% was experienced after aging the titanium sample at 260°C (500°F) for 2000 hr.

Method [27] *Monomer preparation: oxalamidrazone* Dicyanogen (16 g, 0.5 mol) is bubbled through a 120 ml of 95% hydrazine dissolved in 600 ml of ethanol at -10°C with stirring over 15 min. The white suspension is stirred for 1 hr at -10°C and the white solid is isolated, washed with cold ethanol, and dried over P_2O_5. The yield is 27 g (47%) of oxalamidrazone, mp 179-80°C.

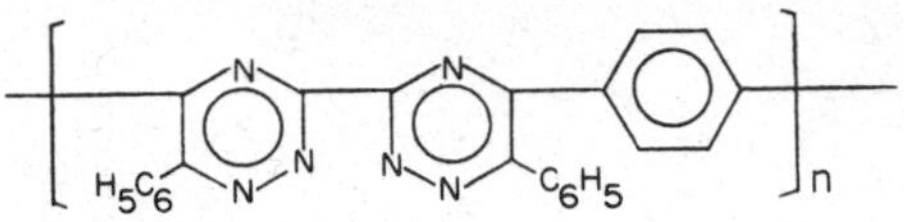

Polymerization Equimolar amounts of oxalamidrazone and p-bis(phenylglyoxaloyl)benzene [35] are dissolved in enough m-cresol at room temperature to make a 10% solution. The solution is stirred at 25°C overnight and poured into well-agitated methanol to give a fibrous, yellow solid. The poly-as-triazine is dried at 130°C in vacuo for 4 hr.

Films are prepared by doctoring a 10% solution of polymer (in 1:1 m-cresol:xylene) onto a glass plate. The film is dried in a forced air oven at 70°C for 8 hr, then at 130°C for 4 hr, and finally at 150°C in vacuo for 18 hr.

7.10 Polybenzoxazinones

Yoda and co-workers have published much since 1966 on their investigations of the incorporation of the benzoxazinone nucleus in polymers [52,53,55,56,75-77]. However, some of the papers repeat earlier data and results together with new ones. This subject has been reviewed [74] and the composition has been patented [69,75]. This polymer has a similarity to the polyquinazolinediones (Section 7.7) in that both are derived from benzidine-3,3'-dicarboxyli acid. In this case, however, it is reacted with a diacid chloride rather than a diisocyanate. Again a two-step process is used involving a tractable poly(amic acid) intermediate which is soluble in DMAc, NMP, DMF, and DMSO (all containing LiCl) or in H_2SO_4. The final cyclization step is accomplished by (1) heating a prepolymer film at 180-360°C under vacuum or (2) treating it with a 1:1 mixture of acetic anhydride/pyridine followed by heating at 130-180°C in a vacuum for 20 min; or (3) heating it in PPA for 1-3 hr at 200-250°C. The postcure is confirmed by the benzoxazinone carbonyl IR absorption at 1760 cm^{-1}.

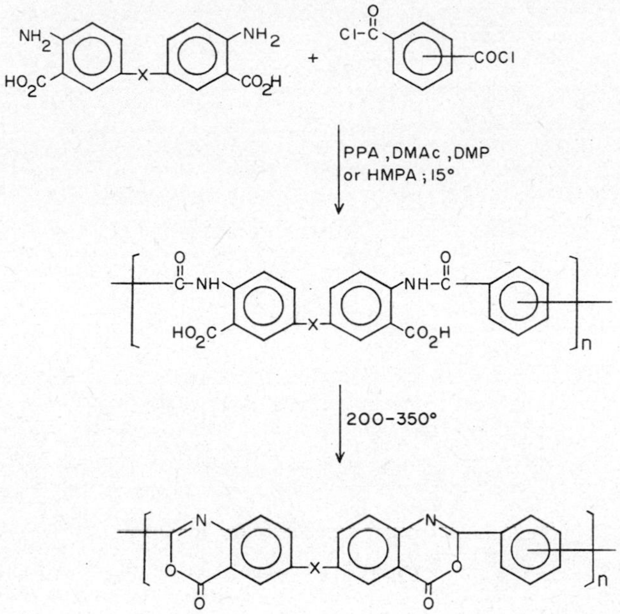

X is: nil, CH_2, or O

It was found that a slight excess (2%) of acyl chloride was beneficial to the molecular weight of the product. It was postulated that this excess served to remove trace amounts of water, which has a disasterous effect on molecular weight. For example, an increase in the water content from 0.001 to 0.5% dropped the inherent viscosity from 1.28 to 0.70 dl/g (in H_2SO_4) [77].

A great deal of the synthetic work on this system has been done in 84% PPA [53]; hence, the mechanism of the reaction in PPA has been studied extensively [52,56]. It was established that the process involves (1) low-temperature phosphorylation of the amine and acid; (2) above 140°C a dephosphorylation of the amine with poly(amic acid phosphate) formation, which can be isolated as the poly(amic acid); and finally (3) cyclodehydrophosphorylation to the polybenzoxazinone. This scheme is represented below in a simplified manner:

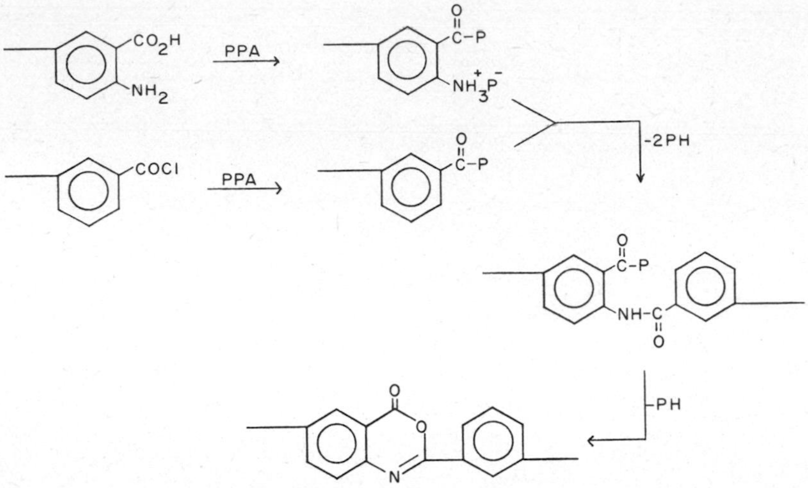

In 84% PPA the poly(amic amide) precursor reached an inherent viscosity of 2.66 dl/g (0.5% in H_2SO_4) [53], approximately twice that attained by low-temperature solution polymerization in NMP [77].

The polybenzoxazinones show excellent hydrolytic and thermal stabilities. Little change in viscosity is effected in strong acid or alkali [74]. TGA shows initial decomposition at 480°C in nitrogen and 440°C in air. The 10% weight loss point is reached at 615°C in nitrogen and 580°C in air [53].

Method [53]

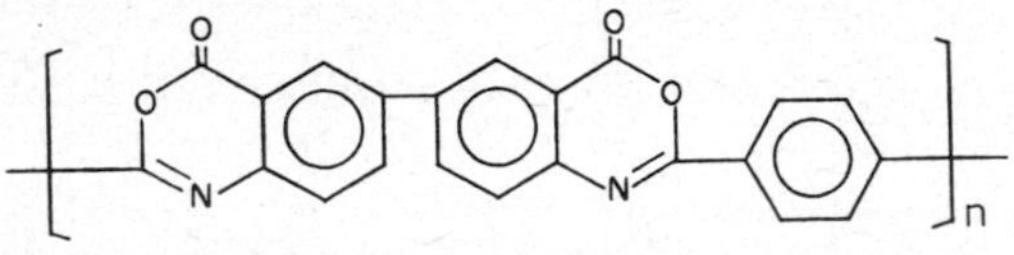

Polymerization Under an atmosphere of nitrogen, 200 g of 84% PPA is heated to 60°C and 2.08 g (0.100 mol) of terephthaloyl chloride is added; and the mixture is stirred and heated to 80-100°C to effect dissolution. Then 2.72 g (0.100 mol) of 4,4'-diamino-3,3'-biphenyldicarboxylic acid is added gradually, and the mixture is heated to 140-160°C for 5-8 hr. The hot mixture is poured into water to precipitate the polymer, which is isolated and thoroughly washed successively with water, 5% aq. sodium carbonate, dilute hydrochloric acid, water, and methanol. The poly(amic acid) is then dried at 80°C in vacuo to give a nearly quantitative yield; inherent viscosity 2.66 dl/g (0.5% in H_2SO_4). The polymer is dissolved in NMP which contains 5% lithium chloride, and the solution is cast on a glass plate and dried at 120°C for 30 min to give a transparent, tough film; IR: 1670 cm^{-1} (C=O) and 3400 cm^{-1} (NH).

Cyclodehydration The poly(amic acid) film is heated under vacuum at 180-360°C; IR: 1760, 1260, and 1060 cm^{-1}.

7.11 Polybenzoxazinediones

A one-step reaction between a diisocyanate and an aryl bis(o-hydroxycarboxylic acid ester) yields polybenzoxazinediones [9]. The reaction is run in DMSO and uses tertiary amines as catalysts. A poly(urethane ester) can be envisioned as an intermediate structure.

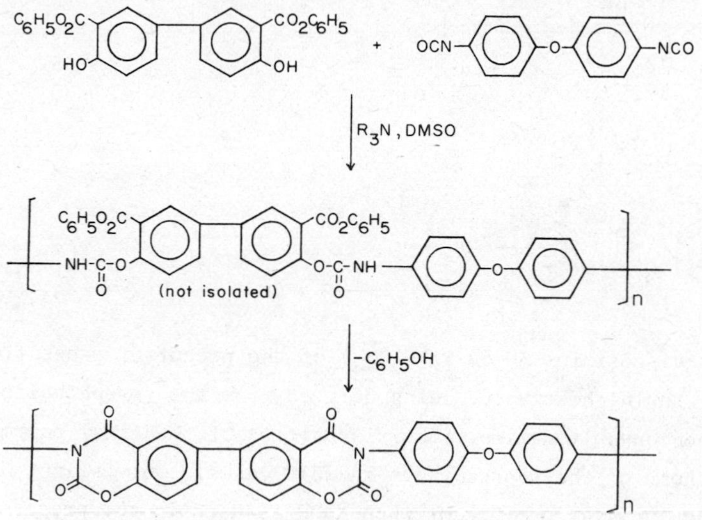

$C_6H_5O_2C$ $CO_2C_6H_5$ HO OH + OCN O NCO

R_3N, DMSO

$C_6H_5O_2C$ $CO_2C_6H_5$ NH-C-O O O-C-NH O (not isolated) O n

$-C_6H_5OH$

O O N N O O O O n

The polymers are soluble in polar solvents and can be cast into films, which can be oriented and crystallized by drawing. They soften above 390°C and are said to be useful as high-temperature insulating films.

7.12 Polybenzothiadiazinedioxide

A copolymer of amide and benzothiadiazinedioxide functions has been reported to result from the condensation of a diacid chloride with a diaminosulfonamide [44,45] followed by chemical cyclodehydration [44]. The first step is a low-temperature solution reaction (in polar solvents), which quantitatively yields a poly(amic sulfonamide). The second, cyclodehydration, step is successful only with organic bases (tri-n-butylamine or γ-picoline) and not with sodium hydroxide or by thermal means, due to the strong base causing chain degradation by amide hydrolysis and to heat effecting cleavage to aryl nitrile and aminosulfonic acid.

Both the poly(amide amic sulfonamide) precursor and the final poly(amide benzothiadiazinedioxide) are soluble in polar solvents such as DMF with 5% LiCl and DMAc, DMSO, and NMP, and from such solutions can be cast to tough films. Inherent viscosities (0.5% in H_2SO_4) of the precursor range from 0.6 to 3.0 dl/g, the highest value being derived from the terephthalic acid and 2,5-diaminobenzenesulfonamide. The viscosities of the final polymers are similar to those of their precursors in most cases. Only slight decreases accompany cyclization, except when NaOH is used in the process. One exception is the highest viscosity precursor (3.03 dl/g), which is reduced by cyclization even with tributylamine (to 2.63 dl/g).

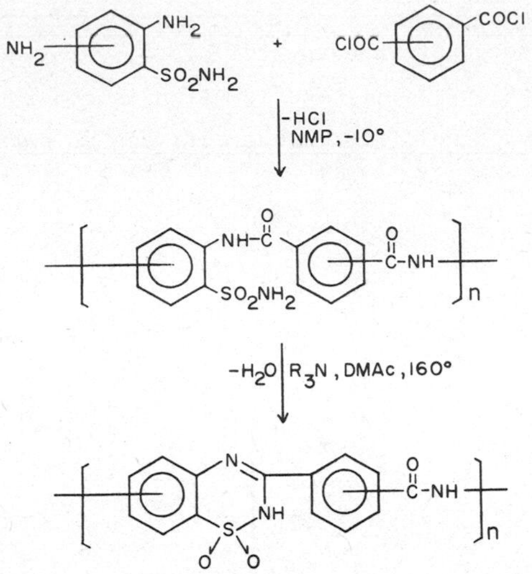

The thermal properties of the cyclized polymers are good and similar in either air or nitrogen. Weight losses of 10% by TGA are observed to occur at 475-485°C. In addition, both precursor and final polymer exhibit self-extinguishing characteristics.

Method [44] *Prepolymerization* A solution is prepared containing 0.94 g (0.005 mol) of 2,5-diaminobenzenesulfonamide [44] in 7 ml of NMP, and it is solidified in a Dry Ice bath. Then 1.03 g (0.005 mol) of isophthaloyl chloride is added, and the mixture is stirred in an ice-methanol bath at -10 to -20°C for 4 hr and then at 25°C for 20 hr. The solution is diluted with 40 ml of NMP and poured into 800 ml of methanol to precipitate the poly(amide amic sulfonamide). The solid is isolated by filtration, washed with water and then acetone, and dried at 100°C under vacuum to yield 1.59 g (99%); inherent viscosity 3.02 dl/g (0.5% in H_2SO_4).

Cyclodehydration A solution of 0.45 g of the above prepolymer dissolved in 10 ml of NMP and 2 ml of tri-n-butylamine is stirred and heated to reflux (160°C) for 12 hr. The mixture is then poured into 500 ml of 5% aq. acetic acid to precipitate the polymer. The solid is collected, washed well with water and acetone, and dried under vacuum at 120°C. The yield is 0.43 g (99%) with inherent viscosity of 2.63 dl/g (0.5% in H_2SO_4).

7.13 Polytetrazines

A self-cyclocondensation has been reported for iso- or terephthaloyl phenylhydrazide chloride to yield a poly(phenylene-1,4-dihydro-1,2,4,5-tetrazine) [63]. This reaction proceeds to high yield in refluxing benzene or pyridine with a tertiary amine catalyst. The reaction goes through a nitrilamine dipole which subsequently undergoes a 1,3-dipole addition. A similar reaction has been described in Chapter 6 for the synthesis of polypyrazoles.

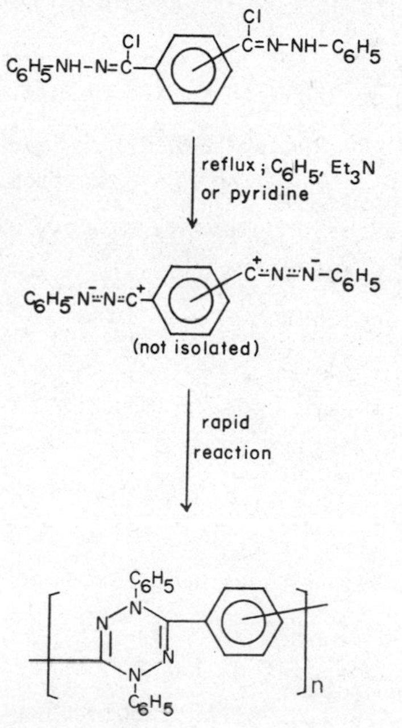

reflux; C_6H_5, Et_3N
or pyridine

(not isolated)

rapid
reaction

7.14 Recent Developments

7.14-1 Polyquinolines

Several recent publications on polyquinolines have emanated from Stille's laboratory [42,101-103]. A higher molecular weight product is possible using di-m-cresyl phosphate in m-cresol solution for the Friedlander synthesis than with crude m-cresol in phosphorus pentoxide [79].

Biphenylene has been used as a crosslinking site for polyquinoxalines as it has for other types of backbones [84,95]. The 2,6-diacetylbiphenylene was incorporated into the backbone and found to open and cure at 340°C for 2-3 hr. A rhodium complex reduced the cure temperature to 170°C.

7.14-2 Polyquinoxalines

Of course considerable research has continued on PQ's and PPQ's and modifications thereof because of promising earlier applications. This work can be divided into three categories: synthesis of new types, crosslinking or curing methods, and applications.

Hergenrother has prepared a concise review of PPQ's and described a new, high molecular weight, soluble poly(carbonate phenylquinoxaline [87]. The materials gave inherent viscosities to 1.28 dl/g, Tg's to 274°C, and thermal stabilities to approximately 475°C by TGA.

Flexibilizing groups such as diphenylether and meta-linked phenyls have been introduced into PPQ's [93]. These products are thermoplastic at 300-315°C and crosslink at 300-500°C. Clear, yellow films can be cast from m-cresol, or the polymer can be molded or used as a composite binder. Flexural lifetimes are 150 hr at 350°C or 24 hr at 400°C.

Ether-PPQ's have also been synthesized and characterized in an effort to bring some flexibility to the backbone [96,97].

Several publications have been directed toward crosslinking PQ's or PPQ's to eliminate high-temperature plasticity [85,86,88,94,100]. One of these studies incorporates a quinoxaline function into a PPQ system to act as a cure site. These polymers were cured at 400°C for 0.5 hr, whereupon the Tg increased from 280 to 360°C. The tensile strength at 400°C of backbones which contained quinoxaline units was twice that of those which did not.

Phenylethynyl pendant groups also are incorporated into PPQ systems [85]. These unsaturated sites provide a possibility of intramolecular cyclization to stiffen the backbone. The cure takes place at 245°C and raises the Tg of the polymer from 215 to 365°C.

A trisbenzil comonomer, 4,4'-bis(4'-oxybenzilyl)benzil was used to provide crosslinking to PPQ [94].

$$\left[C_6H_5{-}\overset{O}{\overset{\|}{C}}{-}\overset{O}{\overset{\|}{C}}{-}C_6H_4{-}O{-}C_6H_4{-}\overset{O}{\overset{\|}{C}}{-} \right]_2$$

Gelation of the initial polymerization was averted by using DMA as solvent and by adding the trisbenzil compound to a prepreg of PPQ oligomer. Postcure at 482-510°C in nitrogen of a Modmor II (graphite fiber) laminate provided a material which retained its mechanical properties even at 371°C.

Other studies have been reported of polyimide (P13H and Skybond 703) PPQ-graphite fiber composites [98]. These investigations encompassed fabrication parameters and mechanical properties and aging effects after 500 hr at 316°C (600°F).

Thermosetting plasticizers for PQ's were also reported [91].

7.14-3 Polyquinazolones

Polyquinazolones and their prepolymers were synthesized and characterized [80] and then cyclization [81] and thermal degradation were studied [82]. Some decarboxylation was discovered which meant that the actual structure deviated from theory. The thermal stability varied from 425 to 510°C, with differences in cyclization method due to structural variations brought on by different cures. The conclusion was that stability sequences cannot be defined because the backbones are not similar. This is only the second reference to a potentially very serious concern in the field of thermally stable polymers. These polymers were evaluated for use in tribological (i.e., friction) applications such as bearings.

A quinazoloneimide block copolymer was synthesized by the condensation of PMDA with an oligomeric (molecular weight 1600-3100), amine-terminated quinazolone [83]. The final product had an inherent viscosity of 2.3-2.8 dl/g and a TGA stability of 480-510°C in air. Isothermal gravimetric analysis showed a 5-7% weight loss at 250°C after 1000 hr.

7.14-4 Miscellaneous

A new class of polymers, polypyridazinones, were discovered from the condensation of a dihydrazine with 4,6-dibenzoylisophthalic acid [89]. One form of this class is the polypyridazinophthalazinediones:

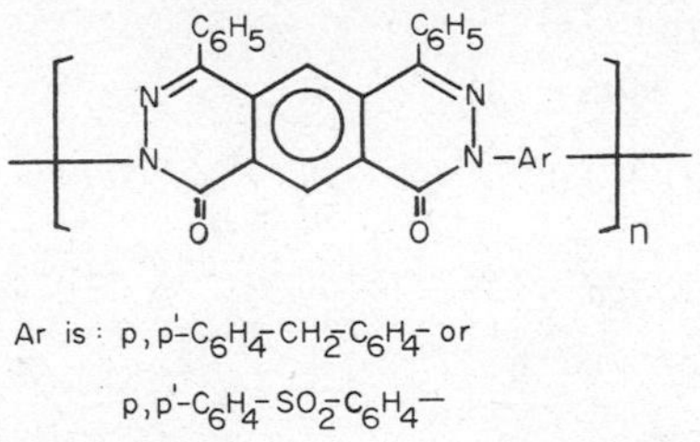

Ar is: p,p'-C_6H_4-CH_2-C_6H_4- or
p,p'-C_6H_4-SO_2-C_6H_4-

Yields were very good (99%), and the cis form shown was soluble in m-cresol and hot dichloroacetic acid. The trans isomer was insoluble, however. Thermal stabilities by TGA were 490-520°C in air or nitrogen.

A similar polymer, polydihydropyridazinone, was also prepared but found to be much less stable (290-350°C) than the above polymer, due no doubt to the lack of aromatic character of the ring [90]:

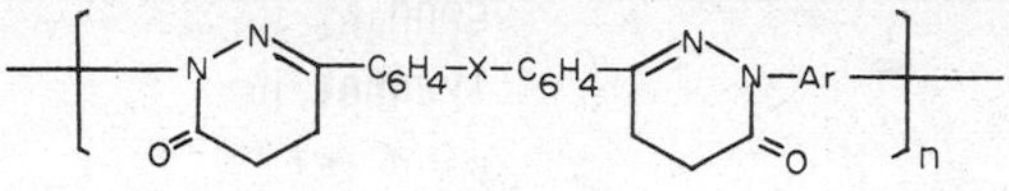

X is : $-O-$, $-CH_2CH_2-$

Ar is : p,p'-$C_6H_4-SO_2-C_6H_4-$
or
p,p'-$C_6H_4-CH_2-C_6H_4-$

Polymers are also possible with naphthalene tetracarboxydiimides and triazine rings as cofunctions in the backbone [99]:

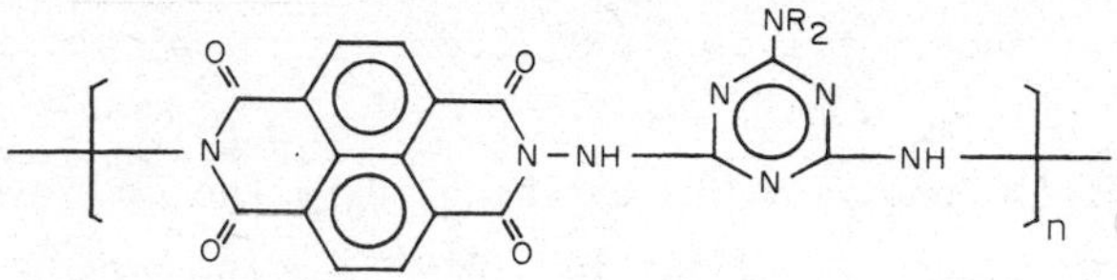

This backbone is made, of course, from the naphthalene tetracarboxylic acid dianhydride and a dihydrazine-substituted triazine, and in good yield (82-96%). Molecular weights range from 13,000 to 16,000 with inherent viscosities of 0.23-0.65 dl/g. Thermal stabilities are 380-410°C. Solubility in THF or dioxane is possible depending on the pendant groups.

References

1. D. R. Anderson and J. M. Holovka, *J. Polym. Sci., A-1, 4,* 1689 (1966).
2. J. M. Augl, *J. Polym. Sci., 10,* 3651 (1972).
3. J. M. Augl, *J. Polym. Sci., A-1, 10,* 2403 (1972).
4. J. M. Augl and H. J. Booth, *J. Polym. Sci.: Polym. Chem. Ed., 11,* 2179 (1973).
5. J. M. Augl and J. V. Duffy, *J. Polym. Sci., A-1, 9*(5), 1343 (1971).
6. J. M. Augl, J. V. Duffy, and S. E. Wentworth, *J. Polym. Sci.: Polym. Chem. Ed., 12*(5), 1023 (1974).
7. L. R. Belohlav and J. R. Costanza, *Ger. Offen.,* 1,806,295 (1969).
8. R. P. Biehle and J. Higgins, *J. Polym. Sci.: Polym. Chem. Ed., 10,* 2919 (1972).
9. L. Bottenbruch, *Angew. Makromol. Chem., 13,* 109 (1970).
10. W. Bracke, *Macromolecules, 2,* 286 (1969).
11. H. Brown, *J. Polym. Sci., 44,* 9 (1960).

12. J. V. Duffy, *J. Polym. Sci.: Polym. Letters, 11*(1), 29 (1973).

13. J. V. Duffy and J. M. Augl, *J. Polym. Sci., A-1, 10*(4), 1123 (1972).

14. G. F. L. Ehlers and W. M. Powers, *Materials Res. Stds., 4*, 298 (1964).

15. G. F. L. Ehlers and J. D. Ray, *J. Polym. Sci., A, 2*, 4989 (1964).

16. G. F. L. Ehlers and J. D. Ray, *J. Polym. Sci., A-1*, 1645 (1966).

17. R. T. Foster and C. S. Marvel, *J. Polym. Sci., A, 3*, 417 (1965).

18. G. P. de Gaudemaris and B. J. Sillion, *J. Polym. Sci., B, 2*, 203 (1964).

19. T. L. Graham, U.S. Air Force Syst. Command, Air Force Mater. Lab Tech. Rep., AFML TR-66-402, 1969.

20. G. L. Hagnauer and G. D. Mulligan, *Macromolecules, 6*(4), 477 (1973).

21. P. M. Hergenrother, *J. Polym. Sci., A-1, 6*, 3170 (1968).

22. P. M. Hergenrother, *J. Polym. Sci., A-1, 7*, 745 (1969).

23. P. M. Hergenrother, Report D1-82-0969, Boeing Scientific Research Laboratories, Seattle, May 1970.

24. P. M. Hergenrother, *J. Macromol. Sci. Rev. Macromol. Chem., C6*(1), 1-28 (1971).

25. P. M. Hergenrother, *SAMPE Quant., 3*, 1 (1971).

26. P. M. Hergenrother, *Polym. Preprints, 13*(2), 930 (1972).

27. P. M. Hergenrother, *Macromol. Sci. Chem., A-7*(3), 573 (1973).

28. P. M. Hergenrother, U.S. Patent 3,778,412 (1973).

29. P. M. Hergenrother, *Org. Coatings and Plastics Preprints, 35*(2), 166 (1975).

30. P. M. Hergenrother, *J. Appl. Polym. Sci., 18*(6), 1779 (1974).

31. P. M. Hergenrother, *J. Polym. Sci.: Polym. Chem. Ed., 12*(12) 2857 (1974).

32. P. M. Hergenrother, *Macromolecules, 7*(5), 575 (1974).

33. P. M. Hergenrother, *Polym. Preprints, 15*(1), 781 (1974).

34. P. M. Hergenrother and D. E. Kiyahara, *Macromolecules, 3*(4), 387 (1970).

35. P. M. Hergenrother and D. E. Kiyahara, *Macromolecules Sci. Chem., A5* (2), 365 (1971).

36. P. M. Hergenrother and H. H. Levine, *J. Appl. Polym. Sci., 14*(4), 1037 (1970).

37. P. M. Hergenrother and H. H. Levine, *J. Polym. Sci., A-1, 5*, 1453 (1967).

38. J. Higgins and Z. Janovic, *J. Polym. Sci., B, 10*(5), 357 (1972).

39. J. Higgins, J. F. Jones, and A. Thornburgh, *J. Polym. Sci., A-1, 9*, 763 (1971).

40. J. Higgins, J. F. Jones, and A. Thornburgh, *Macromolecules, 2*(5), 558 (1969).

41. J. Higgins and C. S. Menon, *J. Polym. Sci., B, 10*(2), 129 (1972).

42. L. C. Hsu and T. T. Serafini, *Org. Coat. Plast. Preprints, 34*(1), 193 (1974).

43. Y. Imai, E. F. Johnson, T. Katto, M. Kurihara, and J. K. Stille, *J. Polym. Sci.: Polym. Chem. Ed.*, *13*, 2233 (1975).

44. Y. Imai and H. Koga, *J. Polym. Sci.: Polym. Chem. Ed.*, *11*, 289 (1973).

45. Y. Imai and H. Okunoyama, *J. Polym. Sci.*, *A-1*, *10*, 2257 (1972).

46. Y. Imai, M. Ueda, and M. Ishimori, *J. Polym. Sci.: Polym. Chem. Ed.*, *13*, 1969 (1975).

47. Y. Imai, M. Ueda, and M. Ishimori, *J. Polym. Sci.: Polym. Chem. Ed.*, *13*, 2391 (1975).

48. Y. Iwakura, K. Uno, and N. Chou, *Polym. J.*, *5*(3), 30 (1973).

49. M. E. B. Jones, D. A. Thornton, and R. F. Webb, *Makromol. Chem. 49*, 62 (1961).

50. R. F. Kovar, G. F. L. Ehlers, and F. E. Arnold, *Polym. Preprints*, *16* (2), 246 (1975).

51. M. Kurihara and Y. Hagivara, *Polym. J.*, *1*(4), 425 (1970).

52. M. Kurihara, H. Saito, K. Nokada, and N. Yoda, *J. Polym. Sci.*, *A-1*, *7*, 2897 (1969).

53. M. Kurihara and N. Yoda, *J. Macromol. Sci. Chem.*, *A-1*, *6*, 1069 (1967).

54. M. Kurihara and N. Yoda, *J. Polym. Sci.*, *A-1*, *5*, 1765 (1967).

55. M. Kurihara and N. Yoda, *Makromol. Chem.*, *105*, 84 (1967).

56. M. Kurihara and N. Yoda, *Makromol. Chem.*, *107*, 112 (1967).

57. R. T. Rafter and E. S. Harrison, *Org. Coatings Plast. Preps.*, *35*(2), 204 (1975).

58. H. K. Reimschuessel, A. M. Lovelace, and E. M. Hagerman, *J. Polym. Sci.*, *40*, 270 (1959).

59. T. T. Serafini, P. Delvigs, and R. D. Vannuccim, *J. Appl. Poly. Sci.*, *17*(10), 3235 (1973).

60. E. M. Shamis and M. M. Dashevskii, *J. Org. Chem.* USSR, *3*, 1005 (1967).

61. J. K. Stille, *Encycl. Polym. Sci. Technol.*, *11*, 389 (1969).

62. J. K. Stille and F. E. Arnold, *J. Polym. Sci.*, *A-1*, *4*, 551 (1966).

63. J. K. Stille and F. W. Harris, *J. Polym. Sci.*, *B-4*, 333 (1966).

64. J. K. Stille and J. R. Williamson, *J. Polym. Sci.*, *A*, *2*, 3867 (1964).

65. J. K. Stille and J. R. Williamson, *J. Polym. Sci.*, *B*, *2*, 209 (1964).

66. J. K. Stille, J. R. Williamson, and F. E. Arnold, *J. Polym. Sci.*, *A*, *3*, 1013 (1965).

67. J. K. Stille, J. E. Wolfe, S. O. Norris, and W. Wrasidlo, *Polym. Preprints*, *17*(1), 41 (1976).

68. S. Toyama, M. Kurihara, K. Ikeda, and N. Yoda, *J. Polym. Sci.*, *A-1*, *5*, 2523 (1967).

69. S. Toyama, N. Dokoshi, K. Ikeda, M. Kurihara, N. Yoda, R. Nakanishi, and M. Watanable, Japan Patent 69 23,108 (1969).

70. S. E. Wentworth and G. D. Mulligan, *Polym. Preprints*, *15*(1), 697 (1974).

71. W. Wrasidlo and J. M. Augl, *J. Poly. Sci., A-1, 7*(12), 3393 (1969).

72. W. J. Wrasidlo and J. M. Augl, *J. Polym. Sci., B, 7,* 281 (1969).

73. W. J. Wrasidlo and P. M. Hergenrother, *Macromolecules, 3,* 548 (1970).

74. N. Yoda, *Encycl. Poly. Sci. Technol., 10,* 682 (1969).

75. N. Yoda and M. Kurihara, Japan Patent 69 20,635 (1969).

76. N. Yoda, M. Kurihara, K. Kieda, S. Tohyama, and R. Nakanishi, *J. Polym. Sci., B, 4,* 551 (1966).

77. N. Yoda, K. Ikeda, M. Kurihara, S. Tohyama, and R. Nakanishi, *J. Polym. Sci., A-1, 5,* 2359 (1967).

78. N. Yoda, R. Nakanishi, M. Kurihara, Y. Bamba, S. Tohyama, and K. Ikeda, *J. Polym. Sci., B, 4,* 11 (1966).

79. W. H. Beever and J. K. Stille, *J. Polym. Sci.: Polym. Symp.,* No. 65, 41 (1978).

80. H. S. O. Chan and R. H. Still, *J. Appl. Polym. Sci., 22*(8), 2173 (1978).

81. H. S. O. Chan and R. H. Still, *J. Appl. Polym. Sci., 22*(8), 2187 (1978).

82. H. S. O. Chan and R. H. Still, *J. Appl. Polym. Sci., 22*(8), 2197 (1978).

83. A. Fukami, *J. Polym. Sci.: Polym. Chem. Ed., 15,* 1535 (1977).

84. J. Garapon and J. K. Stille, *Macromolecules, 10*(3), 627 (1977).

85. F. L. Hedberg and F. E. Arnold, *J. Polym. Sci.: Polym. Chem. Ed., 14* (11), 2607 (1976).

86. P. M. Hergenrother, *Org. Coat. Plast. Preprints, 36*(2), 264 (1976).

87. P. M. Hergenrother, *Polym. Eng. Sci., 16*(5), 303 (1976).

88. P. M. Hergenrother, *J. Appl. Polym. Sci., 21*(8), 2157 (1977).

89. Y. Imai, M. Ueda and T. Aizawa, *J. Polym. Chem.: Polym. Chem. Ed., 14*(11), 2797 (1976).

90. Y. Imai, M. Ueda and T. Aizawa, *J. Polym. Sci.: Polym. Chem. Ed., 16,* 163 (1978).

91. R. F. Kovar, G. F. L. Ehlers, and F. E. Arnold, *Org. Coat. Plast. Preprints, 36*(2), 393 (1976).

92. S. O. Norris and J. K. Stille, *Macromolecules, 9*(3), 496 (1976).

93. G. Rabilloud and B. Sillion, *J. Polym. Sci.: Polym. Chem. Ed., 16,* 2093 (1978).

94. R. T. Rafter and E. S. Harrison, *Polym. Eng. Sci., 16*(5), 318 (1976).

95. A. Recca, J. Garapon, and J. K. Stille, *Macromolecules, 10*(6), 1344 (1977).

96. H. M. Relles, C. M. Orlando, D. R. Heath, R. W. Schluez, J. S. Manello, and S. Hoff, *J. Polym. Sci.: Polym. Chem. Ed., 15,* 2441 (1977).

97. A. K. St. Clair and N. J. Johnston, *J. Polym. Sci.: Polym. Chem. Ed., 15*(12), 2009 (1977).

98. N-H. Sung and F. J. McGarry, *Polym. Eng. Sci., 16*(6), 426 (1976).

99. R. Takatsuka, T. Unishi, I. Honoa, and T. Kakurai, *J. Polym. Sci.: Polym. Chem. Ed.*, *15*, 1785 (1977).

100. S. E. Wentworth, *Org. Coat. Plast. Preprints*, *36*(2), 388 (1976).

101. W. Wrasidlo and J. K. Stille, *Macromolecules*, *9*(3), 505 (1976).

102. W. Wrasidlo, J. E. Wolfe, T. Katto, and J. K. Stille, *Macromolecules*, *9*(3), 512 (1976).

103. J. E. Wolfe and J. K. Stille, *Macromolecules*, *9*(3), 489 (1976).

Chapter 8

Heterocyclic Polymers: Fused Rings

The five polymers of fused ring heterocycles and a brief description of their properties are given in Table 8.1.

8.1 Two Fused Five-Member Rings

8.1-1 Polythiazones

The topic included here as well as in the following fused ring sections could well be placed in the chapter on ladder polymers as pseudoladders. For example, the imidazopyrrolones are discussed as ladder-like systems.

Polythiazones [2] are fused analogs of polyarylenesulfimides (polysaccharins), which were discussed in Section 6.9. This fused system could also be termed poly(benzimidazosulfimides) or poly(sulfimidobenzimidazoles).

The synthetic difference between these polythiazones and the polysaccharins is that for the former, a tetraamine in place of a diamine is condensed with a bis(carboxyl sulfonic) acid derivative. The process is conducted in PPA or as a melt at 170-200°C or at 25°C in DMAc to give a polyamide precursor. Of these the melt process gives the highest viscosity products (1.3 dl/g in DMAc). A thermal postcure brings about cyclodehydration commensurate with insolubility in DMAc and DMSO.

Limited trials were also attempted with 1,2,4,5-tetraaminobenzene and 3,3',4,4'-tetraaminodiphenylether.

Cyclodehydration results in a decrease in viscosity, and the final brown product is soluble only in sulfuric acid. Postcuring solution polymers causes a loss of sulfur by aromatic desulfonation, an expected reaction. However, the melt polymers would cyclize without loss of sulfur.

TGA shows decomposition beginning in the range 380-440°C, with a 10% weight loss occurring at 470-525°C. DTA studies combined with analysis of the evolved gas show that loss of SO_2 occurs at 400-420°C.

Table 8.1 Fused Ring Heterocyclic Polymers

Name	Structure	Comments
Polythiazones	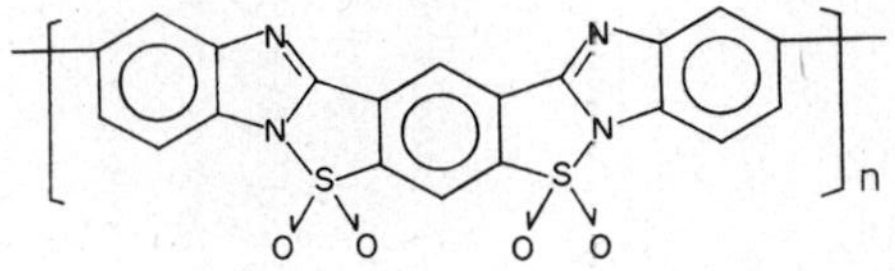	Soluble in H_2SO_4. Brown. TGA stability: 470-525°C. Lose SO_2 at 400-420°C.
Poly(benzopyrroletriazole)	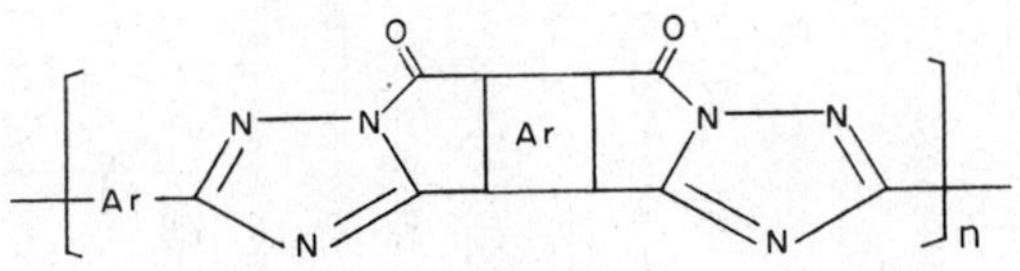	Stable to 410-525°C.
Poly(thioquinazolopyrrolone)	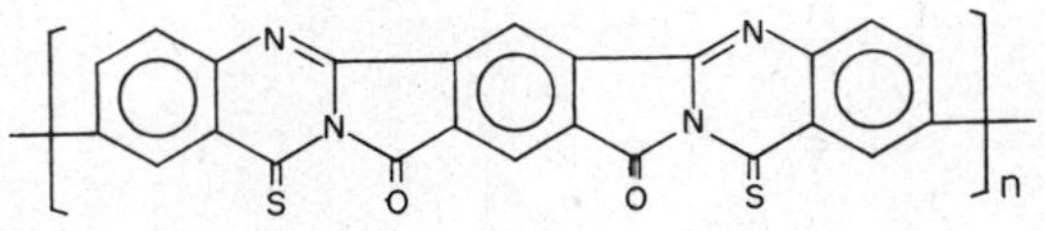	Red, transparent films. Soluble intermediate.

Polybenzimidazoquinazolines	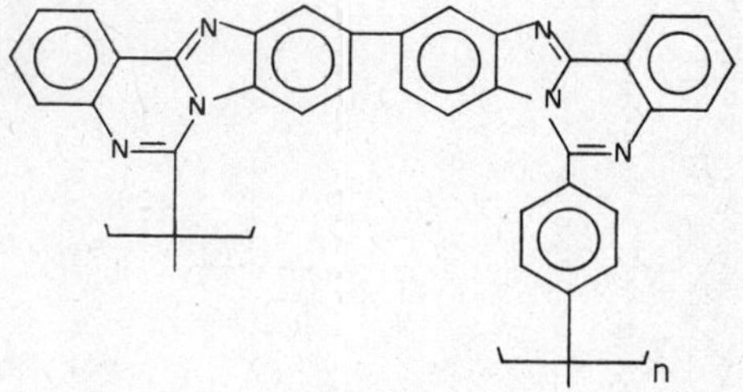 Yellow or brown. High yield. Soluble in H_2SO_4. TGA stability: 500-560°C.
Polyquinazolotrizoles (PQT)	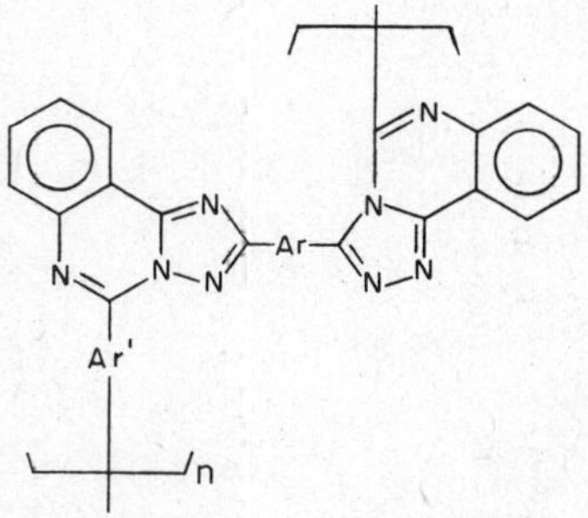 Brown. Soluble in H_2SO_4 and CF_3CO_2H. Softening temperature: 400-450°C. TGA stability: 560°C.

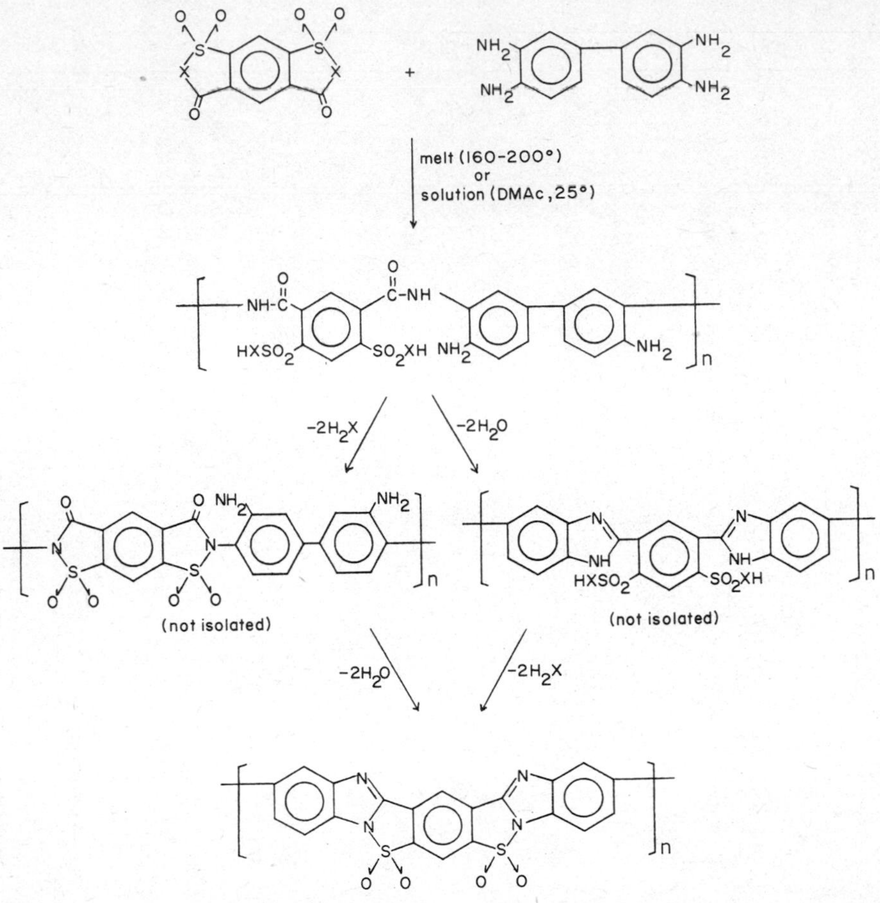

melt (160-200°)
or
solution (DMAc, 25°)

$-2H_2X$ $-2H_2O$

(not isolated) (not isolated)

$-2H_2O$ $-2H_2X$

X is: O or NH

Method [2] A mixture of 2.17 g (0.0075 mol) of 2H,6H-benzo[2,1-d, 4,5-d']-1,2,6,7-diisothiazole-1,1,7,7-tetraoxide-3,5-dione (i.e., the cyclic sulfone monomer where X is -O-) [1], 1.62 g (0.0075 mol) of 3,3'-diaminobenzidine, 1.60 g (0.016 mol) of triethylamine, and 1.31 g (0.073 mol) of water is heated according to the following schedule: 17 hr at 100°C, 3.5 hr at 120°C, 1 hr at 130°C, 2.5 hr at 140°C, 3 hr at 150°C, and 11 hr at 160°C. It is cooled and the red-brown solid is crushed and then heated further at 170°C for 2 hr. The polymer is soluble in DMAc, DMSO, and H_2SO_4 (intrinsic viscosity 0.029 dl/g at 1.17% in H_2SO_4). Then the polymer was heated further at 180°C for 16 hr. Postcuring was effected at 420°C, whereupon an 18% weight loss occurred.

8.1-2 Poly(benzopyrroletriazole)

The second type of two fused, five-member ring system has been called a poly(aroylene-s-triazole) and is represented by fused triazole and benzopyrrole rings. These are synthesized in one- or multiple-step processes by heating bisamidrazones with tetraacid dianhydrides in polyphosphoric acid [6]. These step-ladder polymers show a 10% weight loss by TGA at 410-525°C.

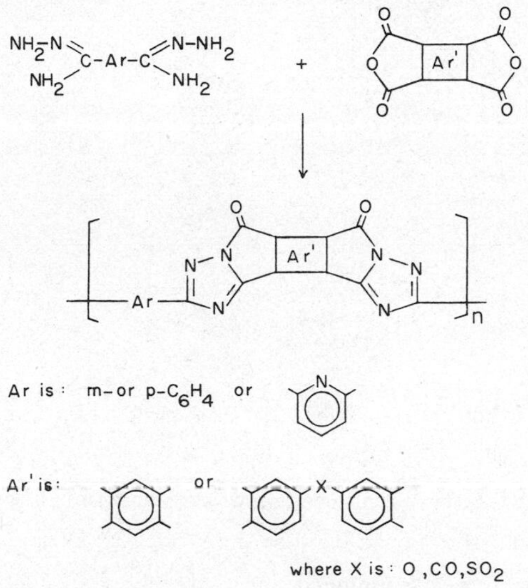

Ar is: m- or p-C_6H_4 or

Ar' is: or

where X is: O, CO, SO_2

8.2 Fused Five- and Six-Member Rings

8.2-1 Poly(thioquinazolopyrrolone) or Poly(isoindolothioquinazolinone)

A polymeric backbone which contains fused five- and six-member rings and which also could be considered a pseudoladder is that which contains the thioquinazolopyrrolone. This system has been patented [9] and is derived from the solution condensation at room temperature of a tetracarboxylic acid dianhydride with a bis(aminothioamide), followed by thermal cyclodehydration at 300-320°C.

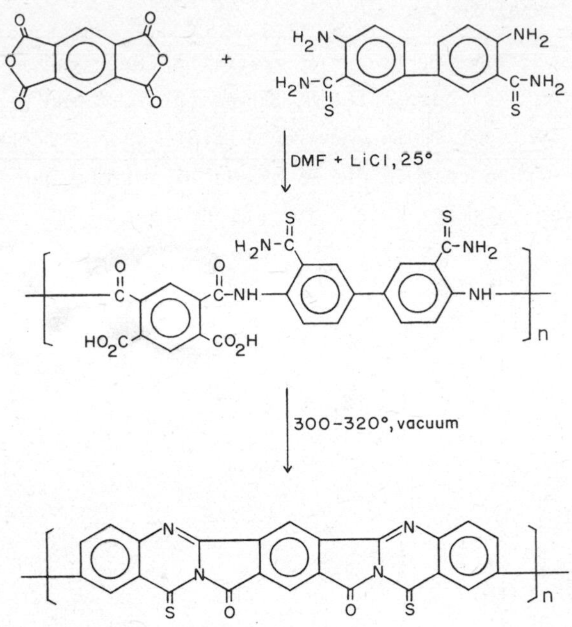

The intermediate polyamide is, of course, soluble in NMP and has an intrinsic viscosity of 1.20 dl/g (0.5% in NMP). It forms a red-brown film.

A postcured film of the final polymer is red and transparent and has a tensile strength of 18 kg/mm^2 and electrical resistance of 100 kV/mm^3. It is said to be useful as films, paints, or adhesives.

Method [9] *Polymerization* A solution is made containing 15.1 g (0.05 mol) of 4,4'-diamino-3,3'-biphenylbis(thiocarboxamide) and 80 g of N-methylpyrrolidone (NMP) and 4 g of LiCl. This solution is added dropwise at room temperature to one containing 10.9 g (0.05 mol) of pyromellitic dianhydride in 80 g of NMP whereupon an exothermic reaction ensues and the viscosity increases. The mixture is stirred 1.5 hr at room temperature and poured into an excess volume of water to precipitate the polyamide as a fibrous solid. The precipitate is collected, washed with methanol and then acetone, and dried for 1 day at 50°C; intrinsic viscosity 1.20 dl/g (0.5% NMP).
To cast a film the yellow viscous reaction solution is poured onto a glass plate, stored at 40°C then heated over 1 hr to 100°C, and held at 100°C for 2 hr to give a clear red-brown film.

Postcuring Cyclodehydration is accomplished by heating the solution-cast film on a frame for 30 min at 100°C under vacuum and then by heating it slowly to 300-320°C and held at that temperature for 1 hr. The resulting film is transparent red.

8.2-2 Polybenzimidazoquinazolines

A fused bicyclic heterocycle with three nitrogens in the system is the benzimidazoquinazoline. The backbone with this moiety is derived from the various routes as follows [8].

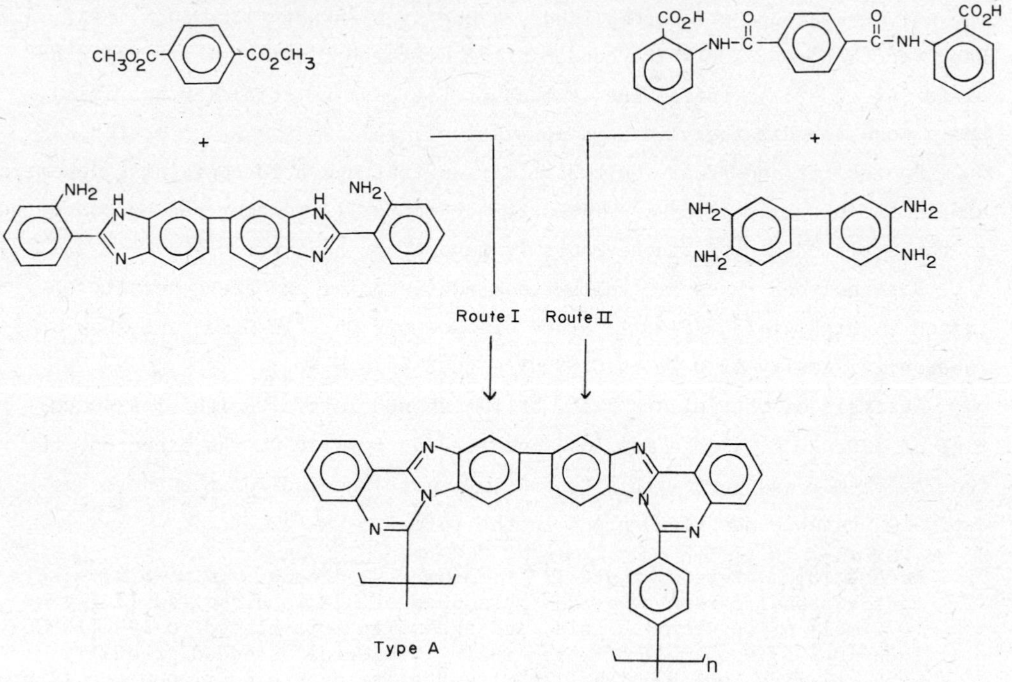

$CH_3O_2C-C_6H_4-CO_2CH_3$

+

Route I Route II

Type A

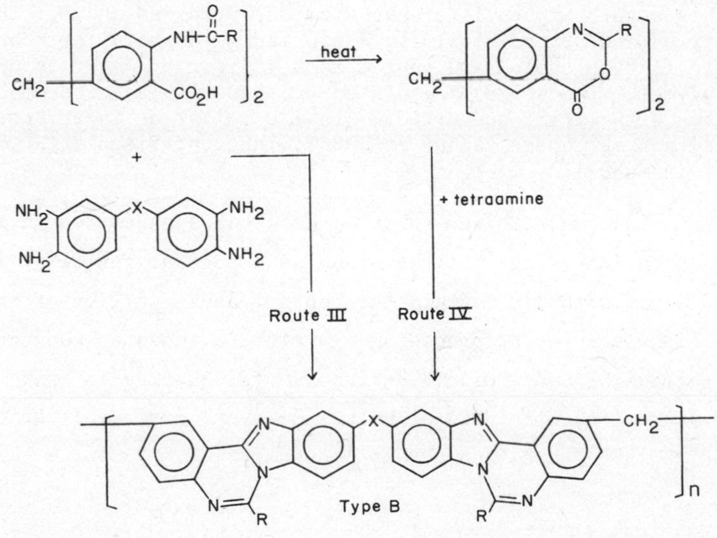

heat

+ tetraamine

Route III Route IV

Type B

X is : nil or CH_2

R is : CH_3 or C_6H_5

The polymerizations are accomplished in one step in hot (230-250°C) PPA. Each reaction represents the condensation of three functional groups in a different order: a tetraamine, a mono or bis(o-amino acid) (anthranilic), and a mono- or dicarboxylic acid derivative.

Routes III and IV are quite similar in that the acid-containing monomers are in the uncyclized forms. In fact, the route III monomer can be converted to the cyclized benzoxazinone route IV monomer by heating.

Both polymer types are yellow to brown in color and are generally obtained in high yield (78-98%). They are soluble only in H_2SO_4 and show an inherent viscosity of 0.10 to 0.58 dl/g (0.25% in H_2SO_4).

Analysis of thermal stability by TGA showed initial weight loss occurring at 380-420°C, with a 10% loss incurred at 500-560°C. As expected, the type A polymer was better than type B, owing to the lack of the thermooxidatively unstable methylene group in the former.

Method [8] *Type A, route II* *Monomer synthesis* *N,N'-bis(o-carboxyphenyl)terephthalamide* In 50 g of PPA is dissolved 13.7 g (0.1 mol) of anthranilic acid, and the mixture is heated to 100-110°C. Then 10.1 g (0.05 mol) of terephthaloyl chloride is added gradually with stirring; and heating (110°C) and stirring are continued for 12 hr. The mixture is cooled and poured into 1 liter of water to yield a white precipitate which is isolated and dissolved in 20% aq. sodium hydroxide. The solution is filtered and neutralized with HCl to produce white crystals, which are isolated and dried. The amide is then recrystallized from DMF, mp 350°C (dec.).

Polymerization Into 50 g of PPA, under nitrogen, are mixed 4.28 g (0.02 mol) of 3,3'-diaminobenzidine and 8.08 g (0.02 mol) of the N,N'-bis(o-carboxyphenyl)terephthalamide, prepared above. The mixture is stirred and heated slowly to 250°C and held there for 4 hr. The mixture is then cooled and poured into 1 liter of water to precipitate the polymer. The solid is isolated, washed with aq. sodium bicarbonate and then with water, and dried. Final cyclization is effected by heating the polymer under vacuum at 300°C; inherent viscosity 0.58 dl/g (0.25% in H_2SO_4).

The benzimidazoquinazoline function has been synthesized in a polymerization which leaves part of the group as a pendant function; i.e., the backbone does not actually contain the benzimidazole portion of the heterocycle. As discussed in the preceding examples, a two-step condensation occurs between a tetraamine and a diacid derivative (dichloride in this case) at high temperature either in HMPA or bulk. However, here cyclization temperatures of 280-370°C are reported as necessary [7].

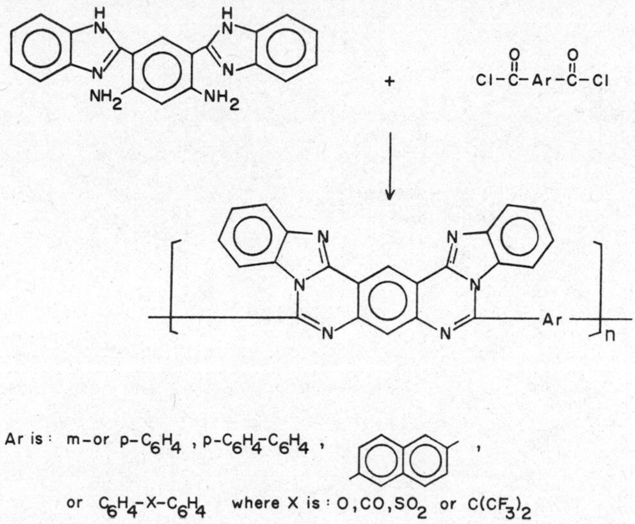

Ar is: m-or p-C_6H_4, p-C_6H_4-C_6H_4, ,

or C_6H_4-X-C_6H_4 where X is: O, CO, SO_2 or $C(CF_3)_2$

The polymers are soluble in sulfuric acid and partly soluble in trifluoroacetic acid. Reduced viscosities are as high as 0.85; but most are in the range 0.2-0.4. Thermal stabilities by TGA (10% weight loss) are from 480 to 620°C, depending on the aryl link, the poorest being sulfone and hexafluoroisopropylene.

A triazole nucleus can be envisioned in place of the benzimidazole function in the above polymer. These types of polymers are made in the same fashion, but are somewhat less stable.

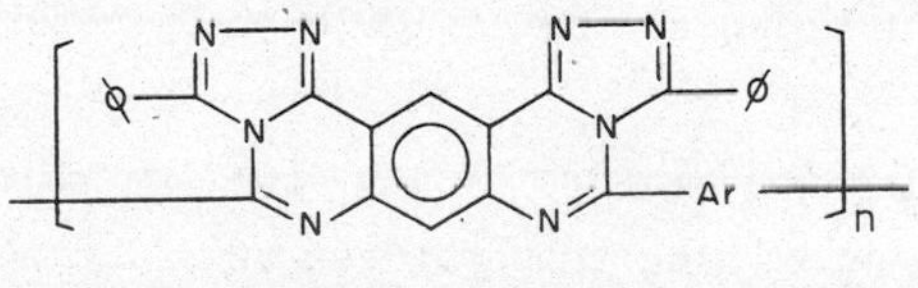

8.2-3 Polyquinazolotriazoles or Poly(triazoloquinazolines)

In 1971 the synthesis of polyquinazolotriazoles (PQT) was reported by both Hergenrother [3] and Korshak et al. [4]. The latter expanded this series in a later publication [5] and referred to the material as a poly(triazoloquinazoline). Both processes comprised the condensation of arylene-bis(o-aminotriazoles) with diacid derivatives, although Hergenrother reported a thermal stability by TGA of some 15-100°C above those of Korshak. Hergenrother also found that polymerizations in PPA were superior to melt or solution methods.

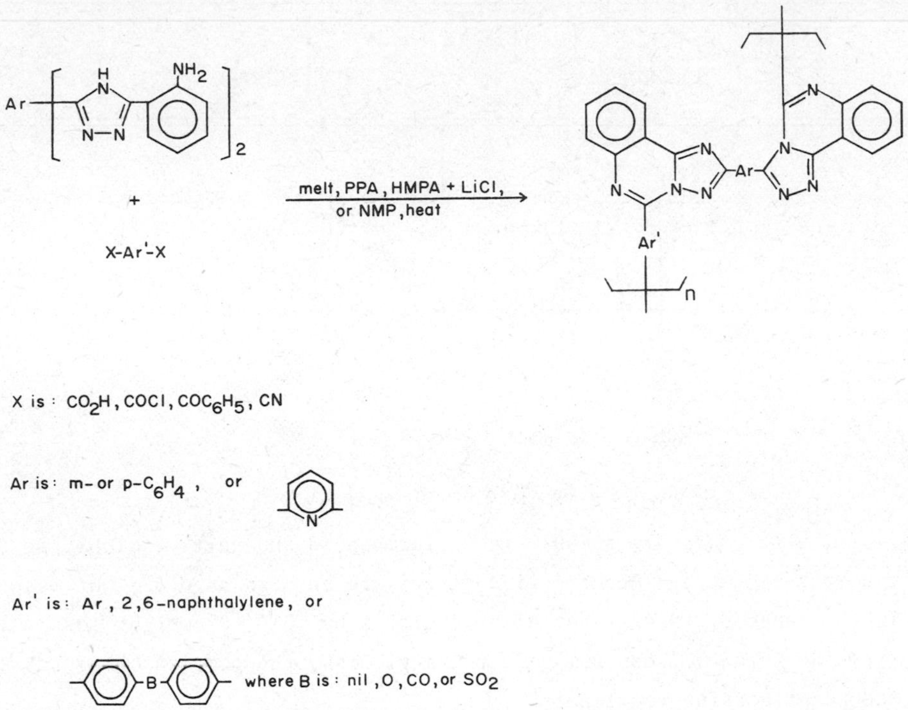

$X{-}Ar'{-}X$

melt, PPA, HMPA + LiCl, or NMP, heat

X is: CO_2H, COCl, COC_6H_5, CN

Ar is: m- or p-C_6H_4, or

Ar' is: Ar, 2,6-naphthalylene, or

where B is: nil, O, CO, or SO_2

The final polymers are brown and soluble in trifluoroacetic or sulfuric acids with inherent viscosities as high as 1.4 dl/g (0.5% in H_2SO_4) [3]. The intermediates, which could be isolated from solution polymerization of the diacyl halide, are white to light brown and are soluble also in formic acid and HMPA [5].

Thermal analyses show a softening temperature of 400-450°C (by thermomechanical analysis) and initial decomposition temperatures of 490°C in either air or nitrogen. A 10% weight loss by TGA is reached at 560°C in air and 530°C in nitrogen [3]. This inverse relationship of stability to atmosphere composition is said to be due to some initial oxidative crosslinking. At 600°C the nitrogen and air TGA curves cross to their usual positions; and by 680°C the weight loss in air reaches 100%, while in nitrogen it is only 22%. Isothermal aging for 200 hr revealed losses of 1.4% at 260°C and 4% at 316°C [3].

Method [3] To 80 g of PPA, which is held at 80°C under argon, is added 1.977 g (0.05 mol) of 3,3'-(2",6"-pyridinediyl)bis[5-(o-aminophenyl)-1,2,4-triazole] (i.e., the triazole monomer where Ar is 2,6-pyridinediyl), followed by 0.831 g (0.005 mol) of isophthalic acid. The mixture then is heated slowly over 3 hr to 250°C and held there for 3 hr, during which time the color changes from tan to brown and the viscosity increases. The brown polymer is precipitated by pouring the hot reaction mixture into water. The solid is isolated, washed well with aq. sodium carbonate and then water, and dried at 250°C for 4 hr in vacuo. The yield is 2.35 g (95%) of light brown polyquinazolotriazole; inherent viscosity 1.41 dl/g (0.5% in H_2SO_4).

8.3 Recent Developments

A report has appeared of a commercial polybenzimidazolone (Teijin, Japan) which is being developed as a reverse-osmosis membrane for large-scale use [10]. The polymer is said to reject 99% of the salt in the feed stream and to be useful at pH values from 1-12 and up to 60°C.

Further work on the known polybenzimidazoquinazolines (PIQ's) resulted in a series of resins for composite materials [11]. These polymers were synthesized by the thermal (230-250°C) condensation of aromatic phenolic di- and triesters with 5,5'-bis[2-(o-aminophenyl)benzimidazole]. The yellow products had inherent viscosities up to 1.6 dl/g in H_2SO_4. The casting of prepolymer solutions (in DMAc or m-cresol) and curing (at 450°C) resulted in films. Glass transition temperatures were in the range 350-495°C, depending on structure and thermal history. Isothermal weight retention was 85-93% after 100 hr in air at 371°C. At least one laminate with Modmor II showed excellent retention of flexural properties after 200 hr at 371°C.

Several backbones with a purine ring system were produced which were stable by TGA to 300-340°C [12]:

$(RO)_3C-Ar-C(OR)_3$ + $NH_2-C(=NH)-C(NH_2)=C(NH_2)-NH_2$ → polymer (–Ar–purine–)$_n$

A new polymer, copoly(pyromellitimidebenzimidazopiperidone), was found to result from the three-step condensation of 3,3'-diaminobenzidine with 2,2'-pyromellitdiimidodiglutaric anhydride [13]:

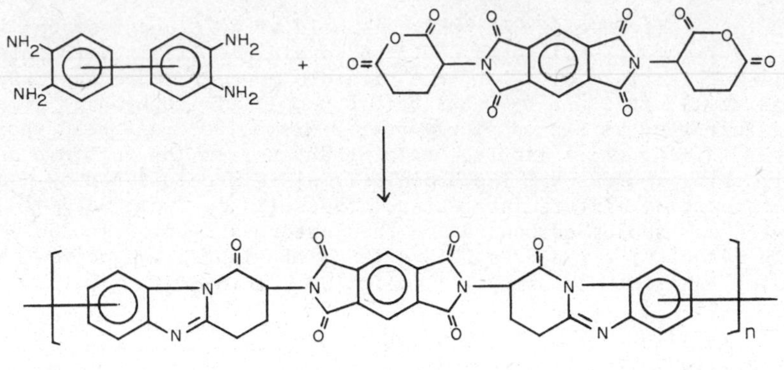

The product was soluble in formic acid, sulfuric acid, and 10% aq. sodium hydroxide. A 10% weight loss by TGA occurred at 420-455°C.

References

1. G. F. D'Alelio, D. M. Feigl, W. A. Fessler, Y. Giza, and A. Chang, *J. Macromol. Sci. Chem.*, *A3*(5), 927 (1969).
2. G. F. D'Alelio, D. M. Feigl, T. Ostdick, M. Saha, and A. Chang, *J. Macromol. Sci. Chem.*, *A6*(1), 1 (1972).
3. P. M. Hergenrother, *J. Polym. Sci.*, *A-1*, *9*(8), 2377 (1971).
4. V. V. Korshak, A. L. Rusanov, E. L. Baranov, Ts. G. Iremashvili, and T. B. Bezhvashrili, *Kokl. Akad. Nauk. SSSR*, *196*, 1357 (1971).
5. V. V. Korshak, A. L. Rusanov, Ts. G. Iremashvili, E. L. Baranov, and I. V. Zhuravleva, *Macromolecules*, *6*(4), 483 (1973).
6. V. V. Korshak, A. L. Rusanov, S. M. Leonteva, and T. K. Jashiashvili, *Macromolecules*, *8*(5) 582 (1975).
7. V. V. Korshak, A. L. Rusanov, L. Kh. Plieva, M. K. Kereselioze, and T. V. Lekae, *Macromolecules*, *9*(4), 626 (1976).
8. M. Saga, Y. Hachihama, and T. Shono, *J. Polym. Sci.*, *A-1*, *8*(8), 2265 (1970).
9. M. Yoda, M. Kurihara, and Y. Hagihara, Japan Patent 69 28,714 (1969).
10. *Chemical and Engineering News*, p. 18, Oct. 3, 1977.
11. R. J. Milligan, C. B. Delano, and T. J. Aponyi, *J. Macromol. Sci. Chem.*, *A10*(8), 1467 (1976).
12. C. D. Rudgeon and O. Vogl, *J. Macromol. Sci. Chem.*, *A11*(11), 1989 (1977).
13. C. Yang, *J. Polym. Sci.: Polym. Chem. Ed.*, *15*, 1239 (1977).

Chapter 9

Ladder Polymers

It was not long after the realization of the field of high-temperature polymers that ladder structures were proposed and their syntheses attempted. In the early and mid-1960s, some success was realized in attaining ladders of double-strand backbones with ester, siloxane, ketal, and quinone units in the backbone [6], with more complex heterocyclic materials becoming known shortly thereafter [8,48].

The reason for the interest in a ladder is obvious; in order for such a structure to sustain a loss in molecular weight, and therefore physical properties, two backbone strands had to be cleaved. Furthermore, this double breakage of the chain had to occur within the same ring before backbone character was destroyed. This is similar to the reasoning used for crosslinked systems; however, unlike crosslinked polymers, ladders should provide essentially a linear and, therefore, soluble product. Finally, it was reasoned, the rigid, ribbon-like backbone will provide a high melting point for property retention to high temperature.

Unfortunately, few of the high hopes have been realized to date for ladder polymers. They are thermally stable, but only in the region of ordinary heterocyclic systems or perhaps slightly higher. It seems then, reasonable enough that once a decomposition temperature is attained both strands of the backbone are readily cleaved. After all at this temperature only a short time is required for sufficient energy input to cause complete decomposition.

The similarities between thermal stabilities of analogous ladder and simple linear structures were demonstrated using a pyrrole-containing backbone in which structure was carefully controlled [32]. It was concluded that (1) protecting oxidizable groups is a primary requirement; (2) increasing chain stiffness contributes to stability; and (3) ring strain decreases stability. The import of these results then is that ladder polymers introduce ring strain to a backbone and thereby decrease thermooxidative stability.

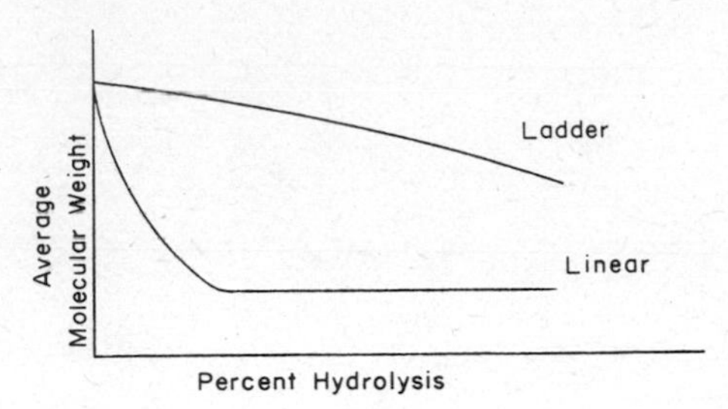

Figure 9.1 Hydrolytic stability comparison of linear to ladder backbones.

Improved hydrolytic stability of ladder polyesters over linear types has been shown to be real. If one plots a graph of molecular weight vs extent of hydrolysis for these systems, the expected classical stability is nicely demonstrated (Figure 9.1). However, this process is not analogous to thermal decomposition since the two proceed by quite different mechanisms. The tendency is to equate the abscissa on the above graph to temperature, i.e., read it as a TGA curve, which is fallacious.

Solubility also proved to be a problem, being slightly less than for simple heterocyclic backbones. Increasing the solubility of ladder systems is more difficult than for linear polymers since such tricks as phenylation do not lessen backbone rigidity or crystallinity.

The synthetic problem of attaining wholly ladder backbones has proven to be an advantage. It is difficult to attain a complete ladder, in that some ring closures do not succeed. One is then left with a partial ladder, referred to as a "pseudoladder" or "step-ladder" or "ladderoid," which is somewhat more tractable than the pure ladder. These structures with occasional single bonds have been purposely designed as a processing aid since the thermal stability of the perfect ladder is not significantly above the pseudoladder.

In this chapter both ladder and ladder-like structures will be discussed. Other step-ladder systems have been referred to in their proper sections in the discussions of heterocyclic backbones. There are six general types of functional groups used for ladder backbones, with variations on each and combinations of them. General examples of these structures are given in Table 9.1. Other backbones which could also be included here are spiro or organometallic bridged types; these, however, are covered in Chapter 5 and Chapter 10, respectively.

Table 9.1 Ladder Polymers

Name	General structure	Remarks
Polypyrrones [also polypyrrolones, poly(biobenzimidazopyrrolones), poly(benzoylenebenzimidazoles, or poly(benzimidazolimides)]		High strength and modulus. Stable to 400°C in air. Tractable intermediates. Films, adhesives, laminates. Moldable.
Polyquinoxalines (PQ or PPQ]		Soluble. Stable to 450-585°C in air (700°C when postcured). No thermal transitions below decomposition. Soluble or partly soluble.
Polyquinolines		Soluble. Stability to 300-440°C in air.
Poly(benzimidazobenzophenanthroline) ladders (BBL) [or poly(imidazo-isoquinoline) ladders]		Slightly soluble. Tg 500°C. Stable to 450-550°C or 200 hr at 370°C in air. Forms supramolecular aggregate films.
Poly(benzobistriazolophenanthrolines)		Soluble in H_2SO_4. Red-black powders. Tough films possible. Stable to 510°C in air.
Graphitic-type ladders		Insoluble (soluble intermediate). Black. Film or fiber possible. Stability: 10% loss at 300°C in 24 hr or at 470°C by TGA. High conductivity (compared to other organic polymers).

9.1 Polypyrrones

Polypyrrones are also known as polypyrrolones, polybisbenzimidazopyrrolones, polybenzoylenebinzimidazoles, or even polybenzimidazolimides and contain the following structural unit, a sort of carbonyl-bridged benzimidazole:

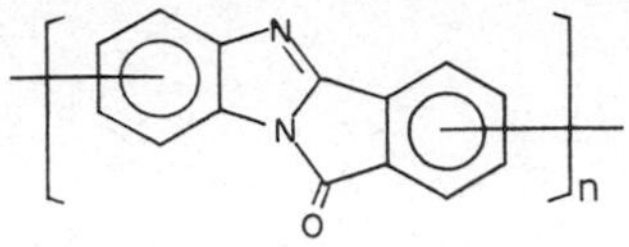

Merely on the basis of functional groups, this polymer should be more stable than benzimidazole systems since it has no N-H or aliphatic C-H bonds.

Since their first synthesis in 1965, the pyrrone polymers have received attention from the aerospace scientists, this effort being recently well reviewed by Nartissov [45], who includes the phenanthroline system. The latter subject is covered under that title in Section 9.4. In reviewing publication dates of pyrrone patents and articles, one observes almost all of them to fall from 1965 to 1969 with very few from 1970 to 1973.

Basically all pyrrones are produced by the thermal condensation of an aromatic tetraacid with an aromatic tetraamine via two intermediates, although a one-step process is possible. Note that the first intermediate, a poly(amide acid amine), is similar to amic acid intermediates available in the polyimide and polybenzimidazole syntheses. Consequently, the second intermediates shown are both the imide and benzimidazole types, each of which has been shown to exist [9] and can undergo further condensation. The polyimide-type intermediate appears to be the predominate one. Of course the first amide intermediate is the more soluble and processable one. It is also obvious that 4 mol of water are lost (one at each step) for each pyrrone moiety formed so that off-gas may be a more serious problem with these types of materials than some of the other heterocycles if the pyrrone is formed in one step. However, the staging process, i.e., the stepwise reaction, serves to remove water incrementally prior to final cure.

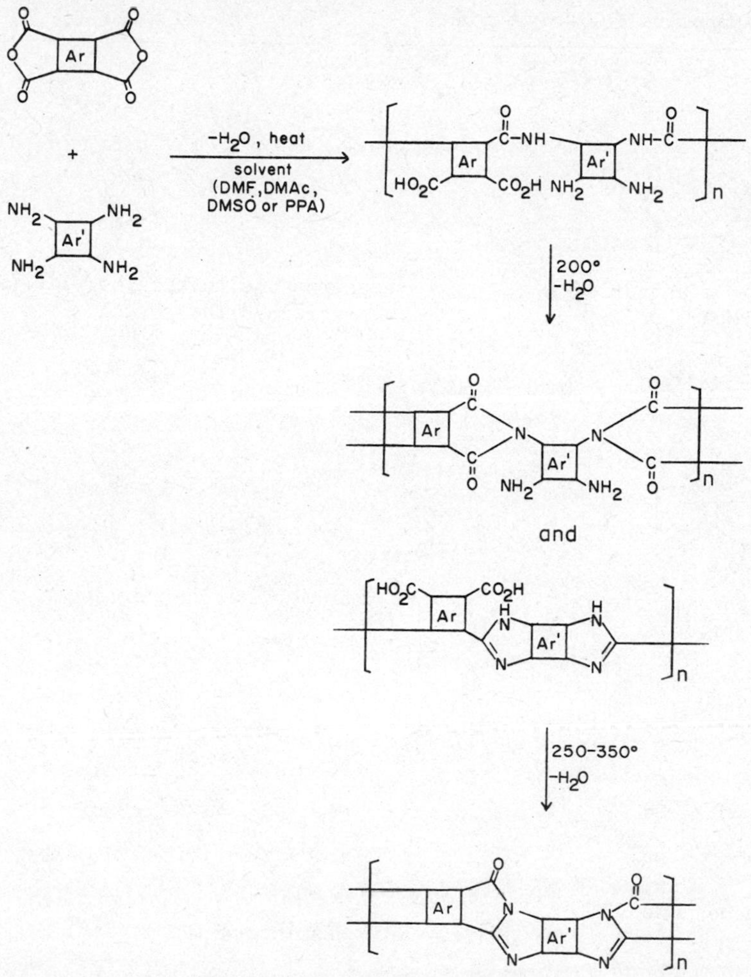

-H_2O, heat
solvent
(DMF, DMAc, DMSO or PPA)

200°
-H_2O

and

250-350°
-H_2O

It is also interesting that a crosslinked benzimidazole polymer is possible by a more rapid heating of tetraamine with the acid [21].

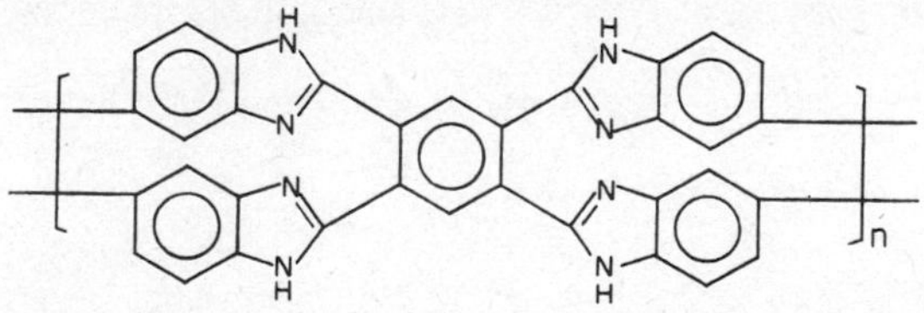

Only a few tetraacids and tetraamines (ortho substitution required) are available for such materials. Further, the only monomers (excluding naphthalene types) which give true ladders are pyromellitic dianhydride

Table 9.2 Monomers for Polypyrrones

Acids or Anhydrides

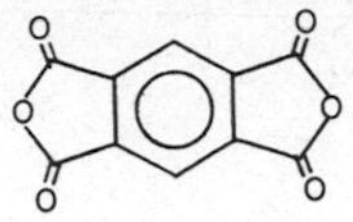

pyromellitic dianhydride (PMDA)

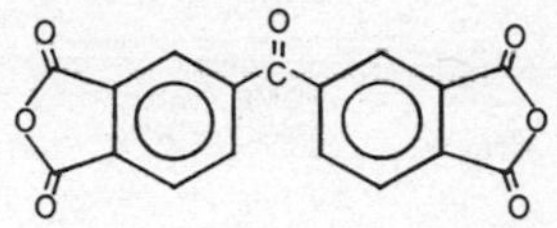

benzophenone tetracarboxylic dianhydride (BTDA)

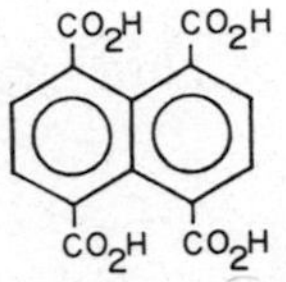

CO_2H CO_2H

CO_2H CO_2H

1,4,5,8-naphthalenetetracarboxylic acid

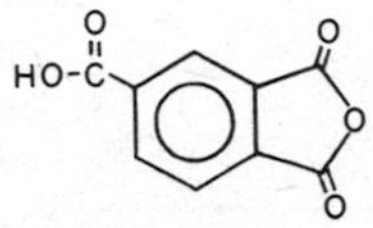

HO-C(=O)-

trimellitic anhydride

Amines

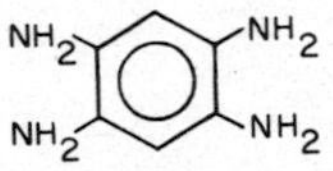

NH_2 NH_2 NH_2 NH_2

tetraaminobenzene (TAB)

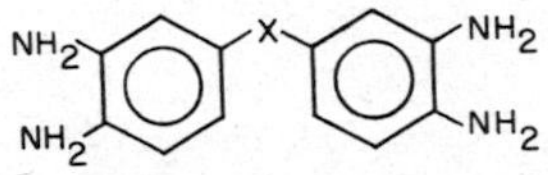

X

nil = 3,3'-diaminobenzidine (DAB)

-O- = bis(3,4-diaminophenyl)ether

-C(=O)- = 3,4,3',4'-tetraaminobenzophenone

$-CH_2-$ = bis(3,4-diaminophenyl)methane

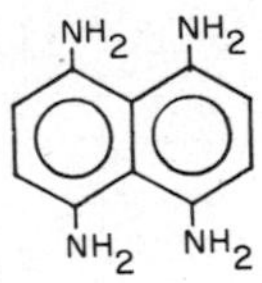

1,4,5,8-tetraaminonaphthalene

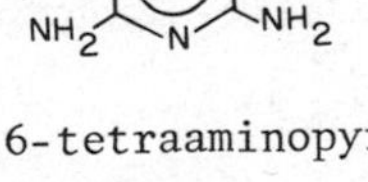

2,3,5,6-tetraaminopyridine

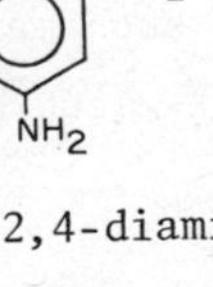

$NHCOCH_3$

acetamido-2,4-diaminobenzene

and tetraaminobenzene. Others give pseudoladder products. A list of those which have been used is given in Table 9.2. The last amine in the table, being only a triamine, gives an imidepyrrolone copolymer. To date no evidence is available for the use of 3,4,3',4'-biphenylethertetracarboxylic dianhydride as a tetraacid in pyrrone synthesis, although a few dianhydride capped, phenylether oligomers have appeared in this synthesis [45].

The black to red-black to red pyrrone polymers are good, flexible film formers and can be used in applications such as films, adhesives, insulating coatings (wire, etc.), and laminating resins for composites. The pyrrone polymers which have been patented for such use are actually pseudoladder systems with imide, amide, and benzimidazole moieties also incorporated in the backbone [22,42,54]. Earlier publications addressed the modification of ladder systems with imide [10,20,23] or benzimidazole [20] cofunctions in regular or random patterns. Table 9.3 lists the range of properties which have been reported for various pyrrones.

The various synthetic routes to pyrrones are quite similar. Regarding forms of monomers, one can use the anhydride, acyl chloride, or methyl or butyl esters of the acid and ordinary amines or amine hydrochloride salts, the latter for less stable amines.

Two general methods of reaction are possible: (1) solution prepolymerization at room temperature followed by initial cyclization at 130-160°C and final cyclization at 350-375°C under vacuum or (2) gradual heating of a monomer mixture or 1:1 salt up to 350°C. These processes are done preferably under an inert atmosphere to prevent oxidation of the very sensitive tetraamines. The latter method, being free from solvent, can be utilized in a

Table 9.3 Pyrrone Polymer Properties

Tensile strength	103-145 MPa (15,000-21,000 psi)
Elongation-to-break	2-7%
Elastic modulus	1.7-6.9 GPa (250,000-1,000,000 psi)
Inherent viscosities	Up to 2 dl/g
Solubility	H_2SO_4, DMSO, and NMP
Thermal data:	
DTA	No transitions below degradation temperatures
TGA	400°C in air, 450-500°C in N_2
Isothermal	3% weight loss at 316°C for 500 hr

direct-molding procedure and is a one-step reaction since processable intermediates are not required.

In the solution process, a few precautions are necessary to prevent gelation and to allow the resulting solution of intermediate to have a viscosity low enough for processing. First, the final solution must be dilute, i.e., below 10% solids. Second, the anhydride or other acid derivative is added very slowly to the amine solution with extensive stirring. Finally, for the highest molecular weight product, the acid derivative should be in excess by 1-5%.

Numerous solution [9,11,12,19,21,26,27,37] and nonsolvent [11,21,43] methods have been detailed in the literature. Given below are one of each which are representative and which use common monomers.

> *Solution method* *Prepolymerization* [37] The initial step in solution polymerization is extremely sensitive to procedure. Under dry, air-free conditions, 0.85 g of 3,3'-diaminobenzidine is placed in a reaction vessel and dissolved in 30 ml of DMAc. To this is added slowly at room temperature and with vigorous stirring a solution of 0.88 g of pyromellitic dianhydride in 30 ml of DMAc. The amounts used are such that the final solution is less than 10 wt % solids and that the anhydride is in 1% excess. The rate of addition decreases considerably as viscosity of the reaction solution increases. Rapid and thorough mixing is essential. The yellow solution is stirred for 10 min and stored cold under nitrogen.
>
> *Cyclization* [21] The solution from the prepolymerization reaction is heated to 160°C under nitrogen for 3 hr. The precipitate is collected, washed with ether, and dried. The polymer at this point is still completely soluble in H_2SO_4.
>
> *Postcure* [21] The final completion of cyclization is brought about by heating the solid at 300°C under vacuum for 3 hr.
>
> *Melt method* [21] A mixture of 1.30 g of pyromellitic dianhydride and 1.28 g of 3,3'-diaminobenzidine is heated at a rate of 2°/min in an oil bath and under nitrogen to 220°C. Vacuum is then applied, and the temperature is raised to 250°C and held there for 2 hr. The product is purified by dissolution in DMSO or DMF and precipitation into methanol. The yield is 75% with inherent viscosity of 0.45-0.50 dl/g taken in DMSO or formic acid.

Pyrrones have also been produced by direct-molding procedures, which are actually modifications of melt polymerizations [19]. By such a process one circumvents solubility problems and goes directly to an intractable product from a monomer mixture. Quite good mechanical properties have been demonstrated by these products and, surprisingly, the water off-gas problem can be resolved nicely by using a porous graphite liner in a stainless steel die. Tensile strength and modulus are ~69 MPa (10,000 psi) and 9.0 GPa (1.3×10^6 psi), respectively, and density is 1.27 g/cm^3.

Either a simple monomer mixture (ballmilled or blended in a nonsolvent) or a salt of the monomers can be slowly heated under pressure to result in a molded block. The molding procedure is critical since a polymerization must be allowed to proceed prior to or concurrently with pressure application. Homogeneity of the monomer mixture also is very important to the speed of the reaction, uniformity of the product, and necessary molding conditions.

TGA of products from molding showed a 10% weight loss at 380°C in air if heated at 0.5°C/min, a much lower heating rate than is normally used. At a rate of 10°C/min, this 10% value went to 580°C, a more realistic or comparable number for this method. At a rate of 0.5°C/min, the TGA approaches isothermal testing in severity.

Method [43] *Monomer mixtures* Equimolar amounts of monomers can be ballmilled; however, a high-speed, nonsolvent liquid blending is preferred. To blend, pure monomers in equimolar amounts are placed in cyclohexane in a high-speed blendor for 15 min. The monomers are isolated by filtration and dried under vacuum.

Salt preparation A solution of 54.5 g of pyromellitic dianhydride in 1200 ml of degassed, hot distilled water is added, under nitrogen, to a well-stirred solution of 54.7 g of 3,3'-diaminobenzidine in 1200 ml of distilled water and 500 ml of methanol. A few seed crystals of the product are also placed in the amine solution before the addition. The fine, orange crystalline product is isolated by filtration, rinsed with water and then methanol, and dried under vacuum to give a 95% yield.

Molding procedure The salt or blended mixture is placed in a porous graphite-lined, stainless steel die (1.9 cm or 0.75 in. in diameter and 1.3 or 0.5 in. long as the final molding size) and heated at a rate of 4-5°C/min at 28-48 MPa (4000-7000 psi). The final temperature of 450°C is held for 1 hr.

9.2 Polyquinoxalines

Following the successful synthesis of polyquinoxalines by Stille and Williamson, it was not long before the value of this system was recognized for ladder backbones. The ladder quinoxalines first appeared in 1967 [51], but then a large number of papers appeared during 1969-1972, work sponsored mostly by Air Force laboratories.

Since the quinoxaline function can be formed by condensation of an aromatic ortho-diamine with a diketone, it is an obvious choice of reactions for ladder preparation [51]:

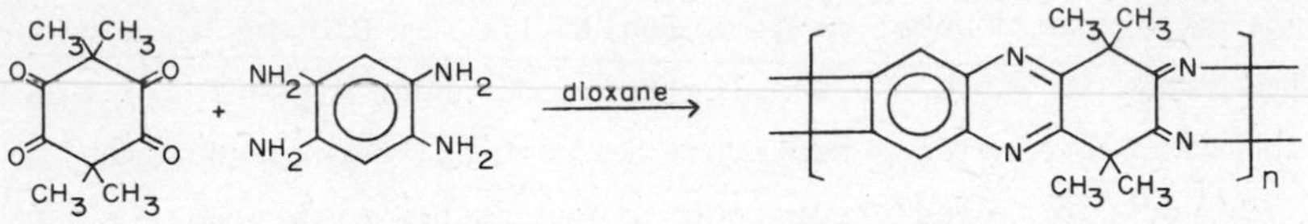

dioxane

Some syntheses start with the quinoxaline nucleus preformed in the monomer and merely couple it through other functional groups [34,41].

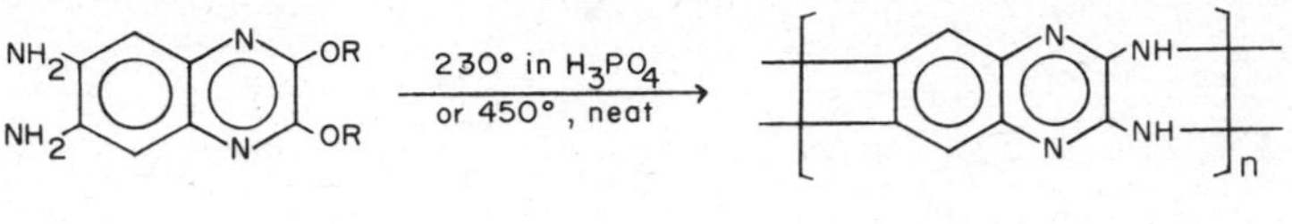

230° in H_3PO_4 or 450°, neat

R is H or C_6H_5

Other tetraketones or diaminodiketones which have been used in actual or attempted preparations are as follows:

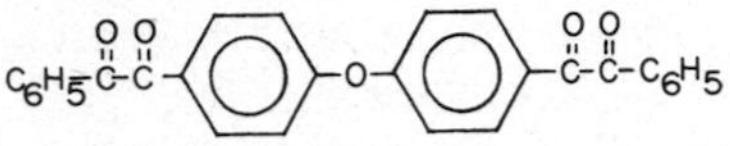

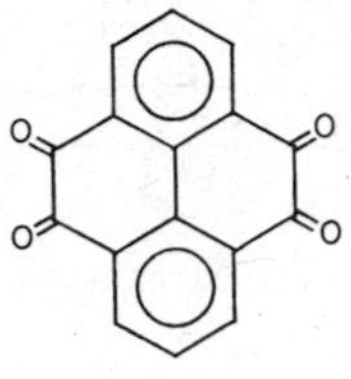

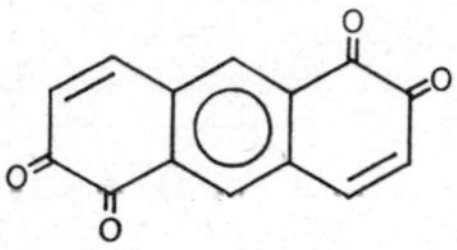

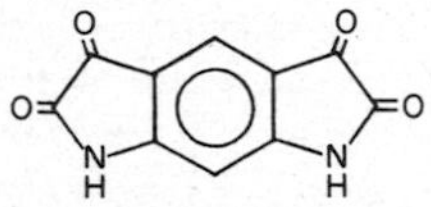

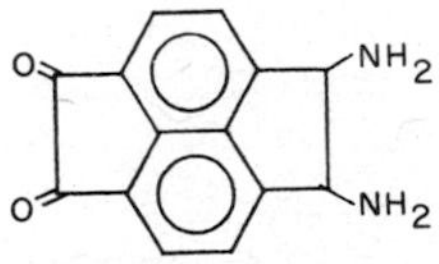

Some tetraamines are the following:

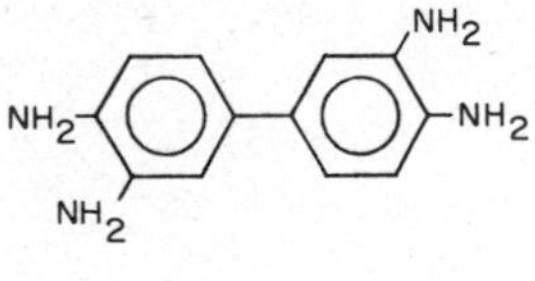

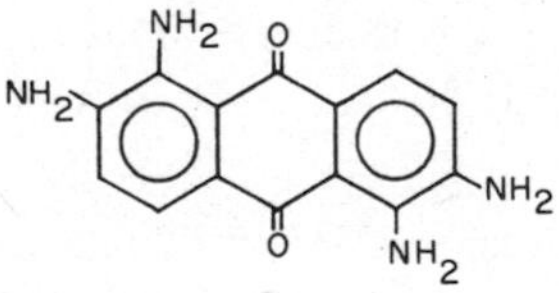

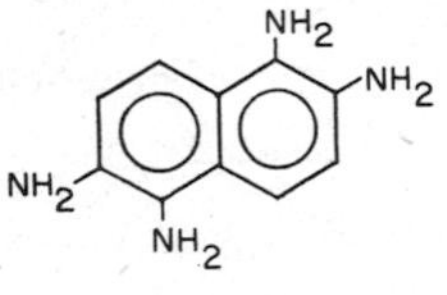

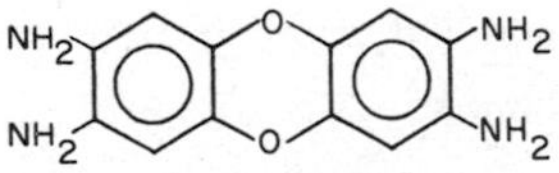

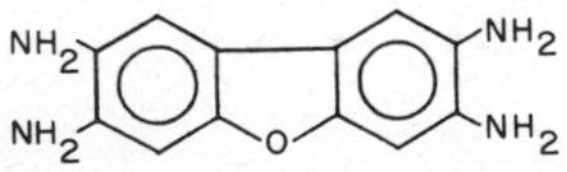

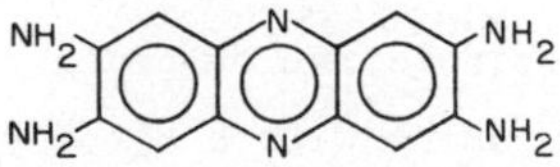

Other more complex, oligomeric quinoxaline tetramines have also been utilized [29] (vide infra).

Inherent viscosities for those products which are soluble range generally from 0.1 to 1.3 dl/g, with one reported as high as 4.7 [52]. Thermal transitions are frequently nonexistent below the decomposition temperature, although a Tg has been reported at 406°C for the polymer with a dibenzofuran unit in the backbone [29].

TGA curves for these materials generally show a gradual decomposition, dropping to only a 25-40% weight loss at 900°C. With a 10% loss as the criterion, thermal stabilities in air show to be in the range of 450-550°C for wholly aromatic systems. In nitrogen these values are 600-650°C. When the polymers are postcured or thermally aged, however, stabilities as high as 775°C have been attained. As expected, some backbones which contain alicyclic portions display less resistance to thermooxidative degradation.

Experimental procedures to synthesize the quinoxaline ladder backbones are generally high-temperature solution processes and yield dark brown to black products. Solvents used in polymerizations are those which are well known to this technology: polyphosphoric acid, m-cresol, pyridine, tetramethylene sulfone, phosphoric acid, and N,N-dimethylacetamide with N,N-diethylaniline. Conditions range from 200 to 350°C for 9-20 hr. Melt or sealed tube procedures are also used, usually at higher temperatures (up to 450°C) than solution reactions. The products are generally soluble or partly soluble in sulfuric, methylsulfonic, and trifluoromethylsulfonic acids, hexamethylphosphoramide (HMPA), and occasionally m-cresol.

A true quinoxaline ladder was synthesized as follows [53]:

NH_2 NH_2 NH_2 NH_2 + O O O O $\xrightarrow[\text{m-cresol}]{200°}$ N N N N n

Products showed a 10% TGA loss of 440°C in air and 585°C in nitrogen. When postcured, this material then goes to 700°C. At 800°C, the total weight loss is 38%.

Method [53] *Polymerization* In a 100-ml round-bottom flask, a slurry is prepared containing 0.410 g (2.18 mmol) of 1,2,5,6-tetraaminonaphthalene in 24 ml of m-cresol, all under nitrogen, and to it is added 0.519 g (2.18 mmol) of 1,2,5,6-tetraketoanthracene with stirring. This mixture is stirred under nitrogen according to the following schedule:

Temp. (°C)	Time (hr)
25	3
65	17
96	4.5
120	18
120-200	4
200	45

The dark black mixture, which includes some polymer precipitate, is poured into chloroform to precipitate all the polymer. The solid is isolated, washed with chloroform, dried, extracted with pyridine, and dried again to yield 0.80 g of poly(benzo[a]phenazino[1,2-h]phenazine-1,2:10,11-tetrayl) with an inherent viscosity of 0.66 dl/g in sulfuric or methane sulfonic acid at a limiting concentration of 0.06 g/100 ml.

Postcuring A secondary postcure or thermal advancement of the polymer serves to complete cyclization. The following schedule, conducted under argon on the above crude polymer before its extraction with chloroform, results in an 8.3% weight loss.

Temp. (°C)	Time (hr)
300	30
400	16.5
455	20

Working in Marvel's laboratory, Banihashemi has taken a somewhat different approach to a full ladder backbone, that of condensing aromatic

NH_2 O NH_2 NH_2 O NH_2 + Cl N N Cl Cl N N Cl → H N N N NH Cl N N N H O O NH_2 n

350°, vacuum

H N N N H N O=C C=O N N N N H H n

amines with aromatic halides, the latter containing a preformed quinoxaline unit [7]. Of course, these materials are not wholly aromatic. The product was soluble in H_2SO_4 and possessed an inherent viscosity of 0.44 dl/g. The 10% weight loss point in nitrogen is 480°C, but the gradual TGA slope tapers to only 40% loss at 900°C.

> *Method* [7] To a mixture of 2.1 g (0.0065 mol) of 2,3,7,8-tetrachloro-1,4,6,9-tetraazaanthracene [35] in 55 ml of pyridine is added 1.73 g (0.0065 mol) of 1,2,5,6-tetraaminoanthraquinone [15], and the whole is heated at reflux with stirring for 2 hr under nitrogen. The black solid is isolated, washed with pyridine and then ether, and dried at 50°C.
>
> *Postcuring* To postcure, the solid is heated at 350°C for 6 hr under vacuum and extracted with ethanol and then benzene to afford a 100% yield.

A recent synthesis of a pseudoladder quinoxaline polymer has been published by Hedberg and Arnold using a solution process [29] with a linear tetraketone and highly fused aromatic amines.

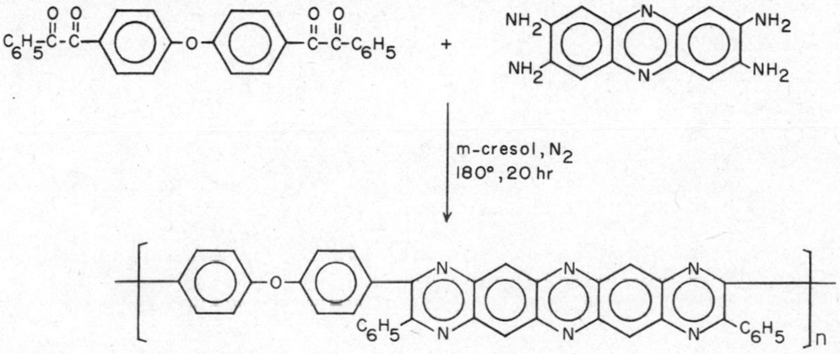

m-cresol, N_2
180°, 20 hr

Other tetraamines:

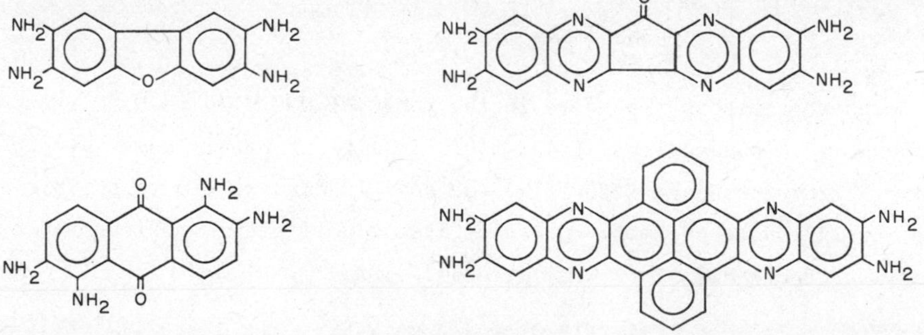

The products are soluble in m-cresol and have inherent viscosities up to 1.8 and Tg's above 400°C. The TGA data show stabilities in air of 460-550°C. TGA's in nitrogen are difficult to classify for many of these polymers. The less highly fused systems show a 10% loss at 590-625°C but a total weight loss (plateau) of only 25-30% at 900°C. The more highly fused products only gradually relinquish their weight to a final value of merely 8-20% at 900°C.

Method [29] Under nitrogen, 0.332 g (0.137 mol) of 2,3,7,8-tetraaminophenazine [46] and 0.598 g (0.137 mol) of p,p'-oxydibenzil [30] are added to 30 ml of freshly distilled m-cresol, and the mixture is stirred overnight at room temperature. The dark red solution is heated to 120°C for 6 hr and then cooled and poured into 500 ml of methanol. The dark red precipitate is isolated, washed with methanol, reprecipitated from m-cresol into methanol, and washed again with methanol. Vacuum drying at 150°C yields a 96% product with an inherent viscosity of 0.32 dl/g in H_2SO_4.

Higgins and Janovic combined quinoxaline and pyrrole functions in a backbone which displayed a typical stability and solubility for ladder or pseudoladder systems [31,36]. Condensation of benzodipyrrole tetraketone with aromatic tetraamines in PPA gave highly colored products whose solubilities (in sulfuric and methane sulfonic acids) were inversely related to total ladder character.

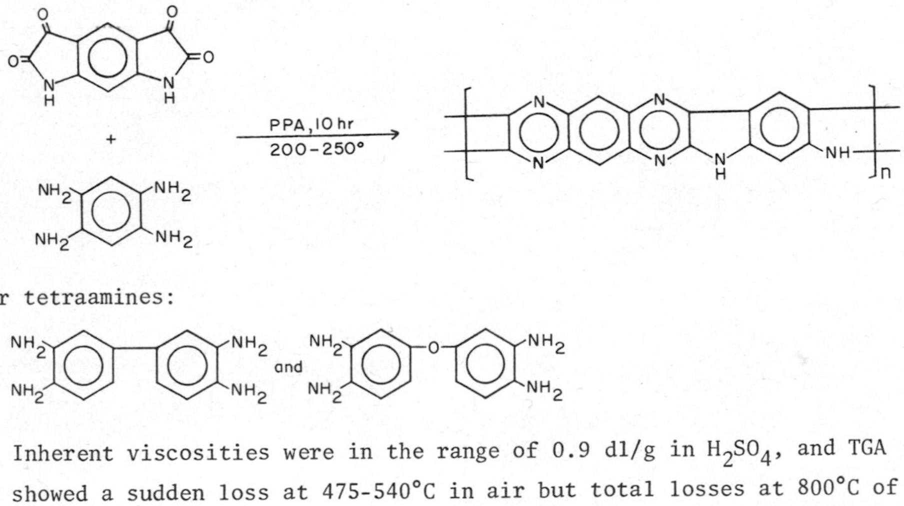

PPA, 10 hr
200-250°

Other tetraamines:

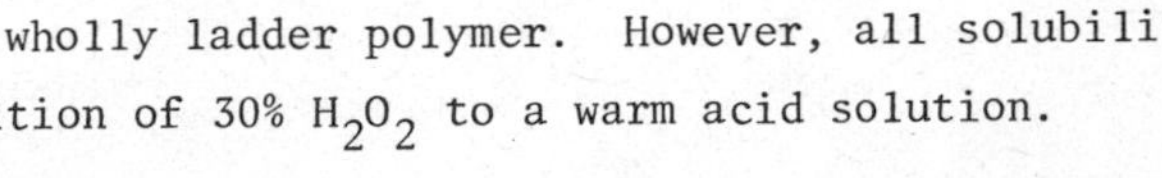

and

Inherent viscosities were in the range of 0.9 dl/g in H_2SO_4, and TGA data showed a sudden loss at 475-540°C in air but total losses at 800°C of only 25% in nitrogen. Naturally, the near-ladders were more soluble than the wholly ladder polymer. However, all solubilities were improved by the addition of 30% H_2O_2 to a warm acid solution.

Method [36] A mixture of 0.415 g (0.003 mol) of 1,2,4,5-tetraaminobenzene and 0.649 g (0.003 mol) of benzo[1,2-b:5,4-b']dipyrrolo-2,3,5,6-tetraone [1] with 100 ml of PPA is heated at 200°C for 10 hr under nitrogen. The resulting blue solution is poured into 800 ml of methanol to precipitate the polymer. The isolated polymer is extracted continuously for 2 days with water and then washed with methanol and dried to give a 95% yield (0.805 g) of green solid. The inherent viscosity is 0.86 dl/g at a concentration of 0.25 wt % in 25 ml of H_2SO_4 which contains four drops of 30% H_2O_2.

Dihydrobenzothiazine and oxazine heterocyclic functions were first incorporated into a backbone with quinoxaline units in Marvel's laboratory in 1967 [47] by the condensation of an aromatic tetrachloride with diaminodithiophenols or diaminodiphenols. Since the aryl halide was based on a quinoxaline structure, the advantage of this function was included.

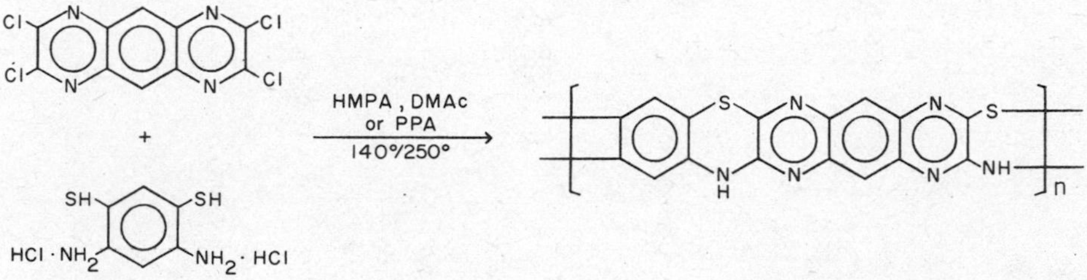

Other amines and chloroquinoxaline nuclei, some of which provided near ladder backbones were used:

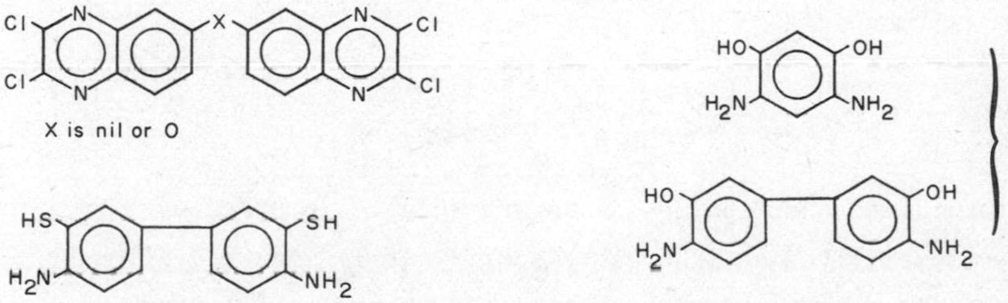

X is nil or O

give oxazine polymers

Yields were commonly high (80-98%), and inherent viscosities were found to be as high as 1.54 dl/g. The brown to black polymers were soluble in sulfuric and methanesulfonic acids. The full-ladder system shown above, when subjected to TGA, showed only a 6.5% weight loss at 500°C and 37% at 900°C.

Method [47] The 4,6-diamino-1,3-dithiophenol dihydrochloride (0.005 mol) is added slowly to a mixture of 0.005 mol of 2,3,7,8-tetrachloro-1,4,6,9-tetraazaanthracene in 60 ml of oxygen-free HMPA, and the mixture is stirred for 2 hr at room temperature under nitrogen. The temperature is then raised slowly to 140°C and held for 24 hr, whereupon the mixture is poured into 800 ml of water which contains 20 ml of conc. NH_4OH. The resulting dark brown precipitate is isolated by centrifugation, washed with water and then alcohol, and dried. The polymer is then heated at 250°C for 6 hr under nitrogen and purified by washing with aq. ammonium hydroxide solution then with methanol and finally by extracting with benzene for 100 hr. Vacuum drying yields 1.34 g (80%) of polymer with inherent viscosity of 0.58 dl/g in methanesulfonic acid.

A rather unique approach to polydihydrobenzothiazines was demonstrated by Augl and Wrasidlo sometime later [5]. Their model compound study showed the feasibility of going through a 2,2-spiropentamethylenebenzothiazoline intermediate which suffers thermal rearrangement:

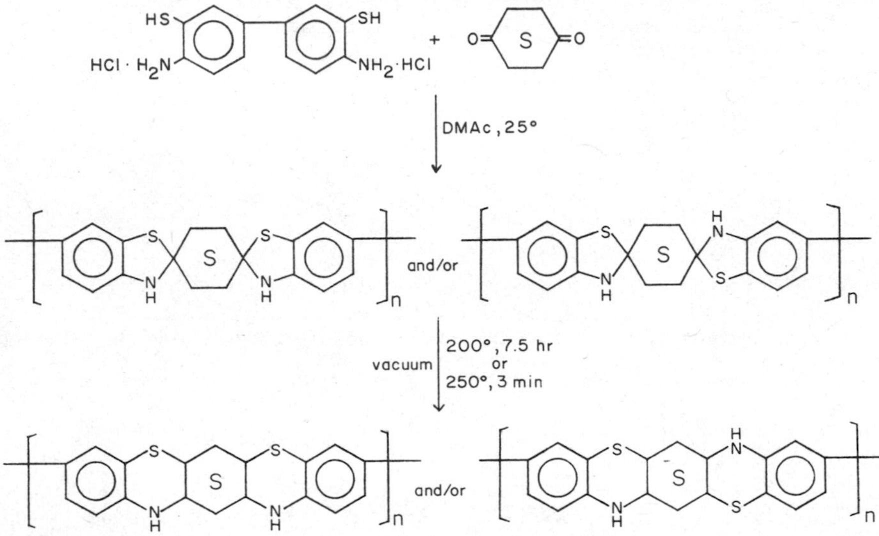

Both intermediate and final products are soluble in DMAc, pyridine, or DMSO, and the former gives an inherent viscosity of 1.01 dl/g. Films can be cast from the intermediate which can then be thermally rearranged.

TGA shows a decomposition temperature at 350°C for either precursor or final product, and the latter has a Tg of 210°C.

Perhaps the most interesting property of these polymers is their sensitivity to a combination of light and oxygen. With a few minutes of exposure to both of these conditions, the polymer films changed from light yellow to purple. The reason proposed for this is a very small amount of photooxidation of the backbone leading to some quinoid structures.

Method [5] *Polymerization* A mixture of 0.867 g (0.003 mol) of 3,3'-dimercaptobenzidine, 0.336 g (0.003 mol) of 1,4-cyclohexanedione, and 8.4 ml of DMAc is stirred for 1 hr at room temperature. Pyridine (0.3 ml) is then added to the viscous solution, and it is stirred for 1 more hour. The solution is poured into an excess of well-stirred deaerated ethanol to precipitate the polymer, which is washed with ethanol and dried (all in the absence of oxygen). Films are cast from a 10% solution of poly[2,2'-(1",4"-dispirocyclohexane)-6,6'-bis(benzothiazoline)] in DMAc.

Thermal rearrangement Films (1 mil) are heated under vacuum either for 7.5 hr at 200°C or for 3 min at 250°C.

9.3 Polyquinolines

A true ladder polyquinoline was reported by Yoda and Kurihara in a 1971 review [57], but referenced only as a private communication. This route was by condensation of a diketal with a bis-o-aminoester.

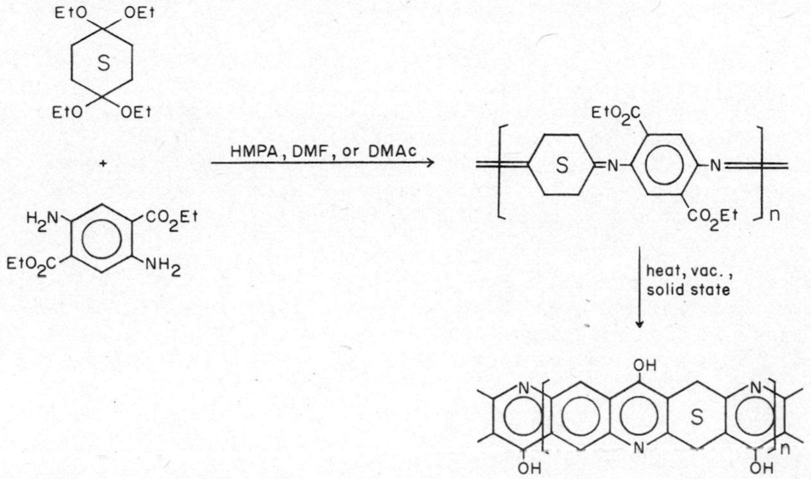

The final product, albeit partially crosslinked, shows a disappointing thermal analysis in a break in the TGA curve at 300°C in air.

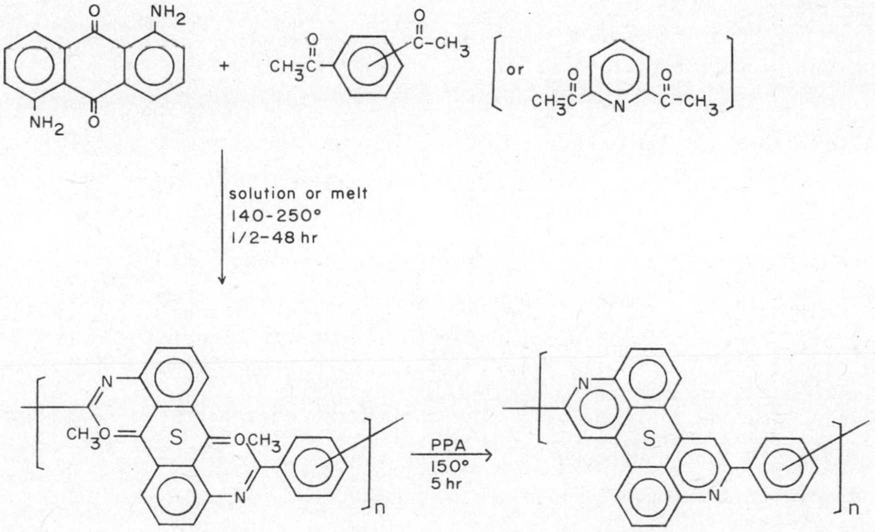

The first confirmed report of a quinoline-containing pseudoladder emanated from Marvel's laboratory [44]. These were brought about by the condensation of 1,5-diaminoanthraquinone with 1,4- and 1,3-diacetylbenzene or 2,6-diacetylpyridine. Generally, melt conditions seemed to provide better (i.e., more soluble) prepolymers, which gave inherent viscosities of 0.21-0.46 dl/g in sulfuric or methanesulfonic acids. The final polymer with para catenation showed an inherent viscosity of 0.56 dl/g in the same solvents and an outstanding stability in nitrogen of 900°C (10% weight loss by TGA). In air, however, this value is reduced drastically to 440°C, where initial decomposition occurred.

> *Method* [44] *Melt polymerization* The monomers of 1,4-diacetylbenzene [0.324 g (0.002 mol)] and 1,5-diaminoanthraquinone [0.476 g (0.002 mol)] are ground together and placed in a polymerization tube [25] under nitrogen. This system is quickly placed in a preheated metal bath at 240°C for 0.5 hr, whereupon some sublimation of monomers occurs. The product is cooled, ground, extracted with ethanol and then DMAc, and dried to yield 89%; inherent viscosity 0.46 dl/g in conc. H_2SO_4 (0.48% solution).
>
> *Cyclization* The above powder is mixed with 10 g of PPA in a 100-ml flask under nitrogen and heated at 150°C for 4 hr with stirring. The mixture is then cooled and poured into water; and the polymer is isolated by filtration, extracted with water and then alcohol, and dried in vacuo at 60°C.

9.4 Phenanthroline Ladder Polymers

9.4-1 Poly(benzimidazobenzophenanthrolines)

The rather complex phenanthroline ladder and near-ladder systems have been investigated almost exclusively by Air Force research laboratories or under their sponsorship [2,4,13,18,40,49,50,55]. Van Deusen termed these BBB or BBL in his original papers [3,56], the latter being a true ladder. Representative structures are the following (only cis isomers shown):

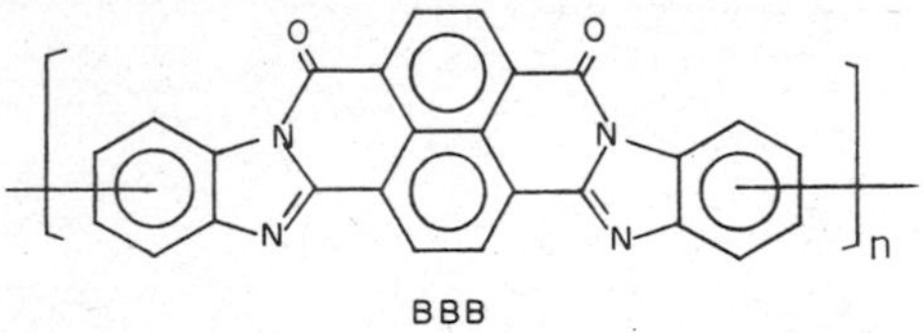

BBB 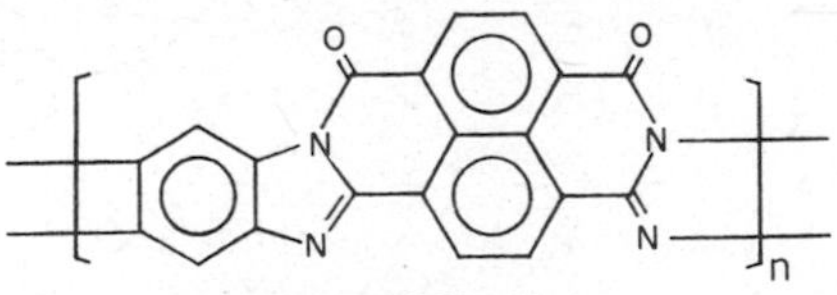 BBL

More recently, these materials have been called imidazoisoquinoline ladders. They are synthesized from starting materials similar to those for pyrrones, namely, a tetraacid or its dianhydride (1,4,5,8-substituted naphthalene) and a tetraamine. The only differences between these and pyrrones are the fused naphthalene ring and the six-member ring in the heterocyclic portion.

The amines which have been successfully condensed with the naphthalene tetraacid or anhydride are as follows:

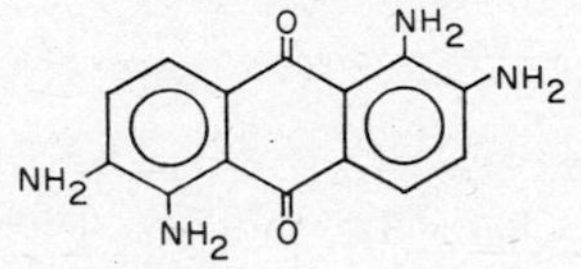

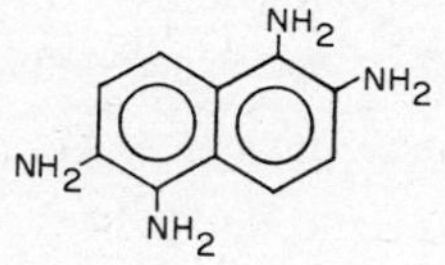

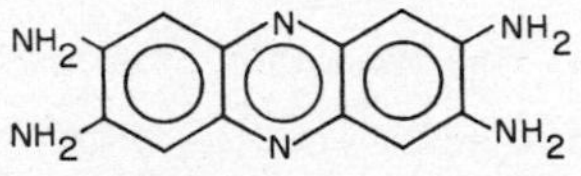

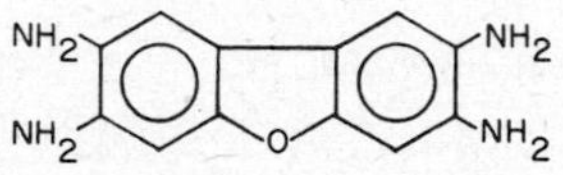

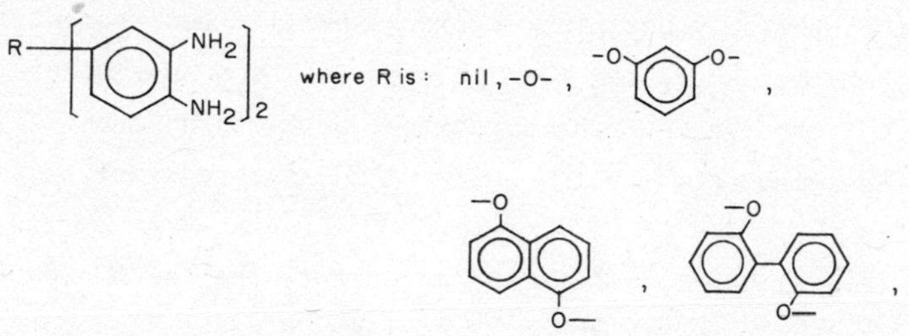

where R is: nil, –O–,

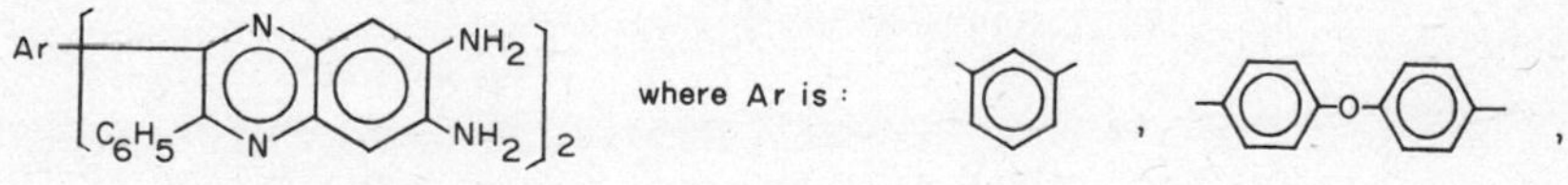

where Ar is:

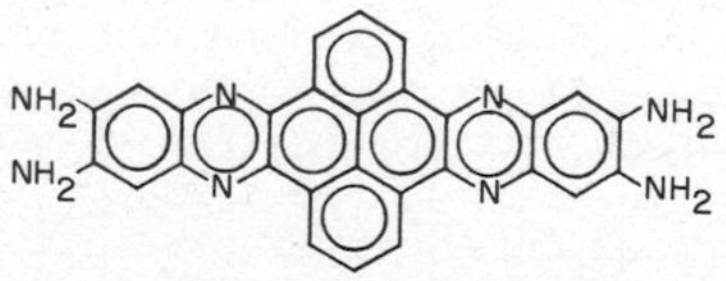

Polymerizations are generally thermal condensations at 150°C in solutions of PPA or molten $SbCl_3$, or at 380°C in the melt. The products are soluble or slightly soluble in methanesulfonic acid. Tg's for pure ladders are above 500°C, if they are present at all, and, of course, are lower when flexibilizing functions are included in the backbone, i.e., near-ladder backbones with phenylether or sulfone. This dependence of Tg to the functional group was nicely shown by Arnold to be in the order BBB > single-ether > phenyl and naphthylethers > phenylether sulfone [4].

Thermal stabilities for these ladders or near-ladders are impressive but not well above other heterocyclic polymers, being 450-550°C in air and 650-775°C in nitrogen. An isothermal study showed minimal weight loss after 200 hr at 370°C in air [49].

A good review of processing and properties has been prepared for the BBB and BBL systems [57].

A peculiar phenomenon has been demonstrated with the BBB system, that of being able to form a quite strong, high-modulus film merely by precipitation and subsequent vacuum filtration of the polymer from solution [4,50]. A development such as this is important to the field of thermally stable polymers in that the necessary but elusive processability requirement is circumvented. It has been suggested that this is due to a supramolecular aggregate structure with high interchain packing, but with no crystallinity; i.e., the planar repeating units, being quite rigid, are regularly stacked in a graphitic array.

There was an earlier indication in the literature that such a packing phenomenon might exist. The pseudoladder from 1,4,5,8-naphthalenetetracarboxylic acid and 3,3'-diaminobenzidine showed flexible coil behavior in dilute solution but a significant amount of interchain complex formation in the solid phase [14]. This complex contributed to the bulk properties which one would expect from a highly crosslinked material.

Any thermal effects below 500°C are said to be due to a rearrangement in this noncrystalline supramolecular order.

Method [50] PPA (100 g) is deoxygenated, and to it under nitrogen is added 1.000 g (2.993 mmol) of 1,2,5,6-tetraaminonaphthalene tetrahydrochloride; and it is heated to 65°C for 8 hr to expel all HCl gas. The mixture is cooled to room temperature, and 0.9104 g (2.993 mmol) of 1,4,5,8-naphthalenetetracarboxylic acid is added; and heat is reapplied to raise the temperature slowly (3°C/min) to 280°C, where it is held for 10 hr. The mixture is then cooled to 25°C and poured into 1 liter of methanol to precipitate the polymer, which is washed with methanol, vacuum-dried, and reprecipitated from methanesulfonic acid into methanol. When washed with methanol and dried again, the polymer is received in 97% yield (1.11 g) with an intrinsic viscosity of 0.8 dl/g in methanesulfonic acid.

9.4-2 Poly(benzobistriazolophenanthrolines)

Poly(benzobistriazolophenanthrolines) are of course very similar to the BBB materials except that the benzimidazole portion of the backbone is replaced with a triazole. This is accomplished by the PPA solution condensation of the naphthalene tetraacid (or its dianhydride) with bisamidrazones (dihydrazidines) [24].

CO_2H CO_2H / CO_2H CO_2H + NH_2-NH-C(=NH)-Ar-C(=NH)-NH-NH_2 $\xrightarrow[\text{20 hr / 200°}]{\text{PPA}}$ [polymer]$_n$

Ar is: m- or p-C_6H_4 or (2,6-pyridinediyl)

The red-black, powdery polymers are quite soluble in sulfuric and methanesulfonic acids and display inherent viscosities up to 2.5 dl/g. Deep red films which were tough and creasable can be cast from the latter solvent. However, the strong films from precipitation as with BBL could not be obtained.

TGA gave initial decompositions of 450°C in air and 475°C in nitrogen, with 10% weight losses at 510 and 520°C, respectively. The materials do not soften under load below their decomposition temperatures.

Method [24] *Terephthaldihydrazidine dihydrochloride monomer* Diethylterephthalimidate [39,58] (11.0 g, 0.05 mol) is slurried at room temperature in a solution of 3.2 g (0.05 mol) of anhydrous hydrazine and 200 ml of acetonitrile. Gradual dissolution gives a clear yellow solution and then (after 48 hr) a pale yellow precipitate which is isolated by filtration and dried under vacuum (92%, 8.9 g, mp dec. 223-225°C). A portion (2 g) of the crude dihydrazidide is dissolved in 30 ml of 4 N HCl at 25°C, and dry HCl gas is bubbled through the solution at 0°C. The white precipitate is collected, washed with cold conc. HCl, and dried over P_2O_5 and KOH at 140°C and 0.1 mmHg for 20 hr to give 2.0 g (74% yield), mp dec. 317-320°C.

Polymerization A mixture is prepared with 1.766 g (0.0067 mol) of terephthaldihydrazidine dihydrochloride and 110 g of oxygen-free PPA and stirred for 3 hr at 80-90°C to release HCl gas. Then 2.026 g (0.0067 mol) of 1,4,5,8-napthalenetetracarboxylic acid is added, and the temperature is raised to 180-185°C over 1 hr and held there for 22 hr to result in a viscous, red solution. The mixture is cooled to 100°C and poured into 500 ml of vigorously stirred water. Upon settling, the polymer is collected by filtration and washed well with water, then DMAc, and finally methanol. Further purification is conducted by dissolving the red-black polymer in methanesulfonic acid, precipitating it into methanol, washing it with ether, and drying it at 180°C and 0.01 mmHg over P_2O_5. The yield is 88% (2.3 g) with inherent viscosity of 2.51 dl/g in methanesulfonic acid.

9.5 Graphitic-Type Ladder Polymers

Perhaps the ultimate in rigid, ladder backbones was discovered in 1972, again in Marvel's laboratory [35]. A two-step process consisting of a solution condensation and a postcure leads to an insoluble, black product through a soluble intermediate.

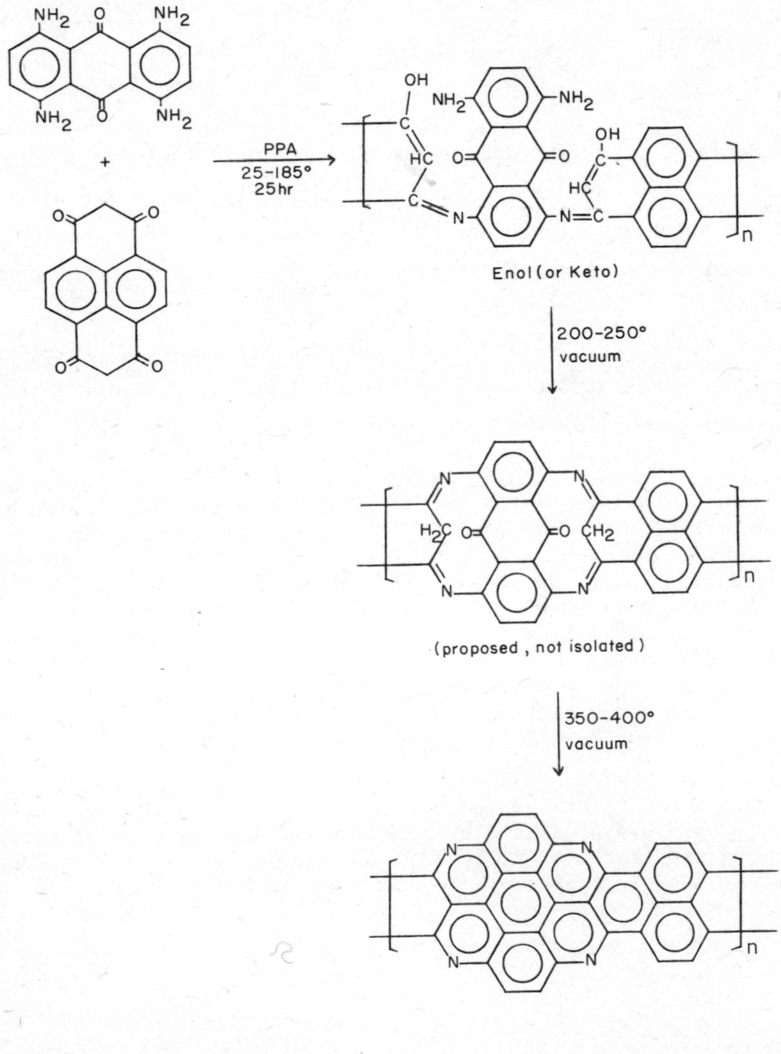

NH2 O NH2
NH2 O NH2
+
PPA
25-185°
25hr
OH
NH2 NH2
OH
HC O O
HC
N N=C
n
Enol (or Keto)
200-250°
vacuum
N N
H2C O O CH2
N N
n
(proposed, not isolated)
350-400°
vacuum
N N
N N
n

Improving the purity of the tetraketone and modifying the reaction conditions in PPA made possible a yield of ~80% for the prepolymer, which is soluble in sulfuric or methanesulfonic acids [38]. The inherent viscosity ranged from 0.1 to 1.6 dl/g, but higher viscosity led to lower solubility. The first prepolymer is better solubilized by treatment with sodium dithionite ($Na_2S_2O_4$) in alkaline, aq. DMAc.

A long (3-5 days), high temperature (>340°C), vacuum postcure brings about final cyclization. This process is conducted on a powder, film, or wet-spun fiber of the prepolymer. Brittle films are cast from methanesulfonic acid, with vacuum evaporation at 60-100°C

Stability analysis by isothermal methods showed that after the final polymeric material is dry it demonstrates at 300°C the following weight loss data, which are not as good as those for BBL (vide supra).

% Loss	Time (hr)
4.8	9
10.4	24
31.8	92

TGA gave an initial break at ~470°C and a 10% weight loss at 560°C. The weight loss reached and held a plateau of 25-30% from 700 to 900°C.

The reported good ring closure of this system, as well as its graphitic structure, naturally makes it considered for electrical conductivity. This measurement has been done on compacted powders of this polymer, and it has been found to display one of the highest conductivities ever found for an organic polymer [28]. The conductivity is still, however, considerably below that of graphite. The prepolymer interestingly demonstrates a high conductivity also, 10% of that of the final product.

Method [38] *Tetraketone preparation* To 24.0 g (0.176 mol) of fused anhydrous $ZnCl_2$ are added 12.0 g (.0447 mol) of 1,4,5,8-naphthalenetetracarboxylic dianhydride and 50 ml (0.313 mol) of diethylmalonate, and the mixture is stirred for 3 hr at 170°C under nitrogen. The reaction is cooled, filtered, and the dark solid washed with 95% ethanol, three 400-ml portions of hot water and similarly with cold water. The solid is then stirred for 30 min in 400 ml of 10% aq. ammonia; and the solution is filtered, treated with charcoal, filtered, and acidified to pH 1 with conc. HCl. It is then heated to reflux for 15 min, cooled, filtered, and the solid residue washed with cold water and vacuum-dried at 75°C. The solid is then extracted (Soxhlet) for 4 days with HOAc and then for 3 days with 95% ethanol and finally dried under vacuum at 75°C to yield 6.2 g (52%) of 1,3,6,8-tetraketo-1,2,3,6,7,8-hexahydropyrene(naphthalene-1,8,4,5-diindandione), mp >500°C.

Polymerization In a 25-ml flask with stirrer, condenser, and nitrogen blanket is placed a fine powder mixture of 0.264 g (1.00 mmol) of the tetraketone and 0.268 g (1.00 mmol) of 1,4,5,8-tetraaminoanthraquinone [16], the latter being recrystallized twice from purified nitrobenzene. To this 15 g of PPA is added, and the system is stirred and heated over 1.5 hr to 120°C, then over 4 hr to 170°C, where it is held for 20 hr. The mixture is cooled to 25°C and poured with stirring into 600 ml of ice and water. The black precipitate is isolated, washed vigorously with two 600-ml portions of methanol and dried. The prepolymer is extracted (Soxhlet) with methylene chloride for 1 day, DMAc for 3 days, water for 1 day, and 95% ethanol for 1 day. The dried, finely ground powder is dried again under vacuum at 100°C for 3 days to yield a shiny, black granular polymer.

Postcuring The powder or film of prepolymer is heated for 5 days at 340-400°C and 0.01 mmHg.

9.6 Recent Developments

Little information is available on recent work on ladder polymers, probably due to the absence of truly exceptional thermal behavior of these materials and to the processing difficulties. The dilute solution properties of BBL (from 1,4,5,8-naphthalene tetracarboxylic acid and 1,2,4,5-tetraaminobenzene) have been investigated by light scattering, intrinsic viscosity, gel-permeation chromatography, and fluoresence depolarization [59]. This was done to search for liquid crystal effects, and a nearly rodlike conformation was concluded.

A siloxane-type ladder was reported and studied in dilute and concentrated solution [60]. cis-Syndiotactic poly(phenylsilsequioxane) or PPSA was found by viscometry and light scattering to possess a wormlike conformation with a persistance length of 74 Å:

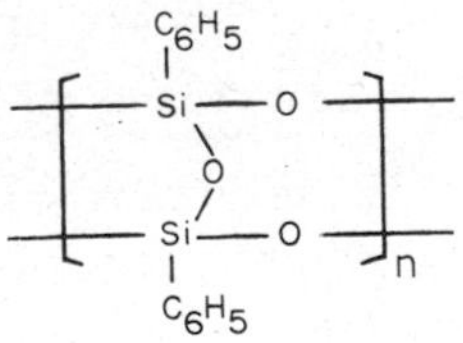

$$\left[\begin{array}{c} C_6H_5 \\ | \\ -Si-O- \\ \backslash \\ O \\ / \\ -Si-O- \\ | \\ C_6H_5 \end{array}\right]_n$$

References

1. Z. J. Allan, *Chem. Listy, 46,* 228 (1952).
2. F. E. Arnold, *J. Polym. Sci., A-1, 8*(8), 2079-2080 (1970).
3. F. E. Arnold, *Polym. Preprints, 16*(2), 251 (1975).
4. F. E. Arnold and R. L. Van Deusen, *Macromolecules, 2,* 497 (1969).
5. J. M. Augl and W. J. Wrasidlo, *J. Polym. Sci., A-1, 8*(1), 63 (1970).
6. W. J. Bailey and A. A. Volpe, *Polym. Preprints, 8,* 292 (1967).

7. A. Banihashemi and C. S. Marvel, *J. Polym. Sci., A-1, 8*(11), 3211 (1970).
8. V. L. Bell, *J. Polym. Sci., B, 5,* 941 (1967).
9. V. L. Bell, *Encycl. Polym. Sci. Technol., 11,* 240-246 (1969).
10. V. L. Bell, *J. Appl. Polym. Sci., 14,* 2385 (1970).
11. V. L. Bell and R. A. Jewel, *J. Polym. Sci., A-1, 5,* 3043 (1967).
12. V. L. Bell and G. F. Pezdirtz, *J. Polym. Sci., B-3,* 977 (1965).
13. G. C. Berry, *J. Polym. Sci.: Polym. Phys. Ed., 14,* 451 (1976).
14. G. C. Berry and S. P. Yen, *Advan. Chem. Series,* No. 91, 734-756 (1969).
15. W. Bracke and C. S. Marvel, *J. Polym. Sci., A-1, 8,* 3177 (1970).
16. W. Bracke and C. S. Marvel, *J. Polym. Sci., A-1, 8* 3177 (1970).
17. E. E. Braunsteiner and H. Mark, *J. Polym. Sci., D, Macromol. Rev., 9,* 83 (1974).
18. M. Bruma and C. S. Marvel, *J. Polym. Sci.: Polymer Chem. Ed., 12*(10), 2385 (1974).
19. J. G. Colson, R. H. Michel, and R. M. Pauflen, *J. Polym. Sci., A-1, 4,* 59 (1966).
20. G. F. D'Alelio and H. E. Kiefer, *J. Macromol. Sci. Chem., A2*(6), 1275 (1968).
21. F. Dawans and C. S. Marvel, *J. Polym. Sci., A-3,* 3549 (1965).
22. N. Dogoshi, S. Toyama, K. Ikeda, M. Kurihara, and N. Yoda, Japan Patent 69 20,111 (1969).
23. W. R. Dunnavant, *J. Polym. Sci., B, 6,* 49 (1968).
24. R. C. Evers, *J. Polym. Sci.: Polym. Chem. Ed., 11,* 1449 (1973).
25. R. T. Foster and C. S. Marvel, *J. Polym. Sci., A-1, 3,* 417 (1967).
26. L. W. Frost and G. M. Bower, *J. Polym. Sci., A-1, 9,* 1045 (1971).
27. A. H. Gerber, *J. Polym. Sci.: Polym. Chem. Ed., 11,* 1703 (1973).
28. J. L. Gilson and W. R. Hertler, *J. Polym. Sci.: Polym. Lett. Ed., 14*(3), 151 (1976).
29. F. L. Hedberg, and F. E. Arnold, *J. Polym. Sci.: Polym. Chem. Ed., 12*(9), 1925 (1974).
30. P. M. Hergenrother, *J. Polym. Sci., A-1, 7,* 945 (1969).
31. J. Higgins and Z. Janovic, *J. Polym. Sci., B, 10*(4), 301 (1972).
32. K. A. Hodd and W. A. Holmes-Walk, *J. Polym. Sci. Symp.,* No. 42, 1435-1442 (1973).
33. S. A. Hurley, P. K. Dutt, and C. S. Marvel, *J. Polym. Sci., A-1, 10*(4), 1243 (1972).
34. W. G. Jackson and W. Schroeder, U.S. Patent 3,484,387, Dec. 1969.
35. H. Jadamus, F. DeSchryver, W. DeWinter, and C. S. Marvel, *J. Polym. Sci., A-1, 4,* 2831 (1966).
36. Z. Janovic and J. Higgins, *J. Polym. Sci., A-1, (10),* 1609 (1972).
37. N. J. Johnston, *J. Polym. Sci., A-1, 19*(9), 2727 (1972).

38. R. Kellman and C. S. Marvel, *J. Polym. Sci.: Polym. Chem. Ed.*, *13*, 2125-2131 (1975).

39. H. Kersten and G. Meyer, *Makromol. Chem.*, *138*, 265 (1970).

40. R. F. Kovar and F. E. Arnold, *J. Polym. Sci.: Polym. Chem. Ed.*, *12*, 401 (1974).

41. L. J. Miller, *Tech. Report AFML-TR-68-249, II*, Dec. 1969.

42. M. Minami, Japan Patent 69 19,878, August 1969.

43. P. E. D. Morgan and H. Scott, *J. Appl. Polym. Sci.*, *16*(8), 2029 (1972).

44. R. M. Mortier, P. K. Dutt, J. Hoefnagels, and C. S. Marvel, *J. Polym. Sci.*, *A-1*, *9*, 3337 (1971).

45. B. Nartisissov, *J. Macromol. Sci. Rev. Macromol. Chem.*, *C11*(1), 143-176 (1974).

46. R. Nietzki and E. Müller, *Berichte*, *22*, 440 (1889).

47. M. Okada and C. S. Marvel, *Polym. Preprints*, *8*(1), 229 (1967).

48. C. G. Overberger and J. A. Moore, *Fortschr. Hochpolym. Forsch.*, 7(1), 113-150 (1969).

49. J. W. Powell and R. P. Chartoff, *J. Applied Polym. Sci.*, *18*(1), 83 (1974).

50. A. J. Sicree, F. E. Arnold, and R. L. Van Deusen, *J. Polym. Sci.*, *Polym. Chem. Ed.*, *12*, 265 (1974).

51. J. K. Stille, and M. E. Freeburger, *J. Polym. Sci.*, *B*, 5, 992 (1967).

52. J. K. Stille, W. Alston, L. Green, K. Imai, L. Mathias, R. Wilhelms, and S. W. Wittmann, *Tech. Report AFML-TR-70-5, II*, Jan. 1971.

53. J. K. Stille, K. Imai, E. F. Johnson, L. Mathias, J. Wittmann, J. Wolfe, and S. Wratten, *Tech. Report AFML-TR-70-5, III*, Jan. 1972.

54. M. Suzuki, E. Hosokawa, S. Hirata, and T. Hoshino, Japan Patent 69 26,310, November 1969.

55. J. Szita, L. H. Brannigan, and C. S. Marvel, *J. Polym. Sci.*, *A-1*, *9*, 691 (1971).

56. R. L. Van Deusen, O. K. Goins, and A. J. Sicree, *J. Polym. Sci.*, *A-1*, *6*, 1777 (1968).

57. N. Yoda and M. Kurihara, *J. Polym. Sci. Macromol. Rev.*, *5*, 109-194 (1971).

58. E. L. Zaitseva, A. Ya. Yakubovich, G. I. Braz, and V. P. Bazov, *Zh. Obsch. Khim.*, *34*, 3709 (1964).

59. G. C. Berry, *J. Polym. Sci.: Polym. Symp.*, No. 65, 143 (1978).

60. T. E. Helminiak and G. C. Berry, *J. Polym. Sci.: Polym. Symp.*, No. 65, 107 (1978).

Chapter 10

Organometallic Polymers

This chapter discusses backbones which contain metal or metalloid atoms in the order silicon, boron, iron, phosphorus with other metals, and a few miscellaneous types. Although these materials have been studied least, perhaps their study is where high-temperature polymers should have begun. After all, inorganic systems are much more suitable than organics to resistance to thermo-oxidative degradation. The advantage of the high bond energies in organometallic compounds is now becoming well recognized. These types of polymeric materials have even been suggested for prolonged use at 800°C [91]. However, a natural barrier presents itself in that polymer chemists receive their training with organic chemicals and are reluctant to venture into "less scientific" areas. Similarly, inorganic chemists dislike dealing with smoking cauldrons of black tar, which are known by them to be the bases of all polymer chemistry. Those bridges which have been made between the two extremes are quite successful.

A large number of the materials discussed in the following sections have been commercialized to give products ranging from hard, crosslinked resins to crystalline thermoplastics to elastomers to viscous liquids. Each commercial material is included under its proper section. A summary table which includes structures and general properties has been prepared (Table 10.1).

The discussion begins with polysiloxanes but, as before in this work, the common silicones will not be discussed since their commercial success is well known as sealants, encapsulants, and elastomers. Rather, special cases of siloxane copolymers with nitrogen functions will be presented. Then in later sections, the copolymers of each of the special functions with siloxanes will be discussed anew.

Some very interesting generalizations have been made by Carraher as a result of his work on numerous organometallic backbones [16,17]. This work

Table 10.1 Structures and Properties of Organometallic Polymers

Name	Structure	Remarks
Arylsiloxane polymers	$-[SiR_2-O-Ar-O]_n-$	Fiber, film, or coating. Stable at 270°C for 96 hr or 600°C by TGA. Soluble.
Arylazine silane polymers	$-[Si(C_6H_5)_2-O-C_{10}H_5(OH)-CH=N-N=CH-C_{10}H_5(OH)-O]_n-$	Stability <360°C. Fragile films.
Siloxane heterocyclic copolymers	$-[(bibenzimidazole)-R-SiAr_2-O-SiAr_2-R]_n-$	Rubbery films. Stable at 425°C for 12 hr. Soluble.
Arylsilane polymers	$-[C_6H_4-Si(C_6H_5)_2-C_6H_4-X]_n-$ where X is: amide, oxadiazole, or benzimidazole	Soluble. Flexible films. Adhesives. Stable to 440°C.
Poly(arylsilane siloxanes)	$-[Si(CH_3)_2-C_6H_4-Si(CH_3)_2-O-(SiRR'-O)_m-SiRR'-O]_n-$	Stable to 355-576°C. Viscous oils to stiff gums.
Polycarbosilanes	$-[Si(CH_3)]-CH_2-[Si(CH_3)]_n-$	Converted to silicon carbide at 1300°C.

Polymer	Structure	Properties
Ten-boron cage-siloxane copolymers	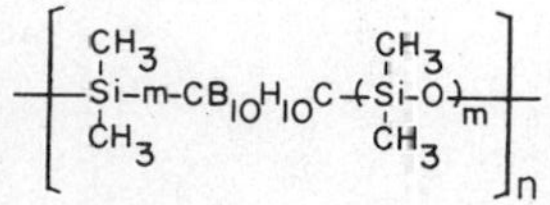 $\left[-Si(CH_3)_2-m-CB_{10}H_{10}C-(Si(CH_3)_2-O)_m- \right]_n$	Rubbery or waxy. Soluble. Tg -60 to -30°C. Stability 570-700°C in argon.
Poly(perfluorophenylene carboranes)	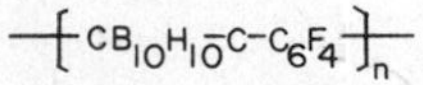 $\left[CB_{10}H_{10}C-C_6F_4 \right]_n$	Soluble. 15% weight loss at 1000°C in nitrogen.
Five-boron cage-siloxane copolymers	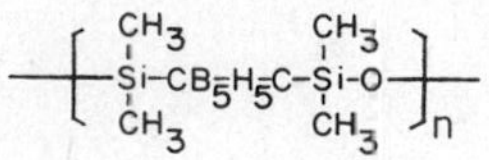 $\left[-Si(CH_3)_2-CB_5H_5C-Si(CH_3)_2-O- \right]_n$	Soluble. Stable to 700°C in argon. Tg -60°C. Tm 71°C.
Poly(ferrocenylene oxadiazoles)	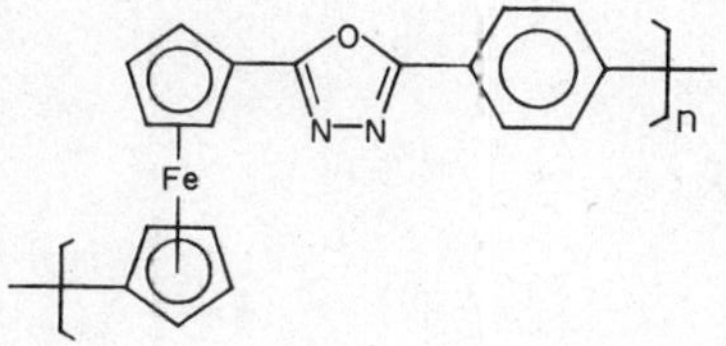 $\left[-C_5H_4-Fe-C_5H_4-(C_2N_2O)-C_6H_4- \right]_n$	Stable to 270-450°C. Soluble intermediate.
Poly(ferrocenyl siloxanes)	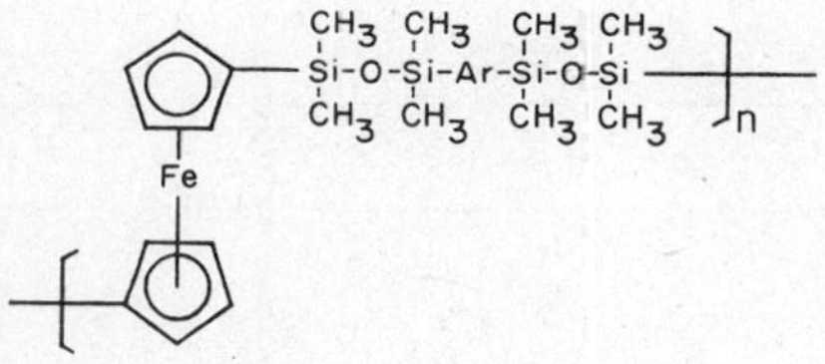 $\left[-C_5H_4-Fe-C_5H_4-Si(CH_3)_2-O-Si(CH_3)_2-Ar-Si(CH_3)_2-O-Si(CH_3)_2- \right]_n$	Tough flexible films and fibers. Soluble. Melting point 80°C. Stable to 440°C in air. Orange.

Table 10.1 (Continued)

Name	Structure	Remarks
Poly(ferrocenylpyrazoles)	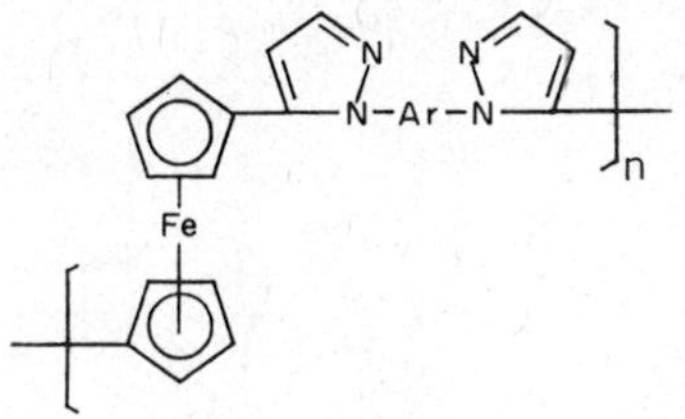	Soluble in H_2SO_4. Stable to 420-490°C. Brown-black.
Poly(ferrocenyl Schiff bases)	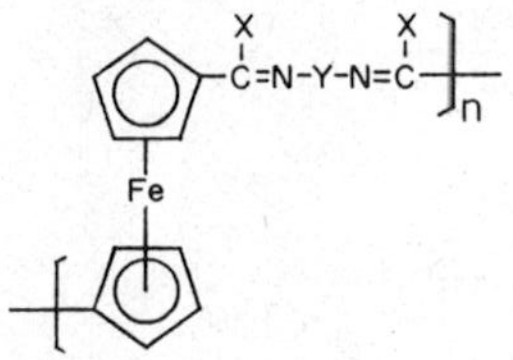	Soluble. Low molecular weights. Stable to 310-420°C.
Polyphosphazenes	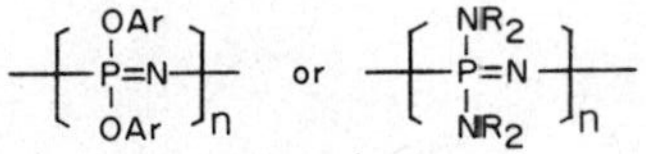 $-[P(OAr)_2=N]_n-$ or $-[P(NR_2)_2=N]_n-$	Elastic. Soluble. Tg as low as -84°C. Tm as high as 390°C. Stable to 310-410°C in argon by TGA. Isothermally stable to 200°C.
Cyclophosphazene-siloxane copolymers	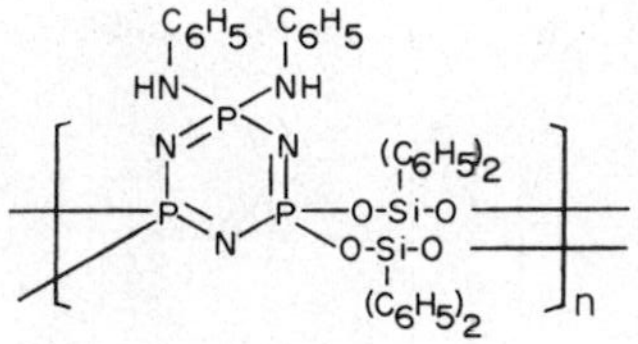	Soluble. Stable to 540°C in air. Amorphous.

Polyphosphonylureas	$\left[-P(=O)(C_6H_5)-NH-C(=O)-N(R)-R'-N(R)-C(=O)-NH- \right]_n$	Soluble in acids. Softening points 165-280°C. Stable to 240-330°C. White to yellow color.
Zinc or cobalt organophosphinate polymers	$\left[M(-O-PR_2-O-)_2 \right]_n$; M is Zn or Co	Soluble in water (R is CH_3). Hydrolysis-resistant. Color: white (Zn), blue (Co). Partially crystalline. Meltable. Oriented fibers. Stable to 480-540°C.
Beryllium organophosphinate polymers	$\left[Be(-O-PR_2-O-)_2 \right]_n$	Soluble. Stable to 410-610°C.
Chromium organophosphinate polymers	$\left[Cr(H_2O)(HO)(-O-PR_2-O-)_2 \right]_n$	Soluble. Stable to 430°C. Hydrolysis-resistant. Can form triple bridge (film former).
Iron organophosphinate polymers	$\left[Fe(CO)_2(-O-P(C_6H_5)_2-O-)_2 \right]_n$	Intractible. Insoluble.

encompassed metal-containing backbones where the metal was Ti, Zr, Hf, Sb, V, or Sn (and occasionally Co as cobaltnocene or Fe as ferrocene), all of which were bonded to organic species through oxygen and sometimes nitrogen.

Certain trends are seen which involve thermal stability, solubility, and viscosity. As can be seen throughout this chapter, thermal stabilities cannot be interpreted the same for organometallic polymers as for simple organic backbones: (1) Thermal properties (stability Tg and Tm) are generally independent of molecular weight once the backbone is above a degree of polymerization of 3, a quite low value. (2) TGA data show distinct plateaus owing to the different bond energy regions of a backbone. Further, the slope of these plateaus is an indication of the kinetic dependency of the process which is occurring, a flat slope being independent of kinetics. (3) With increasing temperature, the colors usually progress to a brown-black (oxidized aromatics) at 300-600°C and then to white (metal oxide) at 800°C. (4) The stabilities in either air or nitrogen are equivalent up to 300-500°C because to this point the degradation is nonoxidative. (5) Ferrocene-containing polymers lose ferrocene quantitatively at 200-300°C. This confirms the general disillusionment with ferrocene moieties in many high-temperature polymers. (6) Owing to the formation of metal oxides and subsequent gain of weight in the face of degradation, weight retention in air is not necessarily a valid evaluation of stability. (7) Due perhaps to the weakening of secondary, interchain bond forces with larger interatomic distances, the solubility of these systems is inversely proportional to the size of the metal atom. In general, the solubility of many of these metal-organic backbones is poor due to crystallinity, strong interchain secondary bonds, and the combination of nonpolar and polar bonds in the backbone. The last factor means that there are few solvents which can interact with all portions of the backbone.

10.1 Silicon-Containing Backbones

10.1-1 Arylsiloxane Polymers

Because of the commercial success of arylsiloxanes, which are usually methyl- or phenyl-substituted, attempts have been made to improve their behavior by changing pendant groups or incorporating other functions in the backbone.

Generally, the silicones are arrived at by allowing the hydrolysis of a dichlorodialkylsilane to the dihydroxysilane, which self-condenses with loss of water:

$$Cl-\underset{R}{\overset{R}{Si}}-Cl \xrightarrow[-2HCl]{+2H_2O} HO-\underset{R}{\overset{R}{Si}}-OH \xrightarrow{-H_2O} \left[\underset{R}{\overset{R}{Si}}-O \right]_n$$

(unstable)

R is CH_3 or C_6H_5

Accordingly, other organic protonic species will react to form a copolymer:

$$X-\underset{R}{\overset{R}{Si}}-X + HO-Ar-OH \xrightarrow{-2HCl} \left[\underset{R}{\overset{R}{Si}}-O-Ar-O \right]_n$$

X is: Cl, OR, NHR, NHAr

Ar is: m- or p-C_6H_4, n-C_6H_{12},

(naphthylene), (phenylene–Y–phenylene)

where Y is nil, O, $C(CH_3)_2$

The biphenylene polymer can be melt drawn to a fiber, forms a coating or film, and is stable to 270°C for 96 hr and at 500°C for 1 hr (by TGA 10% loss at 600°C).

> *Method* [26] Equimolar amounts of bis(anilino)diphenylsilane [8] are placed in a resin kettle equipped with a condenser and vacuum takeoff, and the mixture is heated slowly with mixing. In 30 min vacuum is applied slowly, and the conditions are made more stringent over 6 hr to reach 300-325°C and 1 torr. Products are soluble in THG, DMF, and DMSO.

Such a material has been investigated as a protective coating [80] where Ar is biphenylene [26], which had the best properties in the above series. This hard solid was brown or black, had molecular weights of 56,000 and 84,000, and was soluble in aromatic solvents.

A solution (THF) of the siloxane was tested isothermally after casting on metal sheets and curing for 1 hr at 300°C (572°F). Adhesion was lost after 1 hr at 425°C (800°F), and some discoloration resulted. Weight measurements showed a 20% loss after just 1 hr at this higher temperature, a discrepency from the above data ascribed to the difference between laboratory and pilot plant production. The addition of TiO_2 pigment decreased this weight loss to less than half of that for the nonloaded material.

10.1-2 Arylazine Siloxane Polymers

Similar to the above process, an azine function can be incorporated simply by using it as a bridge between two hydroxyarenes. Molecular weights as high as 38,000 and conversions to 85% are possible by careful control of the melt condensation conditions [33].

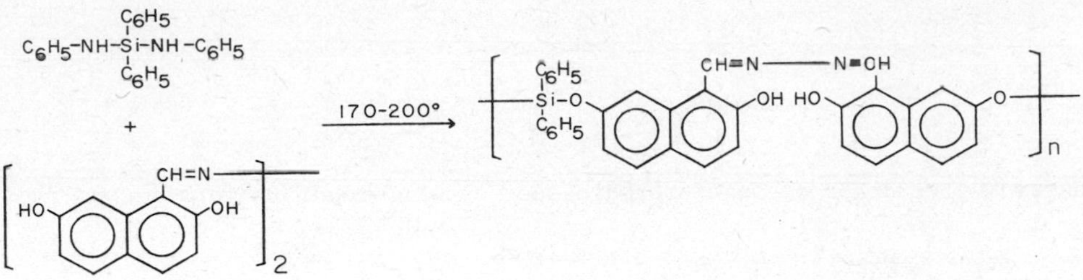

Numerous other azines were used which included phenylenes, methoxy, and acetoxy groups. Strangely, no crosslinking was experienced via the 2,2'-hydroxy groups in the above formula when the reaction was run below 200°C.

Fragile films can be cast from solutions of THF or chloroform. Thermal stability data by TGA are not available since decomposition of the azine function occurs at 360°C.

Because of the limited stability and poor physical properties, the rather complex synthetic method [33] will not be presented here.

10.1-3 Siloxane Heterocyclic Copolymers

With the success of heterocyclic functions--particularly benzimidazoles--in thermally stable systems, the desirability of coupling of these groups to silanes was obvious. The association proved to be an advantage to the heterocyclic systems in terms of greater solubility (in THF, DMSO, DMF, formic and sulfuric acids) and lower softening points. However, the thermal stability did suffer somewhat [48,49].

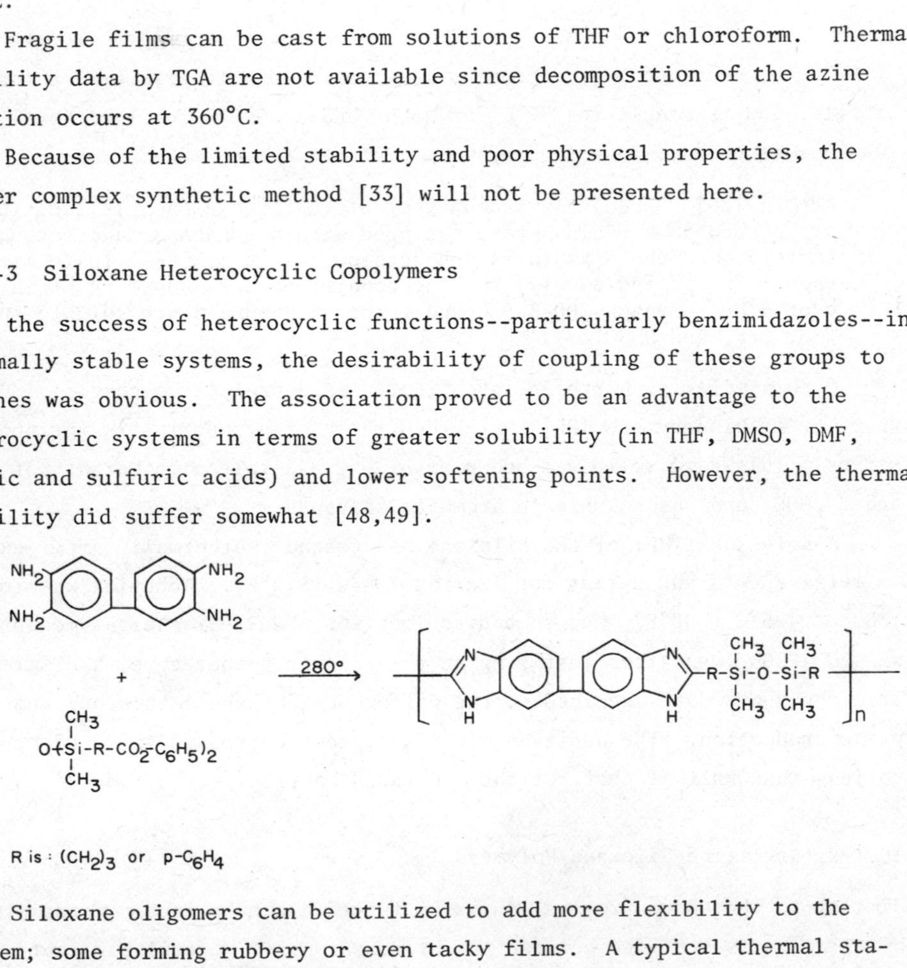

R is: $(CH_2)_3$ or $p\text{-}C_6H_4$

Siloxane oligomers can be utilized to add more flexibility to the system; some forming rubbery or even tacky films. A typical thermal stability is 8.3% weight loss after 12 hr at 425°C [49].

Method [49] Equimolar amounts of monomers, 0.162 g of 3,3'-diaminobenzidine, and 1.899 g of bis(phenylcarboxy-n-propyl)dimethylpolysiloxane are stirred under nitrogen and heated to 120°C for 3 hr and then at 300°C for 20 min. Alcohol is added to the cool mixture to yield 1.5 g of rubbery polymer which is soluble in THF.

10.1-4 Arylsilane Polymers

Poly(amide silanes) are possible from solution or interfacial condensation methods using aromatic diamines with the dicarboxysilane derivatives. Note the slight structural difference here in that no oxygen adjoins silicon. Similarly, one can obtain benzimidazole, hydrazide, and oxadiazole analogs [42].

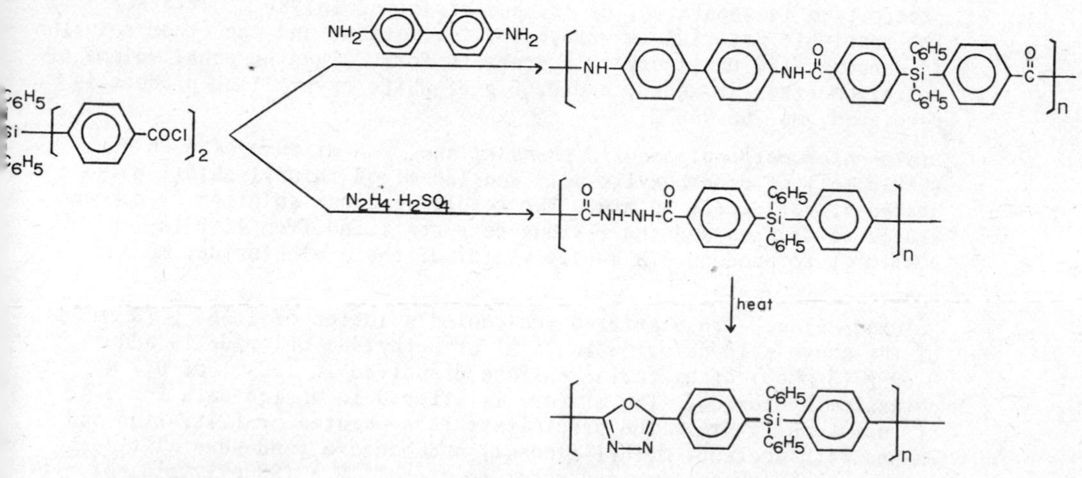
NH2
NH2
NH-
NH-C
C6H5
Si
C6H5
C
n
C6H5
Si
C6H5
COCl
2
N2H4·H2SO4
C-NH-NH-C
C6H5
Si
C6H5
n
heat
O
N—N
C6H5
Si
C6H5
n

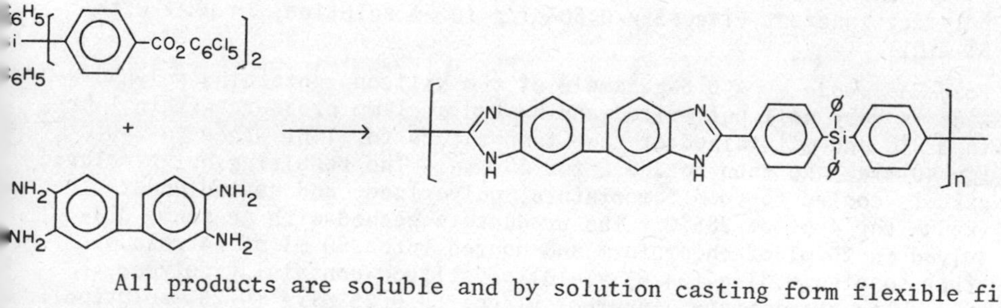
C6H5
Si
C6H5
CO2C6Cl5
2
+
NH2
NH2
NH2
NH2
N
N
H
N
N
H
ø
Si
ø
n

All products are soluble and by solution casting form flexible films which demonstrate good adhesion to glass and metal.

Regarding thermal stability, flexibility and adhesion are maintained even after 100 hr at 300°C plus 3.5 hr at 400°C in air. TGA's of these systems show a 10% weight loss at 410-440°C with the hydrazide being lowest. Considering the variables in the method and structures, however, these data are essentially identical for all these phenylsilane polymers [28].

Given below are experimental details for syntheses of the polyoxadiazole and its monomer and precursors. The polyamide is synthesized in a standard interfacial fashion in methylene chloride and water with Na_2CO_3 at 0°C.

> *Method* [42] *Monomer preparation* *Bis(p-carboxyphenyl)diphenylsilane* To a well-stirred suspension of 10.92 g (0.03 mol) of diphenyldi(p-tolyl)silane [45] in a mixture of 450 ml of glacial acetic acid, 150 ml of acetic anhydride, and 18 ml of conc. sulfuric acid is added 120 g (1.2 mol) of chromic acid over 55 min while maintaining the temperature at 15°C. The reaction mixture is stirred for an additional 10 min, poured onto ice, and stirred vigorously for about 30 min. After filtering, the residue is washed thoroughly with water and air-dried to produce 12.3 g (16% yield) of colorless, crude product. The crude material is dissolved in acetone, treated with activated charcoal, filtered, and reprecipitated with water. The precipitate is separated, dried, and dissolved in 250 ml of ether. The insoluble material is removed by filtration, and the ether solution is concentrated until crystals start to form. Then an equal volume of petroleum ether is added, and 9.65 g of white crystalline product is recovered, mp 266-268°C.
>
> *Bis(p-chlorocarbonylphenyl)diphenylsilane* A mixture of 6.36 g (0.015 mol) of dicarboxylic acid and 120 ml of thionyl chloride is heated at reflux for 40 min. The resulting clear solution is concentrated to dryness and the residue recrystallized from ligroin (bp 90-120°C) to produce 6.3 g (91% yield) of the acylchloride, mp 183-185°C.
>
> *Polyhydrazide* To a stirred ice-cooled solution of 1.384 g (3 mmol) of the above acid chloride in 75 ml of methylene chloride is added 0.39 g (3 mmol) of hydrazine sulfate dissolved in 120 ml of 0.1 N potassium hydroxide. The mixture is stirred in an ice bath for 1 hr at room temperature. The precipitate is separated by filtration and washed with acetone, methyl alcohol, and benzene, and several times in sequence with acetone and water to give 1.18 g (93.6% yield) of polymer; inherent viscosity 0.56 dl/g (0.5% solution, in DMAc with 5% LiCl).
>
> *Polyoxadiazole* A 0.5-g sample of the silicon-containing polyhydrazide is very well pulverized and heated at 1-mm pressure within 1 hr to 170°C and maintained at that temperature for 1 hr, then to 240°C for 40 min, and then to 285°C for 35 min. The resulting beige-colored cake is cooled to room temperature, pulverized, and heated under vacuum for 4 hr at 285°C. The product is washed with acetone, dissolved in 25 ml of chloroform and poured into 250 ml of petroleum ether to give 0.31 g (64.6% yield) of silicon-containing polymer, as a white precipitate; inherent viscosity 9.23 dl/g (0.25% solution).

10.1-5 Poly(arylsilane siloxanes)

An arylsilane siloxane copolymer, termed oxysilane, has been produced by the solution condensation of a bis(hydroxysilyl)benzene with bis(dimethylamino)silanes [63]. This then is an alternate method to the usual dichlorosilane (disilanol) condensations and produces backbone functions where silicon is bonded only to phenylene.

$$(CH_3)_2N{-}Si(CH_3)_2{-}C_6H_4{-}Si(CH_3)_2{-}N(CH_3)_2 \; + \; HO{-}Si(CH_3)_2{-}C_6H_4{-}Si(CH_3)_2{-}OH \longrightarrow \left[Si(CH_3)_2{-}C_6H_4{-}Si(CH_3)_2{-}O \right]_n$$

or

$$HO{-}Si(CH_3)_2{-}C_6H_4{-}Si(CH_3)_2{-}OH \; + \; (CH_3)_2N\left[SiRR'{-}O \right]_m SiRR'{-}N(CH_3)_2 \xrightarrow[100^\circ]{toluene} \left[Si(CH_3)_2{-}C_6H_4{-}Si(CH_3)(CH){-}O\left[SiRR'{-}O \right]_m SiRR'{-}O \right]_n$$

R or R' is CH_3 or C_6H_5
m is 0, 1, or 2

The ratio of silane to siloxane units can be regulated by the diamine used. For example, incorporating an oliogomeric, amine-terminated disiloxane (m = 1 or 2) increased the siloxane to silane ratio in the backbone. Comparative studies showed that an increase in the relative number of dimethylsiloxane units lowered the Tg and the thermal stability (by approximately 30-40°C per siloxane in both cases). A plateau was reached, however, in the lowering of Tg. As dimethylsiloxanes increased to 10, the Tg approached -125°C, the value for commercial poly(dimethylsiloxane). The known increase in pendant or backbone silylphenylene groups was reaffirmed. The addition of each phenyl group (as R or R' or both) caused the Tg to rise by 25-30°C above the comparable methyl-substituted analog. Also, it was not surprising that the use of ferrocenylene backbone functions in place of phenylene lowered stability by 35°C while it increased Tg by 41°C.

Thermal stabilities of these materials by TGA (10% loss) range from 355° to 475°C in nitrogen. The aforementioned inverse relation between stability and dimethylsiloxane content of the backbone is explained by the principle degradation made in nitrogen: Si-O bond scission resulting in cyclic trimers and tetramers. In air, pendant methyl groups allow formation of free radicals.

The products ranged from colorless, viscous oils to stiff gums. They are soluble in aromatic solvents and THF. Number average molecular weights were as high as 3×10^5, and highest inherent viscosities were about 1.2 dl/g in THF.

Method [63] *Monomer preparation* *Bis(dimethylamino)dimethylsilane* Freshly distilled dimethyldichlorosilane (129.0 g, 1.0 mol) is placed in a dry, nitrogen-purged, equilibrating funnel with anhydrous ethyl ether (100 ml). A 1-liter, three-necked flask is dried, purged with nitrogen, and fitted with a mechanical stirrer, Dry Ice condenser, and mercury overpressure value. Anhydrous ether (200 ml) is added to the flask, and cooled to below 0°C. Anhydrous dimethylamine (200.0 g, 4.44 mol) is added to the cooled solution under a nitrogen purge. The chlorosilane is added dropwise while the reaction temperature is maintained close to 0°C. The mixture is heated at reflux for 1 hr following the addition, and the hydrochloride salts are quickly filtered off to minimize exposure of the filtrate to moisture. The salts are washed quickly with fresh ether (200 ml), and the combined filtrates are concentrated to remove the ether. The residue is fractionated on a 35-cm vacuum-jacketed, silvered Vigreux column. Bis(dimethylamino)dimethylsilane is collected in 75% yield at 118-119°C/760 mm and stored by sealing in a dried, nitrogen-purged bottle equipped with a rubber septum for syringe withdrawal.

Poly[1,4-bis(oxydimethylsilyl)phenylenedimethylsilane] A three-necked, 300-ml round-bottomed flask is thoroughly dried, purged with nitrogen, and fitted with a magnetic stirrer, condenser, and nitrogen gas inlet tube. The neck of the flask is fitted with a rubber septum and a slight positive pressure of nitrogen is maintained. Bis(dimethylamino)dimethylsilane (12.754 g, 0.090 mol) is withdrawn into a dried, preweighed syringe, the new weight is determined, and the needle tip immediately sealed by sticking into a soft rubber plug. A stoichiometric quantity of 1,4-bis(hydroxydimethylsilyl)benzene (cf. also W. R. Sorenson and T. W. Campbell, *Preparative Methods of Polymer Chemistry,* John Wiley and Sons, New York, 1961, p. 32), (20.381 g, 0.090 mol) is transferred into the polymerization flask along with toluene (30 ml, distilled from Na). Additional bis(dimethylamino)dimethylsilane equivalent to 5 mol % (0.6377 g, 0.0045 mol) of the stoichiometric quantity is weighed into a second dry syringe. Dry toluene is pulled into the syringe to form a solution which could be accurately dispensed incrementally, and this syringe is similarly plugged. The polymerization mixture is stirred, and heating is initiated to bring the solution to mild reflux (100-105°C). At about 50°C there is a perceptible exotherm as dimethylamine evolution becomes significant. Heating is continued for 1 hr, followed by addition of an incremental amount of aminosilane monomer equivalent to 0.5 mol % of the original stoichiometric quantity. The reaction is continued for 0.5 hr, and a small aliquot is removed for inherent viscosity determination. An additional 0.5 mol % aminosilane is added to the polymerizing mixture. The sequence of monomer addition, 0.5 hr polymerization period, and inherent viscosity check is continued until the polymer achieves a desired molecular weight. The resulting polymer solution is slowly added to vigorously stirred methanol (2000 ml). The resulting off-white, highly viscous oil is precipitated a second time from methanol and evacuated at 100°C/1 mm for 24 hr to yield 78%; [η], 128.5 ml/g; $\overline{M}_w$, 255,000; $\overline{M}_n$, 199,000 (GPC). The IR spectrum (NaCl plates) shows bands at 3060, 2960, 2890, 1405, 1360, 1260 ($Si\text{-}CH_3$), 1220, 1140 ($Si\text{-}C_6H_4$), 1040-1080 (Si-O-Si), 1020, 925, 805, and 710 cm^{-1}.

10.1-6 Polycarbosilanes

The ultimate in the simplification of the silicon-containing backbone has been accomplished with the preparation of a crosslinked silicon-carbon repetition termed carbosilane. It is synthesized from dimethyldichlorosilane via the poly(dimethylsilane) [94]. Such a polymer is said to occur via the homolytic cleavage of the Si-Si bond. The Si free radical abstracts a hydrogen atom to give a $-CH_2$ which subsequently couples with another Si radical [75,76,85]. There must be considerable chain cleavage because the final product is said to have a molecular weight near 2000.

$$Cl-Si(CH_3)_2-Cl \xrightarrow{Na} -[Si(CH_3)_2]_n- \xrightarrow[30\ atm]{500°} -[Si(CH_3)-CH_2-Si(CH_3)]_n-$$

The polycarbosilane is converted to granular silicon carbide (SiC) at 1300°C under vacuum or inert gas. Because of this ultrahigh temperature behavior, the polycarbosilane is used as the matrix material in composites of SiC or Si_3N_4 (silicon nitride) powders. The polymer and filler are mixed, shaped, and heated to produce a low-density, high compressive strength, porous, inorganic composite. It is of interest that high-pressure techniques are not necessary. Such compositions retain strength and modulus to 1250°C.

10.2 Polycarboranes

10.2-1 Ten-Boron Cage-Siloxane Copolymers

The carborane nucleus is another peculiar structure to the organic polymer chemist. However, in 1966 both Olin Mathieson and Princeton Laboratories published reports on polymers which contained the carborane and siloxane moieties. The first commercial endeavor in these materials was by Olin under the product name Dexsil. That effort was terminated, however; and the polymers now can be purchased in experimental quantities for approximately $8800 per kilogram ($4000 per pound) under the name Ucarsil from Union Carbide. The value of n for the siloxane block is from 2 to 4, and occasionally a phenyl substituent is found on silicon in some products. [As n increases, thermal stability decreases [41].] The materials are known as SiB polymers,

$$\left[-\overset{\displaystyle CH_3}{\underset{\displaystyle CH_3}{Si}}-m-CB_{10}H_{10}C\left(\overset{\displaystyle CH_3}{\underset{\displaystyle CH_3}{Si}}-O\right)_m \right]_n$$

or more specifically SiB-2 or SiB-4 where the number is the value of m in the above structure. Actually, the polymers which are synthesized for use are end-capped with vinyl carboranes. This vinyl function provides a site for crosslinking with a peroxide, the final product being an elastomer [83, 84,90]. The final commercial product is filled with talc or mined silica (25-50 phr) to give values of tensile strength of 2.5-3.1 MPa (330-450 psi), elongation 100-200%, and Shore A hardness of 60-85. At 260°C (500°F) the tensile strength drops to around 0.69 MPa (100 psi) and elongation to 40-100%. For low temperature (-60°C or -80°F) applications, the SiB-4 system is better.

This type of polymer is also referred to occasionally as a 10-SiB type. The designation of $-m-CB_{10}H_{10}C-$ in the above formula is a molecular formula for the meta-carboranylene (decaborane) functional group which is an icosahedron (20 triangular sides) in which two carbons are nonadjacent by one position. The ortho isomer has the carbons on adjacent vertices.

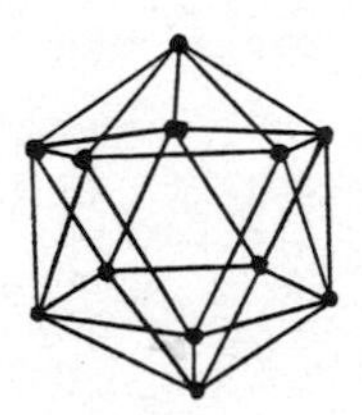

The most recent carboranes are those which contain a 5- or even an 8-boron cage which can also be included in polymers (vide infra).

The polymer can be a rubbery substance generally soluble in NMP if not crosslinked, but unaffected by DMAc, DMSO, and DMF [80]. The low molecular weight waxes (16,000-20,000) or viscous liquids (20,000-30,000) are soluble in common organic solvents [41]. Glass transition temperatures are from -60° to -30°C for commercial materials. Properties of filled and unfilled commercial products have been reported [46,84].

In one test as a film a polysiloxane carborane gave poor results as a coating until it was heated to 425-540°C (800-1000°F), at which point it formed a glassy residue. It showed only a 1.4% weight loss after 25 hr at 540°C, a figure somewhat fallacious since the oxidation of the carborane causes a gain in weight beginning at 275°C [80]. This means that thermal stability tests in an inert gas are probably more meaningful for this type of material than the data taken in air.

Some TGA data under argon show 10% weight losses at 570° to above 700°C, depending on phenyl content (directly proportional) or siloxane content (inversely proportional) [41,69]. This inverse relationship of siloxane content to thermal stability is demonstrated nicely by comparative TGA curves in argon showing 10% weight losses at 390°C for polydimethylsiloxane, but at 550-560°C for the carborane copolymers. Further the partially phenylated versions begin decomposition only at 580-630°C, while the comparable phenylmethylsiloxane goes at 465°C.

An interesting comparative isothermal aging study has been completed with a polycarborane (D-202, Olin-Mathieson), a polyamide (BK692, duPont), a polyimide (RC #5081, duPont), and a polysulfone (360, 3M). These materials were all pretreated from 150°C (300°F) to 315°C (600°F) and aged at 288°C (550°F). The carborane polymer reached a constant weight loss of 1.4% by 200 hr and held there to 300 hr. The others showed an increasing loss of 2.6-3.6% at 300 hr [29].

Two methods have been successful in synthesizing the polycarborane siloxanes: a ferric chloride condensation of methoxysilane with chlorosilane [58] or a water-catalyzed self-condensation of chlorosilanes [41], the former of which can induce crosslinking to insoluble gums.

$$CH_3O\text{-}Si(CH_3)_2\text{-}CB_{10}H_{10}C\text{-}Si(CH_3)_2\text{-}OCH_3 + Cl\text{-}Si(CH_3)_2\text{-}Cl \xrightarrow[\text{heat}]{FeCl_3} \left[Si(CH_3)_2\text{-}CB_{10}H_{10}C\text{-}Si(CH_3)_2\text{-}O\text{-}Si(CH_3)_2\text{-}O \right]_n$$

$$CB_{10}H_{10}C(Si(CH_3)_2\text{-}O\text{-}Si(CH_3)_2\text{-}Cl)_2 \xrightarrow[\text{ether}]{H_2O} \underset{\text{(unstable)}}{CB_{10}H_{10}C(Si(CH_3)_2\text{-}O\text{-}Si(CH_3)_2\text{-}OH)_2} \xrightarrow{-H_2O} \left[Si(CH_3)_2\text{-}O\text{-}Si(CH_3)_2\text{-}CB_{10}H_{10}C\text{-}Si(CH_3)_2\text{-}O\text{-}Si(CH_3)_2\text{-}O \right]_n$$

In the second process the secret to forming siloxane bonds is having more than one silicon atom between the carborane and the hydrolyzable function. The minimum allowable formula (two silicons) is shown. It is possible to isolate the silanol intermediate if one treats the chlorosilane with wet ether and $NaHCO_3$; then it can be polymerized with acid catalysis.

Method [41] Under nitrogen in a reactor fitted with stirrer, dropping funnel, and thermometer, a mixture of 2100 ml of diethylether and 502.6 g of bis(chlorodimethylsiloxydimethylsilyl)-m-carborane, $CB_{10}H_{10}C[Si(CH_3)_2OSi(CH_3)_2Cl]_2$, is cooled in an ice bath. To this is added 26.527 g of water in 400 ml of THF over 30 min and the mixture is stirred 30 min longer. The flask is then warmed to 25°C and stirred for 2.5 hr and the solvents stripped at 0.5 torr at 30-35°C for 1 day and 25°C for 3 days. The wax (95% yield) melts at 30-40°C and has a number average molecular weight of 10,000.

10.2-2 Perfluorophenylene Carborane Copolymers

One of the newest modifications to the carborane backbone is the incorporation of a perfluorophenylene group (1:1 with carboranylene). This is accomplished by heating the dilithiocarborane (o or m) with hexafluorobenzene in xylene [9].

$$HCB_{10}H_{10}CH \xrightarrow{BuLi} LiCB_{10}H_{10}CLi \xrightarrow[\text{heat, xylene}]{C_6F_6} \left[CB_{10}H_{10}CC_6F_4 \right]_n$$

The purpose was to overcome the thermal instability of the xylylene analog [24,27] and in so doing have a system with no hydrogen as an oxidation point. (A similar goal was achieved as discussed earlier with pyrazine-containing polyimides.)

Conversions were 75% for the ortho and 93% for meta carboranes. Both ether-soluble (mp 170-240°C) and insoluble (mp >300°C) fractions were obtained, the insoluble portion being about 16% for the ortho but 32% for the meta carboranes.

TGA showed that, indeed, the fluorophenylene moiety increased thermal stability. The ortho isomer weight vs temperature curve is nearly linear to a 15% weight loss at 1000°C in nitrogen, passing the 10% point at 450-500°C. The 10% loss point is not really significant here, however, since there is never any break or definite point of decomposition. The meta isomer is poorer, but still showing no particular decomposition temperature. A small break occurs at about 400°C and again at about 600°C but with a total weight loss of only 35% at 850°C.

Method [9] Equimolar amounts of hexafluorobenzene and dilithiocarborane [36] are heated at reflux in xylene for 10 hr under nitrogen. Water is then added to the mixture to precipitate the polymer, which is isolated, dried, and extracted with diethylether to obtain ether-soluble and ether-insoluble fractions.

10.2-3 Five-Boron Cage-Siloxane Copolymers

Other carborane cages are possible with five borons which have also been incorporated into polymer backbones [30,39,40,69] or with eight borons which can be utilized in backbones or as pendant side chains [39]. The difunctional five-boron two-carbon carborane, closo-carboranylene, molecular formula CB_5H_5C, has a pentagonal bipyramidal geometry with the carbons separated by one boron and with a hydrogen on each boron.

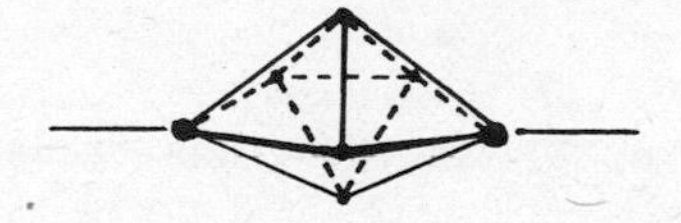

The polymers are actually copolymers with the siloxane groups and are synthesized in the same fashion as the decarborane materials, i.e., by a ferric chloride catalyzed condensation of a methoxysilylcarborane with a chlorosilyl derivative [39,40] under melt polymerization conditions.

$$\begin{array}{c} Cl-\underset{CH_3}{\overset{CH_3}{Si}}-CB_5H_5-\underset{CH_3}{\overset{CH_3}{Si}}-Cl \\ + \\ CH_3O-\underset{CH_3}{\overset{CH_3}{Si}}-CB_5H_5C-\underset{CH_3}{\overset{CH_3}{Si}}-OCH_3 \end{array} \xrightarrow[\substack{185^\circ \\ 3hr}]{FeCl_3} \left[\underset{CH_3}{\overset{CH_3}{Si}}-CB_5H_5C-\underset{CH_3}{\overset{CH_3}{Si}}-O \right]_n$$

The polymer is referred to as a 5-SiB-1 type where the 1 indicates the number of siloxane oxygens. Of course, copolymers are easily prepared by mixing the above pictured monomers with the 10-SiB variety.

The 5-SiB polymers are received in high yield (>95%) and in moderate molecular weight (12,500 number average) [40] and are soluble in aromatic solvents except after heating above 350°C.

Thermal stabilities of 5-SiB and copolymers with 20% of 10-SiB are very similar showing a familiar pattern for the carboranes. The weight loss in argon by TGA is gradual to 5% at 480°C and to 8% at 700°C where it levels out. Homopolymers of 5-SiB are slightly more stable than the copolymers.

By DTA the 5-SiB-1 homopolymer shows a secondary transition at -140°C, Tg at -60°C, and Tm at +71°C. The copolymer, which is largely or wholly amorphous, also shows the -140°C secondary transition and has a Tg of -60°C and a weak Tm peak at 51 or 57°C (for 20 or 5% $C_2B_{10}H_{10}$, respectively) [40, 69]. The chain-stiffening effect of the 10-boron cage on the siloxane backbone is greater than that of the five-boron system as is evident from Tg and torsional braid analysis [70]. This rigidity is to be expected in view of the larger volume and subsequent increased steric hindrance of the

$-CB_{10}H_{10}C-$ cage. The minimal chain stiffening of the boron cages indicates that the carborane-silicon bond is quite flexible. The Tg can be accurately calculated from the Tg/composition copolymer equation:

$$\frac{1}{Tg} = \frac{W_1}{Tg_1} + \frac{W_2}{Tg_2}$$

where W_1 or W_2 is the weight fraction and Tg_1 or Tg_2 is the glass transition of the homopolymeric species.

Further, mixtures of the copoly(carborane siloxanes) with silicone polymers or with copoly(carborane·siloxanes) of different composition were shown to exist in separate phases as incompatible materials by the observation of thermomechanical transitions of only individual components in the mixture [71].

Thermomechanical testing reveals that these polymers are linear to 400°C, rubber to 500°C, and thermoset resin to 625°C. Owing to its improved mechanical properties in air, its virtual lack of crystallinity, and its equivalent thermal stability, the copolymer is a better elastomer than the CB_5H_5C homopolymer.

> *Method 20% Copolymer* [69] The monomers are mixed in the ratio of 1.00 mol of $C_2B_5H_5[Si(CH_3)_2Cl]_2$, 0.80 mol of $C_2B_5H_5[Si(CH_3)_2OCH_3]_2$, and 0.40 mol of $C_2B_{10}H_{10}[Si(CH_3)_2OCH_3]_2$ with 2 mol % of anhydrous $FeCl_3$ under vacuum and heated to 185°C. After 3 hr an additional 2 mol % of $FeCl_3$ is added, and the reaction is continued until methyl chloride is no longer evolved. The polymeric mass is dissolved in hot xylene, filtered, and precipitated into methanol. Purification is accomplished by washing the solid with aq. acetone, dissolving in xylene, and reprecipitating into methanol to yield 98% of a hard, light brown wax, $\overline{M}n$ 12,500. This process can also be accomplished by using the hydroxysilane monomers in a manner discussed in Section 5.2.1.

10.3 Ferrocene-Containing Polymers

Ferrocene is too convenient an organometallic function not to have some commercial use. It is a fairly normal aromatic compound, is readily synthesized and reacted, and has a unique, reversible oxidazability to

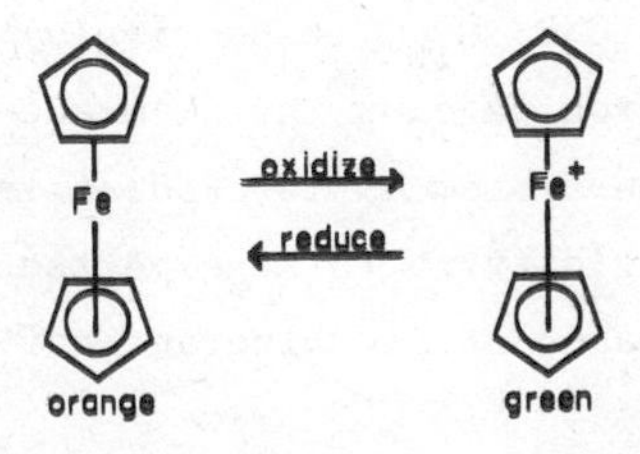

ferrocinium ion, as well as an interesting sandwich structure and moderate thermal stability.

As a result of its structure, considerable effort has been expended in polymer synthesis since the 1950s in the attempt to bring the ferrocene moiety to some application, the effort being well reviewed by Neuse [54] and Pittmann [61] and abstracted by Almond [7], the latter including patents. Among the areas in which it has been found to have useful properties are antiknock additives for fuel (if one can dispose of an annoying deposit of iron oxide), extraterrestial shield due to its ultraviolet radiation resistance, pigments, electron transfer resins, stabilizers [54], catalysts, vulcanization aids, and weathering improvement additive [79]. Ferrocene polymers are generally poor semiconductors unless impurities are present; neither do they protect against corrosion, as one might expect. An early review of the potential uses of ferrocene polymers was made by Pittmann [60].

The ferrocenyl analog of styrene has a melting point of 280-285°C, which probably was responsible for encouraging its use in high-temperature polymers. In fact, when ferrocene is incorporated into a benzimidazole backbone, it permits only the same stability as the aliphatic unit, as do, by the way, even perfluoroaliphatic moieties [47,64]. So, in thermal stability it is considerably below its benzene analog.

Polyferrocenylene of molecular weight 3500-7000 has much better solubility than polyphenylene but poorer thermal stability: 2% loss by TGA at 400°C, 15% at 600°C; mp 250-300°C [56].

The silane ferrocene polymer (molecular weight to 8700) has also been reported in the patent literature in 1965 [54]. Although it possessed thermal stability to 400°C and was touted as a high-temperature hydraulic fluid, lubricant, and heat transfer fluid, there is still no commercial evidence of its use or acceptance.

In ferrocene-containing polymers, the beneficial results have evidently not justified the cost and effort required to use these organometallics, at least to date. Frequently the ferrocenyl nucleus is seen to lower thermal stability of otherwise acceptable polymers.

Other metallocenes have been given cursory investigation: ruthenocene [44], titanocene, zirconocene [15], rhodocene, and cobalticene [57]. Of these, only the ruthenocene systems show a stability superior to ferrocene. Titanocenes have been the subject of recent research and are discussed in the last section of this chapter.

10.3-1 Oxadiazole Copolymers

Oxadiazole functions are incorporated with ferrocenes in the usual manner of cyclization of polyhydrazides or of reaction of acid chlorides with aromatic bistetrazoles [44]. Production of the oxadiazole is difficult due to the problem of chain decomposition proceeding simultaneously with ring formation [52]. Another factor in this problem is the irregular structure (bridging, branching, etc.) of the hydrazide prepolymer. However, even the hydrazide polymers gave thermal stabilities to 300°C by TGA.

Inherent viscosities for the hydrazide systems are low (0.08-0.10 dl/g) compared to the benzene analogs, again indicating an irregular polymerization with intramolecular bridging.

$Fe(COCl)_2$ + N_2H_4 or $Fe(CONHNH_2)_2$ ⟶ $-[Fe-\overset{O}{\overset{\|}{C}}-NHNH-\overset{O}{\overset{\|}{C}}]_n-$ —heat→

$Fe(COCl)_2$ + (bistetrazole) —pyridine→

Fe is 1,1'-Ferrocenylene nucleus

Thermal decomposition of these systems was found to occur over the range 270-450°C with the evolution of aromatic cyano compounds. This indicates a cleavage of the oxadiazole function at the C-O and N-N bonds as the thermal weakness in this system.

Method *Polyhydrazide* [52] In a 500-ml flask equipped with stirrer and dropping funnel is placed a solution of 2.78 g (0.087 mol) of anhydrous hydrazine in 200 g of dry HMPA. The solution is cooled to 3°C, and to it is added dropwise over 1.5 hr a cold solution of 27.0 g (0.087 mol) of 1,1'-di(chlorocarbonyl)ferrocene in 96 g of dry HMPA. The mixture is stirred vigorously at 305°C for 14 hr, and the crude polymer is precipitated by pouring it into 1500 ml of water. The hard brown precipitate is isolated, washed with water, washed with hot methanol, and dried over P_2O_5 in vacuo to yield 10.5 g (73%). As a purification step, the solid is dissolved in 200 ml of concentrated H_2SO_4 containing 10 g of stannous chloride (to reduce any ferricenium ions) and rapidly reprecipitated into water. After being dried at 140°C, the solid (15.1 g) is tan-colored and partially soluble in HMPA containing 3% LiCl; inherent viscosity 0.13 dl/g in H_2SO_4.

Polyoxadiazole Partial success is realized by heating the above polyhydrazide (0.4 g) at 10^{-2} torr at 180°C for 15 hr, then 225°C for 3.5 hr, and finally at 250°C for 1.5 hr. The product is a black, infusible solid which is soluble only in concentrated H_2SO_4; inherent viscosity 0.12 dl/g.

10.3-2 Siloxane Copolymers

Again the siloxane backbone function allows facile incorporation of a flexibilizing group in an aromatic backbone. This subject has been extensively reviewed [55]. The hydrolyzable precursor to siloxane condensation is -OH or $-N(CH_3)_2$ as seen earlier [62].

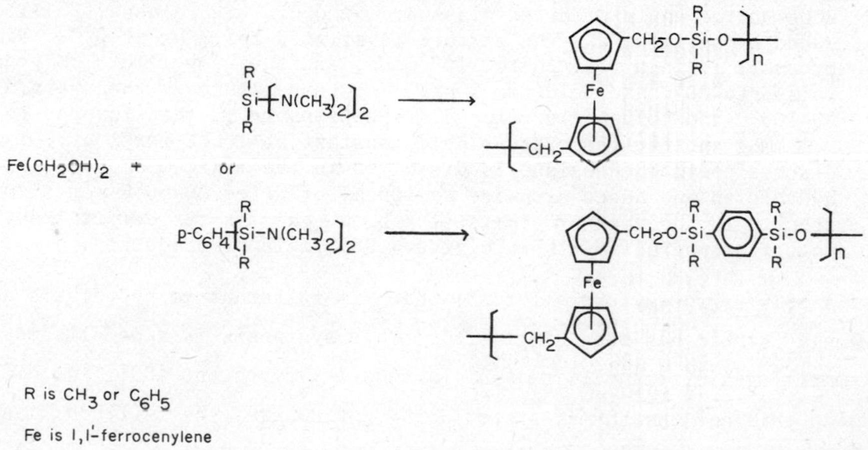

R is CH_3 or C_6H_5

Fe is 1,1'-ferrocenylene

This ferrocenylene-siloxane combination was accomplished by Pittmann and his co-workers, who have done a large amount of work on ferrocenes, by a low-temperature condensation in toluene to avoid cyclization [62]. The products are low-melting, amorphous, waxy solids which are soluble in THF, benzene, and DMF and can be cast into films or melt-drawn to fibers.

Yields from the reaction are generally only moderate (42-62%) with one exception, 93% for the phenylene siloxane copolymer; but intrinsic viscosities in THF are very high (3.2-7.6 dl/g), with the phenylene system again outstanding at 22.0 dl/g. However, number average molecular weights are only as high as 10,000. The polymers all have low melting points of 50-65°C.

The most negative data on these materials are those regarding hydrolytic stability. Owing to the C-O-Si bond, this inherent weakness is unavoidable, and these backbones are readily degraded by nucleophilic attack of water on the silicone. The phenylene isomer is better than the simple siloxane, which hydrolyzes completely in 10% aq. THF in 1 hr at reflux.

Method [62] *Silane monomer* The 1,4-bis(chlorodimethylsilyl)benzene [65] precursor (1 mol) is purified by fractional distillation (bp 108-110°C at 1.5 torr) and converted to 1,4-bis(dimethylaminodimethylsilyl) benzene by treatment with a 5-mol excess of dimethylamine, in 300 ml of ether at 0 to -5°C under nitrogen (all in a dry system with dry reagents). The mixture is then warmed and stirred at reflux for 1 hr. The hydrochloride salts are then isolated by filtration under nitrogen, and the salts are washed twice with excess ether. Ether is removed from the combined filtrates, and the residue is carefully fractionated by vacuum distillation, bp 73-74°C at 0.06 torr or 100-101°C at 0.5 torr to yield 40-60%.

Polymer The above aminosilane (2.7827 g, 0.0099 mol) and 1,1'-bis(hydroxymethyl)ferrocene (2.4412 g, 0.0099 mol) are added together under nitrogen in predried glassware at 0°C. Dry toluene (25 ml) is added by syringe, and the mixture is stirred for 1 hr at 0°C. The pressure is then reduced to ~100 mmHg, and the temperature is gradually raised to 30°C and held for 1 hr. The temperature is then raised to 90-100°C and toluene is removed. The pressure is then lowered to 0.1 mmHg and the temperature held constant at 100°C for 1 hr. The viscous polymeric residue is dissolved in the minimum amount of tetrahydrofuran and added dropwise to 500 ml of dried 30-60°C petroleum ether. The polymer precipitate, a brown gum, is reprecipitated twice more from petroleum ether to give a 93% yield of product.

A poly(ferrocenylene siloxane) which is resistant to hydrolysis and thermally stable has been discovered. This synthesis is accomplished by melt-polymerizing a disilanol with a bis(aminosilyl)ferrocene [59], the difference between this and the former system being, of course, that silicon is attached

$$Fe[Si(CH_3)_2-N(CH_3)_2]_2 + HO-Si(CH_3)_2-OH \xrightarrow{100^\circ} -[Fe-Si(CH_3)_2-O-Si(CH_3)_2-O-Si(CH_3)_2]_n-$$

or

$$Fe[Si(CH_3)_2-N(CH_3)_2]_2 + Ar[Si(CH_3)_2-OH]_2 \longrightarrow -[Fe-Si(CH_3)_2-O-Si(CH_3)_2-Ar-Si(CH_3)_2-O-Si(CH_3)_2]_n-$$

Fe is 1,1'-ferrocenylene

Ar is $-C_6H_4-$ or $-C_6H_4-C_6H_4-$

directly to ferrocene or other aromatic nucleus and never to an alkoxy group, thereby precluding degradation by hydrolytic attack on silicon.

The polymers are obtained in very good yield (76-95%) and intrinsic viscosities (up to 32.1 dl/g in THF corresponding to a number average molecular weight of 20,250). Melting points are all below 80°C and TGA data (10% loss at 440°C in air and 460°C in nitrogen) show them to be comparable to a poly(phenylene siloxane) (420 and 480°C in air and nitrogen, respectively).

Tough, flexible films can be cast, and fibers can be melt-spun.

Method [59] Since melt polymerization affords superior properties over the solution process in toluene, only the former will be given. 1,4-Bis(hydroxydimethylsilyl)benzene [89] (02.9892 g, 0.0132 mol) and 4.8733 g (0.0125 mol) of 1,1'-bis(dimethylamodimethylsilyl)ferrocene [38,80,92] are mixed and heated at 50-60°C at 1 torr for 30 min. The system is then flushed with nitrogen at atmospheric pressure, an additional 0.136 g (0.00035 mol) of the ferrocene monomer is added, and the reaction is continued at 50°C for 30 min. After raising the temperature to 100°C for an additional 30 min, still another like portion of the ferrocene compound is added, and the reaction is continued for 30 min more. The viscous polymer is then dissolved in 50 ml of toluene and precipitated into 2 liters of methanol to yield 94% of fibrous, orange solid.

10.3-3 Poly(ferrocenylpyrazoles)

The disappointing behavior of ferrocene-containing polymers was improved upon considerably by the inclusion of a pyrazole unit in the backbone [53]. The improvement in stability over the benzimidazole-ferrocene systems is 100-200°C. In comparison of the ferrocene to the benzene-analogous pyrazole polymer it is seen that weight loss onset for ferrocenes is 50°C lower; however, the total loss at 800°C is not so severe.

The synthesis is carried out by condensing bis-β-diketoferrocenes with aromatic dihydrazines in HMPA, NMP, or DMF at room temperature to give polyhydrazones. More vigorous conditions produced side reactions and chain irregularities, although these were accompanied by improved yield and molecular weight. Molecular weights of these polyhydrazones were low (number average 1400-2000), and inherent viscosities were 0.09-0.18 dl/g. Yields were only moderate (60-75%).

The next step was to cyclodehydrate the polyhydrazone to a polypyrazole at 180-240°C (stepwise) in a vacuum after oxygen removal. The products, although infusible, were essentially soluble in formic and sulfuric acids. It is not surprising to see better solubility when biphenylsulfone segments are in the backbone or when methyl or phenyl pendant functions are available to decrease crystallinity. Inherent viscosities increased over the polyhydrazone precursors to 0.21-0.47 dl/g in H_2SO_4, a result which is said to be due to chain extension. However, at least a part of this viscosity increase must be attributed to increased chain stiffness.

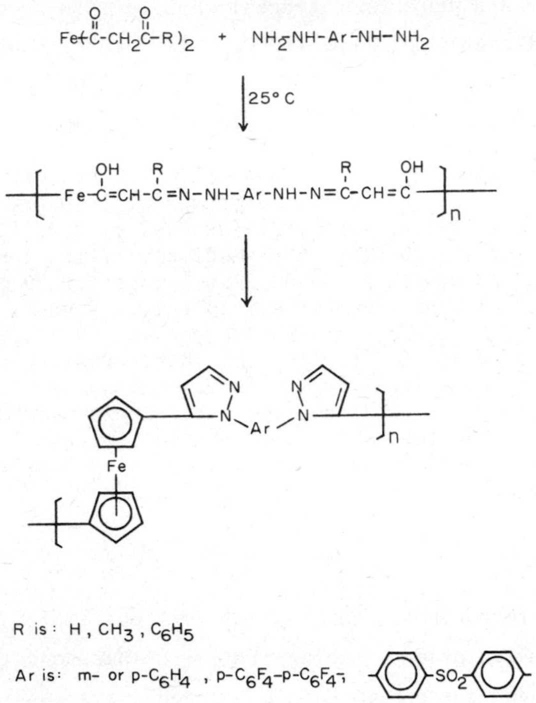

R is: H, CH_3, C_6H_5

Ar is: m- or p-C_6H_4, p-C_6F_4-p-C_6F_4-,

Thermal analyses by TGA show initial weight losses beginning from 350-400°C with 10% losses at 420-490°C, quite respectable for a ferrocene-containing backbone.

Method [53] *Polyhydrazones* In a 750-ml flask equipped with stirrer and gas blanket system are mixed 0.01 mol of the bis-β-diketone [14] and the dihydrazine [82] with 100 ml of DMF. (If dihydrazine dihydrochlorides are used, 0.02 mol of sodium methoxide is also added to neutralize the HCl.) Then 3.0 g (0.05 mol) of HOAc are added, and the solution is stirred in the dark, under nitrogen, for 3 days, and finally heated at 90°C for 1 hr. The mixture is filtered and precipitated into a well-stirred solution of 100 ml of isopropyl alcohol, 200 ml of water, and 1.0 g of ascorbic acid reducing agent. The tan to brown solids are washed with water and methanol and dried at 100°C over P_2O_5 in vacuo.

Polypyrazoles About 0.3-0.5 g of the polyhydrazone is made into a paste with 0.3 to 0.7 ml of DMF or formic acid, and the paste is subjected to solvent removal under vacuum at 50°C in a tube. The solid is then flushed well with nitrogen and heated under vacuum in the dark to 180°C over 4 hr and held there for 12 hr. Heating is then continued at 200°C for 15-30 hr and finally at 240°C for 2-8 hr to yield black-brown solids.

10.3-4 Poly(ferrocenyl Schiff Bases)

The incorporation of C=N bonds in conjugation with ferrocene in polymer backbones results in a thermally stable system, albeit not as stable as the pyrazole ferrocene polymer. A simple condensation between a diketoferrocene and a diamine gives an orange to red solid in moderate to high yield (56-90%). The polymers are soluble in chlorinated solvents and have low molecular weights (<2000) [32].

$Fe(C(=O)-X)_2$ + $H_2N-Y-NH_2$ → (Benzene/heat, Acid catalyst) → $[-C_5H_4-Fe-C_5H_4-C(X)=N-Y-N=C(X)-]_n$

X is: H or CH_3

Y is: nil, C_2H_4, m- or p-C_6H_4, p-C_6F_4 or biphenylene

Although initial weight losses by TGA begin below 300°C (200-280°C), 10% losses are realized from 310 to 420°C. The material of lowest initial stability results when Y is C_2H_4, and the highest is for p-phenylene diamine. However, the most drastic decomposition is witnessed for the polyazine (Y is nil), which drops to a 70% weight loss at 350°C. Others demonstrate only a 25-30% loss to 650°C. The fluorinated phenylene function in the backbone did not contribute to thermal stability. This result is not unexpected since oxidation of benzene is not the anticipated primary decomposition mechanism for this structure. Further, the presence of either hydrogen or methyl on the keto function (as X) did not significantly affect the thermal stability of the product.

[Cyclic structure: two ferrocene units linked by $-C(CH_3)=N-Y-N=C(CH_3)-$ bridges]

Method [32] To a suspension of aluminum oxide in 50 ml of toluene are added 1.35 g (0.005 mol) of 1,1'-diacetylferrocene and 0.54 g (0.005 mol) of p-phenylene diamine. The mixture is heated to reflux with stirring under nitrogen for 48 hr. The mixture is cooled to room temperature and filtered. The solid residue is extracted with hot chloroform, and the solution is evaporated to dryness to yield 0.25 g (15% yield) of poly-1,1'-diacetylferrocene-p-phenylene diamine.

10.4 Phosphorus-Containing Backbones

10.4-1 Polyphosphazenes

Polyphosphazenes, composed of a completely inorganic (P + N) backbone, were synthesized at Pennsylvania State University and developed as a broad-temperature range elastomer and fire-resistant plastic [67,68] by Horizons, Inc. Commercial processes and applications are being pursued by Firestone. Comprehensive reviews have appeared which cover all aspects of this area of research and development [1-3,87].

The starting material is hexachlorocyclotriphosphazene, which will undergo ring opening in the bulk by heating with or without catalysts (metals, sulfur, water, alcohols, ketones, and carboxylic acids). Catalysts tend to induce crosslinking. The broad variety of catalysts indicates the complexity of the as yet unknown mechanism.

$$(Cl_2P{=}N)_3 \xrightarrow[\text{vacuum}]{250^\circ} \left[PCl_2{=}N \right]_n \rightleftharpoons \left[PCl_2{=}N \right]_n$$

This simple, polymeric, inorganic rubber hydrolyzes readily, however, owing to the P-Cl bond, to produce phosphoric acid, hydrochloric acid, and ammonia, far from desirable byproducts in any situation. Therefore, to stabilize the backbone, one must replace chlorine with another nucleophilic group such as R-O-, ArO-, and R_2N-.

$$\left[PCl_2{=}N \right]_n \xrightarrow{RO^-Na^+} \left[P(OR)_2{=}N \right]_n$$

$$\left[PCl_2{=}N \right]_n \xrightarrow{ArO^-Na^+} \left[P(OAr)_2{=}N \right]_n$$

$$\left[PCl_2{=}N \right]_n \xrightarrow{R'_2NH} \left[P(NR'_2)_2{=}N \right]_n$$

R is:	Ar is:	R' is:
CH_3-	C_6H_5-	H, CH_3, C_2H_5-
C_2H_5-	m-or p-$CH_3C_6H_4-$	CF_3CH_2-
CF_3CH_2-	$C_2H_5-C_6H_4-$	C_6H_5-
$C_3F_7CH_2-$	$CH_3O-C_6H_4-$	$CH_3O-C_6H_4-$
$CH_2F-(CH_2)_5-CH_2-$	m-or p-ClC_6H_4-	$(C_2H_5)_2NCH_2-$
$CH_2F-CF_2-CH_2-$		C_5H_{10}(piperidinyl)
$CH_2F-C_3F_6-CH_2-$		m-or p-XC_6H_5, where X is: CH_3, C_2H_5, Cl, F, C_4H_9

Copolymers can be produced by exposing the alkoxy-substituted polymers to conditions of a metathetical reaction with another nucleophile. In this way not only can substituents be exchanged, but crosslinking can be effected by utilizing a difunctional nucleophile:

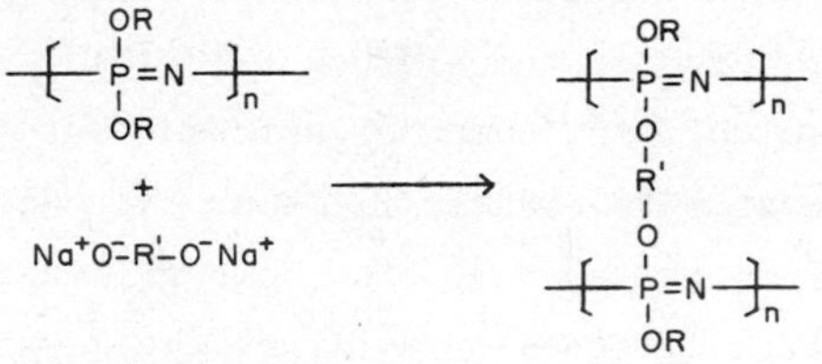

$$\left[P(OR)_2{=}N \right]_n + Na^+\,{}^-O{-}R'{-}O^-\,Na^+ \longrightarrow \left[P(OR){=}N \right]_n{-}O{-}R'{-}O{-}\left[P(OR){=}N \right]_n$$

The difluoro and dibromo polyphosphazenes are also known and have been studied with the dichloro, alkoxy, and aryloxy analogs to elucidate conformation of the backbone, flexibility, and wide temperature range of elasticity [4,6]. The difluoro polymer showed the lowest temperature range for elasticity, -95° to +270°C, with dichloro next higher, -63° to +350°C, and dibromo highest of the series, -15° to +270°C. The diiodo polymer is not known. These polymers were found to crystallize in a cis-trans planar array.

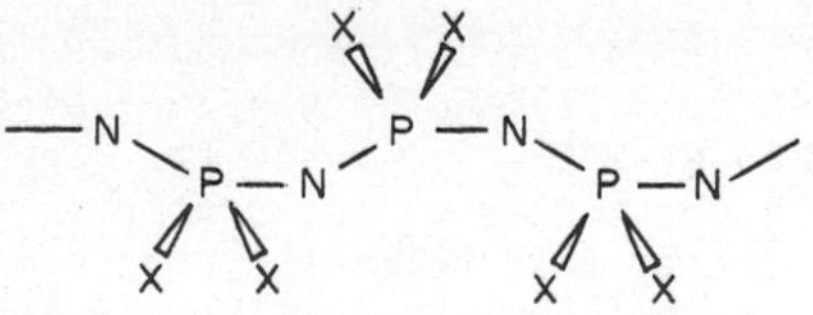

$$-N{-}PX_2{-}N{-}PX_2{-}N{-}PX_2{-}N-$$

Thermal analytic data for all the alkoxy, aryloxy, and aliphatic amino systems are interesting in two respects: very low Tg values and quite similar Tm and TGA decomposition values. Glass transitions for alkyloxy substituted backbones are -66 to -84°C and for aryloxy systems -25 to +5°C, the latter data indicating a restricted degree of rotation of the backbone. All Tg values are below room temperature except when aniline is used as a nucleophile (Tg 91°C), Tm's for aryloxyphosphazenes are from 340 to 390°C, and decomposition temperatures in air or nitrogen or argon are from 310 to 410°C. Lower thermal data result when an alkyl substituent is on the aromatic nucleus. In fact, two first order transitions have been found for the trifluoroethyl-substituted backbone, the lower one (90°C) a crystalline to mesomorphic transition, and the second (240°C) a true Tm [81].

Arylamino groups cause some incongruity in thermal data. The Tg values are high compared to other polyphosphazenes (+53 to 105°C), and thermal stabilities are low (243° to 266°C by TGA) [93].

Isothermal data may conflict with those from TGA, however [87]. Some materials show a loss of molecular weight on aging at 200°C, during which cyclic species as byproducts are formed.

Thermomechanical testing (rigidity and loss modulus by torsional braid analysis) shows two phenomena in addition to the expected losses near the Tg and Tm, one effect considerably below the Tg at 60-70°C attributed to a crystal-crystal transition and one effect at very low temperature (-180°C) [23].

Most phosphazene polymers are soluble in common organic solvents (THF or chloroform) and can be solution-cast into elastic films or extruded to fibers. The fluorinated aliphatic-substituted polymers, however, must be dissolved in chlorofluoroalkanes. With compression molding, the physical properties of the polymers suffer and become brittle or at best slightly flexible and weak. For example, elongation goes from 280% to 4% [87]. The chlorophenoxy-substituted backbones offer flame retardancy, of course, and have an oxygen index rating of 51. They have been proposed as wire coatings and foam flame retardants [34]. Other suggested uses for some cursory tests which show promise are biodegradable polymers, blood-compatible surgical implants, and carriers for biological active species [3].

Method [86] *Polymerization* Hexachlorocyclotriphosphazene (50 g) is sealed in a Pyrex tube (28 x 125 mm), flushed with argon, evacuated to less than 0.1 mm, and heated at 245°C for 15-50 hr. The polymerization is terminated by cooling the tube when the contents cease to flow. Termination of the reaction at this point prevents extensive crosslinking and hence gelation. The polymer is dissolved in 100 ml of benzene and precipitated into 400 ml of n-pentane to yield 15-50% of polydichlorophosphazene. The isolated polymer is immediately dissolved in 100-150 ml of toluene. The isolation and handling of polydichlorophosphazene is performed in a dry box to minimize hydrolysis and crosslinking.

Substitution A solution containing 250 ml of bis(2-methoxyethyl)ether, 100 ml of benzene, and 53.4 g (0.44 mol) of p-ethylphenol is prepared and dried by removal of 30 ml of benzene using a Dean-Stark trap. After the mixture is cooled to room temperature, 9.7 g (0.42 mol) of sodium is slowly added under argon and allowed to react overnight. A solution containing 22.1 g (0.19 mol) of the polyphosphazene prepared above in 100 ml of toluene is slowly added under argon over a period of 1 hr to the sodium p-ethylphenoxide solution at 90°C. The temperature of the viscous suspension is raised to 115°C by further removal of benzene, and maintained at 115°C for 26 hr. The reaction is cooled and added dropwise to several liters of methanol to precipitate the polymer, which is washed with water and methanol to remove sodium chloride and excess p-ethylphenol. The polymer is dissolved in 1 liter of tetrahydrofuran, filtered, and precipitated into several liters of methanol. The solid product is redissolved in 600 ml of chloroform, filtered, and reprecipitated into several liters of methanol. After a final wash with methanol, the polymer is dried under vacuum (25°C, 0.1 mmHg) to give 15 g of polybis(p-ethylphenoxy)phosphazene (27% yield, based on polydichlorophosphazene).

Recent work [5] has sought to polymerize phenylhalogenocyclotriphosphazenes. This was done to shed some light on the reaction mechanism to understand why hexaarylcyclotriphosphazenes will not polymerize. Of the phenylated trimers, only the monophenylpentafluoro material will homopolymerize; i.e., two or more phenyl substituents deactivate the trimer. In copolymerization with hexachlorocyclotriphosphazene, only diphenyltetrafluoro or chloro and triphenyltrichloro will react. Four or more phenyls on a trimer will prevent copolymerization and will deactivate any polymerization when present.

$$[P(F)(C_6H_5)=N-P(F)_2=N-P(F)_2=N]_{cyclic} \xrightarrow[2.\ Na^+O^-R]{1.\ 300°,\ 42\ hr} \left[-P(OR)_2=N-P(OR)_2=N-P(C_6H_5)(OR)=N- \right]_n$$

R is CH_2CF_3

Final polymers are white, adhesive thermoplastics which are soluble in THF or acetone with molecular weights to 300,000 (intrinsic viscosity 0.32 dl/g). The copolymers with the hexachloro monomer are brown, soluble film formers with number average molecular weights to 500,000 (intrinsic viscosity 0.47 dl/g).

10.4-2 Cyclophosphazene Siloxane Copolymers

Another type of backbone can be synthesized from the cyclic, trimeric phosphonitrilic chloride, one in which the ring is retained in a ladder structure. In this case a modified cyclotriphosphazene is condensed directly with a dichlorosilane without a prior ring opening polymerization [2,37].

$2C_6H_5NH_2$; Na^+O^-n-Bu, Toluene, 80°, 25 hr ; $2(C_6H_5)_2SiCl_2$, 100-160°

The polymeric product is soluble in polar organic solvents, amorphous, and obtained in 90% yield. If a 1:1 ratio of silane to phosphazine is used, the products are white, whereas a 2:1 ratio yields a brown material. The latter case gives a higher molecular weight if the reaction temperature is below 120°C. For either material, the molecular weight is increased by two to six times (to 60,000 to 180,000) by heating the polymer to 190°C. This is attributed to further condensation between proper end groups.

Although one might expect crosslinking in the polymerization or post-curing process, there is no reported evidence of an insoluble fraction being formed.

Thermal stability data (by TGA in air) on these systems show weight losses commencing below 300°C, with 10% losses occurring from 300 to 540°C. However, it is interesting that no catastrophic drop in the TGA curve is realized even up to 800°C. The best systems show a leveling of the curve at 15-25% weight loss. Composite materials of these types of condensation polymers with asbestos fillers have even demonstrated a use temperature, by isothermal testing, of 500°C [31].

Method [37] *Monomer* A solution of 3.68 g (0.16 mol) of sodium in 30 g (0.405 mol) of n-butanol is prepared, and the excess butanol is removed under vacuum. Dry toluene 100 ml is added to the residue, and to this is added with vigorous stirring a solution of 18.44 g (0.039 mol) of tetrachlorodianilinocyclotriphosphazene [43] in 200 ml of toluene. After 25 hr at 70-80°C, the mixture is filtered; and the filtrate is washed with dilute HCl, dried over sodium sulfate, and the solvent is removed under vacuum. The residue is recrystallized from low-boiling petroleum ether to yield 85% of tetrabutoxydianilino-cyclotriphosphazene (mp 180-5°C).

Polymer The cyclotriphosphazene prepared above and dichlorodiphenyl-silane in a 1:1 ratio (6.11 g and 2.53 g, respectively, for 0.01-mol quantities) are heated under nitrogen at 120°C. The butyl chloride byproduct is removed in the nitrogen stream, and isolated in a cold trap of Dry Ice temperatures. The white, amorphous solid obtained (7.35 g) is then postcured at 190°C under nitrogen to give another white amorphous solid in 89% yield overall, with a final molecular weight (by end-group analysis) of 180,000.

A polymer analogous to the above siloxane phosphazenes has been suggested where the connecting link between cyclophosphazene units is a divalent metal oxide (Ca, Mg, Pb, Sn, Co, Ni, Ba, Zn, or Cu) [37].

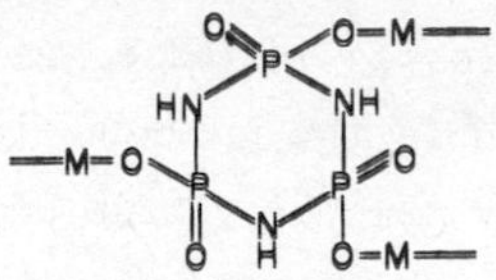

O=P—O—M—, HN, NH, —M—O—P(=O), N, H, P=O, O—M—

This metaphosphomic structure was derived from the reaction of the hexachlorocyclotriphosphazene with metal acetates in water at 30°C.

Only the barium-containing polymer demonstrated any thermal stability (310°C by TGA).

10.4-3 Polyphosphonylureas

The phosphorus counterparts of polyureas have been synthesized and found to have thermal stabilities (>300°C) superior to the more common polyurea backbone [18,20,21]. This inclusion of phosphorus is brought about by the addition of a diamine across a diisocyanate of phosphoric acid.

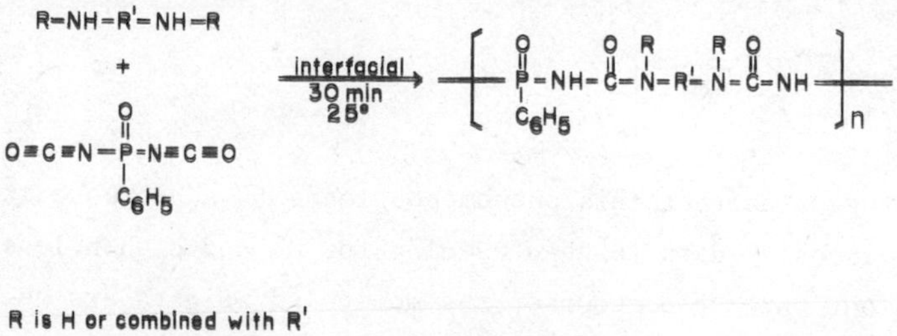

R—NH—R′—NH—R + O=C=N—P(=O)(C_6H_5)—N=C=O $\xrightarrow[\text{30 min, 25°}]{\text{interfacial}}$ —[P(=O)(C_6H_5)—NH—C(=O)—N(R)—R′—N(R)—C(=O)—NH]$_n$—

R is H or combined with R′

R with R′ is piperazine or 1,3-di-4-piperylpropane

R′ alone is m-or p-C_6H_4 or $-(CH_2)_6-$

If one replaces the diamine in the above equation with a dihydrazide, the product is a poly(phosphonylhydrazide), which, although not particularly thermally stable, retains some interest as adhesives, flame retardants, and biodegradable products [19,66].

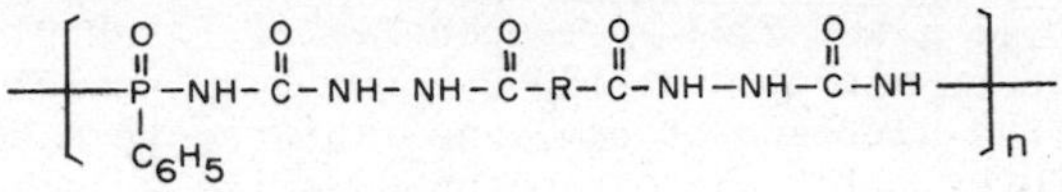

$$\left[-P(=O)(C_6H_5)-NH-C(=O)-NH-NH-C(=O)-R-C(=O)-NH-NH-C(=O)-NH- \right]_n$$

The phosphonylureas are obtained in yields of generally 77-96%, with intrinsic viscosities of 0.05-0.12 dl/g in formic acid, and possess softening points of 165-280°C. As could be predicted, the aromatic backbone functions (p- and m-phenylene) provide the highest softening point products, 270 and 250°C mean values, respectively.

Weight losses by TGA in nitrogen commence below 200°C and reach 10% at 240 to 330°C. At higher temperatures, the weight losses increase gradually to 80% at 800°C. Cyclic secondary amines were used in the backbone in an effort to improve the thermal stability of these systems by virtue of lower oxidazability and the pseudoladder structure. However, little or no improvement is realized by TGA methods.

The polymers are polar and consequently are soluble in formic acid, DMSO, acetic acid, and sometimes acetone (for aromatic backbones). The colors of the powders range from white to light yellow to gray-white. The solubility and color characteristics have been linked to a double-bond character proposed for the backbone through resonance.

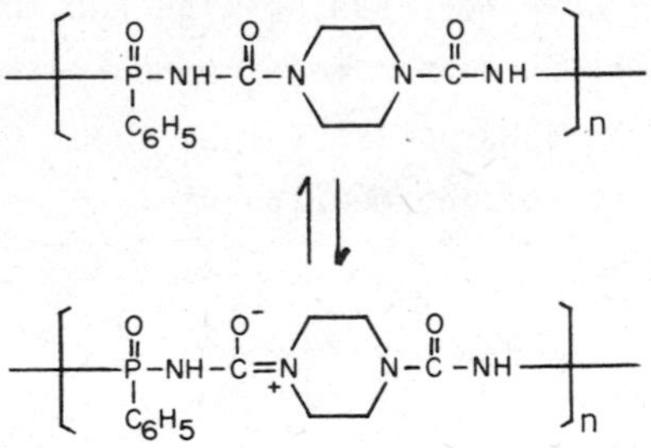

$$\left[-P(=O)(C_6H_5)-NH-C(=O)-N\langle\text{piperazine}\rangle N-C(=O)-NH- \right]_n$$

$$\rightleftharpoons$$

$$\left[-P(=O)(C_6H_5)-NH-C(-O^-)=N^+\langle\text{piperazine}\rangle N-C(=O)-NH- \right]_n$$

Also it is suggested that this phenomenon leads to a backbone stiffness. However, viscosity data (0.05-0.09 dl/g for m- and p-phenylene units) do not bear this out unless, of course, the molecular weights are quite low.

Method [18] In a semimicro stainless steel blendor jar are placed 50 ml of water and 0.01 mol of the diamine. To the vigorously stirred aqueous solution at room temperature is added 0.01 mol of phenylphosphonic diisocyanate [66] in 50 ml of CCl_4, and stirring is continued for 30 sec. The polymer is isolated by suction filtration, washed with water, and dried.

10.5 Metal Organophosphinate Polymers

Coordination polymers of double phosphinate ($-O-PR_2-O-$) bridges have received extensive study from 1962-1968 primarily by Rose and Block and their co-workers; however, the polymeric nature was not immediately recognized. Wide ranges of solubilities, tractabilities, and thermal stabilities are available in these systems. The general approach is to react a dialkyl or diaryl or alkylaryl phosphinic acid with the metal salt (acetate or acetylacetonate) to produce coordinated backbones with two phosphinate bridges between each metal ion:

$$MX_2 + HO-\overset{R}{\underset{R}{P}}=O \longrightarrow \left[M \begin{matrix} O-PR_2-O \\ O-PR_2-O \end{matrix} \right]_n$$

Metals which have been incorporated include beryllium [10-12], zinc [10,12, 25,73,74], chromium [13,65,77,78], cobalt [22,25,73], iron [65], and copper [10,11]. It was first realized in 1962, by a molecular weight analysis, that the materials were indeed polymers. It was further postulated that the square planar conformation for the central metal atom, such as Cu, leads to more ready decomposition.

In this ladder or spiroladder system, a considerable thermal stability is expected and indeed is found vis-a-vis backbone integrity. However, the point of decomposition seems to be the P-C bond. In investigations of thermal degradation volatile byproducts, benzene is found to emanate from phenyl systems and toluene from methylphenyl substituted phosphinates (vide infra). This free radical process is soon followed by backbone cleavage to yield phosphinic acids and phosphine. With some wholly phenylated systems, crosslinking may occur through a p-phenylene link.

10.5-1 Zinc and Cobalt Systems

By a surprisingly simple procedure, quite stable backbones containing zinc, oxygen, and phosphorus as double-bridged coordination polymers can be synthesized [72]. The analogous cobalt system has also been prepared [22,25,50, 51,73].

The white, zinc, and blue cobalt, partially crystalline powders are received usually in greater than 90% yield. The dimethyl-substituted polymers are soluble in water but are insoluble in organics (benzene, chloroform), while the diphenyl analogs are insoluble in all solvents. Diphenylsulfone

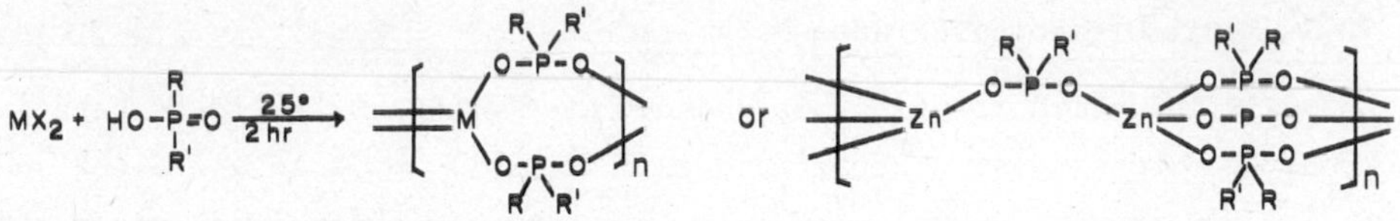

MX_2 is: $Zn(OAc)_2$ or $CoCl_2$

R and R' are: CH_3, C_6H_5, n-C_4H_9, C_6H_5-SO_2-C_6H_4-,
polyphenylene, or polyoxyphenylene

substituents cause the polymer to be soluble in DMF, THF, and chloroform. The methylphenyl zinc polymer is soluble in organics as are the cobalt hybrid and dibutyl cobalt polymers [25]. The soluble polymers also melt to viscous liquids, and on cooling form glassy solids. Hybrid copolymers are possible by mixing phosphinic acids so that a mixture of R and R' groups are present in the final polymer. These hybrids can be prepared in varying ratios of phosphinates and are random so that crystallinity of the original homopolymer is decreased and solubility is increased.

The polymers had molecular weights to 35,000 (degree of polymerization ~100). Quite low intrinsic viscosities of 0.02 were measured for those materials in the range of 10,000 molecular weight.

Highly oriented fibers can be drawn from a gel which is prepared by wetting the dibutyl/zinc polymer with benzene [25].

TGA shows very sharp decomposition points for all materials at 480-560°C for the zinc polymers and from 520-570°C for cobalt polymers, the diphenyl and diphenylsulfone versions being highest. It is important to note, however, that all TGA curves level off after a weight loss of 22-25% and maintain that value to over 800°C, and that the only volatile product from phenylated materials is benzene [73]. It is postulated that cross-linking occurs with loss of benzene and production of a phenylene inter-chain link:

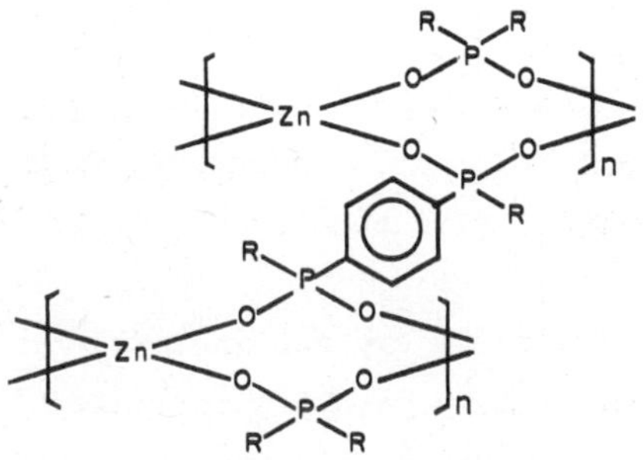

A 25% weight loss corresponds to 1.5 phenyls per repeating unit for the diphenyl cobalt. This indicates a rather high crosslink density.

In the studies of polyphenylene substituents [50], a decrease in thermal stability was noted when more than a diphenyl unit was attached. Further, ortho and meta substitutions were less stable than para. This phenomenon was attributed to the inability of the polymer to absorb energy except by bond cleavage. Phenylsulfone side groups, therefore, imparted greater stability because of energy absorption by rotation [51].

Isothermal testing by monitoring melt viscosity showed about a 13% decrease after 220 hr at 316°C. Weight monitoring gave a 12% loss under vacuum at 316°C for 267 hr. Other more severe and more applications-oriented testing [80] demonstrated an isothermal weight loss of a coating on metal to be 49% after 10 hr at 540°C (1000°F). If the polymer was compounded with a pigment, this loss dropped to only 14% after 100 hr; but in either case film hardness was retained.

The materials also exhibit good resistance to hydrolysis and ^{60}Co radiation but poor dielectric and corrosion prevention properties.

Method [72] *Poly(zinc methylphenylphosphinate)* A solution of 7.718 g (35.5 mmol) of powdered zinc acetate dihydrate and 11.088 g (71.02 mmol) of methylphenylphosphinic acid in 600 ml of ethanol is stirred at room temperature for 2 hr to produce a white precipitate. The solid is isolated, washed with ethanol, and dried at 60°C in vacuo.

Method [73] *Hybrid copoly(cobalt diphenylphosphinate methylphenylphosphinate)* A solution is made using 20 mmol of hydrated cobalt acetate, 20 mmol of diphenylphonphinic acid, and 20 mmol of methylphenylphonphinic acid in 600 ml of ethanol, and it is heated to reflux to give a blue solid. The 1:1 hybrid is extracted with chloroform, and the extract is evaporated to yield 5.2 g of polymer.

10.5-2 Beryllium Systems

The most thermally stable of the metal phosphinates is the beryllium polymer, which is also the least soluble and tractable. In an effort to combine the best of these properties, a series of beryllium polymers was prepared with alkyl substituents [10,88]. In this case, the metal acetylacetonate is used as starting material.

Molecular weights range up to 50,000 for the solid products, most of which are soluble in common organic solvents, except for the dimethyl and diphenyl systems. It is of interest that the methylphenyl polymer is soluble and not, therefore, intermediate in behavior to the dimethyl and

$Be(AcCHAc)_2 + 2R_2P(O)OH \xrightarrow[25-215°]{heat/N_2}$

AcCHAc is acetylacetonate:

R is: CH_3, C_6H_5, C_6F_5, n-C_4H_9, n-C_5H_{11}, n-C_7H_{15} and tetramethylene as R_2.

dipnenyl systems. As seems to be the fate of things, this most soluble isomer is also least thermally stable in air or nitrogen.

Thermal stabilities by TGA (10% weight loss) range from 410 to 610°C in nitrogen in respective order of substituents: n-butyl, n-heptyl, n-pentyl, C_6F_5, cyclic tetramethylene, methyl/phenyl, methyl, phenyl. The cyclic tetramethylene substituent is more stable than the similar di-n-butyl system since it requires the cleavage of two P-C bonds to yield the 1-butene byproduct.

Method [88] Since the methyl phenyl phosphinate polymer is the most stable of the soluble materials and since it is obtained at a significant molecular weight (30,000), its synthesis is given here.

A mixture of 1 mol of triply sublimed beryllium acetylacetonate and 2 mol of phenyl(methyl)phosphinic acid [35] is heated slowly under nitrogen to 165°C for 8 hr while the acetylacetone is removed by distillation. Then the mixture is heated under vacuum to 180°C for 3 hr to yield an amber, glassy solid soluble in benzene.

10.5-3 Chromium Systems

A benzene- or chloroform-soluble polymer with molecular weight above 30,000 is possible with chromium in the backbone [77].

$Cr(OAc)_2 + KOPOR_2 \xrightarrow[H_2O]{heat/N_2}$ … $\xrightarrow{air,\ H_2O}$

R is CH_3 or C_6H_5

The synthesis from chromium hexacarbonyl leaves CO as ligand and results in lower molecular weights (7000) [65].

$Cr(CO)_6$ + $HO-P(O)(C_6H_5)_2$ → (heat/UV, Toluene) → $[Cr(CO)_2\{O_2P(C_6H_5)_2\}_2]_n$

The former example is soluble in benzene and chloroform, obtained in up to 95% yield, and has an intrinsic viscosity of 07 dl/g. A 10% weight loss in TGA is experienced at 430°C, with no melting below this point (lower stability than the Be analog but much better solubility). Further, it is resistant to hydrolysis.

> *Method* [77] A suspension of 2.7 g of $Cr(OAc)_2H_2O$ is heated at reflux under nitrogen in 125 ml of deoxygenated water for 1 hr. To this is added a deoxygenated aqueous solution of 8 g of diphenylphosphinic acid neutralized with KOH, and reflux is continued for 3 hr. The precipitate is isolated and washed with water in the absence of air. It is then suspended in 500 ml of water, and air is bubbled through. The precipitate is collected, washed with water, and dried at 100°C. The polyphosphinate is then dissolved in benzene and filtered, and the filtrate is evaporated to give the pure polymer, which is dried at 120°C.

Still another example of a polymeric chromium phosphinate uses a triple bridge between chromium(III) centers. An advantage is the random mixing of types of organic substituents, particularly octyl, on the phosphinate bridge to provide plasticization and subsequently a flexible film [53].

$CrCl_3 \cdot 6H_2O$ + $CH_3-P(O)(OH)-C_6H_5$ → (1. heat, THF, H_2O; 2. K_2CO_3) → $[Cr(H_2O)(OH)\{O_2P(CH_3)(C_6H_5)\}_2]_n \cdot xH_2O$

→ (1. Dry at 110°; 2. $(C_8H_{17})_2P(O)OH$) → $[Cr\{O_2P(CH_3)(C_6H_5)\}_2\{O_2P(C_8H_{17})_2\}]_n$

The model of preparation of the polymeric intermediate determines whether the final product will form a flexible or brittle film. The flexibility is evidently dictated by the degree of order of the backbone phosphinate bridges. Interestingly, flexibility is inversely related to randomness of placement of the backbone. That is, more crystalline (i.e., more ordered systems are more flexible).

The key to the precursor for flexible film formation is in equilibration of structures prior to addition of K_2CO_3 and isolation of the prepolymer, which is essentially all double bridged. Without this equilibration time, there exist many single and triple bridges.

Both brittle and flexible films will swell in chloroform, and some will dissolve; so crosslinking is evidently at a minimum. They have similar tensile strengths (up to 48 or 55 MPa, or 7000 or 8000 psi); but, of course, elongations are quite different (3-6% compared to 17-25%). Intrinsic viscosities range up to 18 dl/g, suggesting microgel formation.

Thermal stability is quite similar for either film form by TGA with 70% weight losses at 330°C in air and 460°C in nitrogen. Weight loss reaches a plateau in either atmosphere at 60% at 600°C and is due to oxidative cleavage of the alkyl side groups, giving largely 2-octanone, acetic acid, and n-butyric acid. Higher temperatures yielded aromatic compounds (benzene, toluene, and phenol).

In static thermal analysis at 200°C, the films darken and embrittle rapidly with a 13% weight loss in 18 hr.

Method [14,53] *Prepolymer* A solution of 11.04 g (0.04143 mol) of $CrCl_3 \cdot 6H_2O$ in 200 ml of THF is heated at reflux for 30 min after the addition of 12.94 g (0.0829 mol) of methylphenylphosphinic acid in 100 ml of THF and 100 ml of water. Then a solution of 8.6 g (0.062 mol) of K_2CO_3 in 50 ml of water is added dropwise with stirring. The clear, green solution is heated to boiling to remove the THF while water is added to maintain the volume. The powdered solid which forms is collected, washed with water, and air-dried to yield 97% of the hydrated bisphosphinate polymer, $[Cr(H_2O)(OH)[OP(CH_3)(C_6H_5)O]_2]_n$, intrinsic viscosity 0.4 dl/g in $CHCl_3$. The yield may be somewhat optimistic since weight loss will occur by loss of water in a low humidity atmosphere, and therefore the exact composition (% hydration) is not known. The powder is dried at 110°C for 24 hr to yield the anhydrous prepolymer.

Flexible polymeric film Equimolar amounts of the anhydrous prepolymer, $[Cr(OH)[OP(CH_3)(C_6H_5)O]_2]_n$, and dioctylphosphinic acid, $(C_8H_{17})_2P(O)OH$, are shaken in o-dichlorobenzene for a few hours and allowed to stand for 24 hr to complete the dissolution process. Film is produced by evaporating the solution on a Teflon surface at 110-120°C. A cover of perforated aluminum foil is used during film formation to control solvent evaporation rate.

10.5-4 Iron Systems

An insoluble, intractable polymer, stable to 300°C, is obtained by heating to reflux a solution of iron pentacarbonyl and diphenylphonphinic acid in THF with simultaneous UV light (250-W sunlamp) irradiation. The light radiation is said to minimize crosslinking [65].

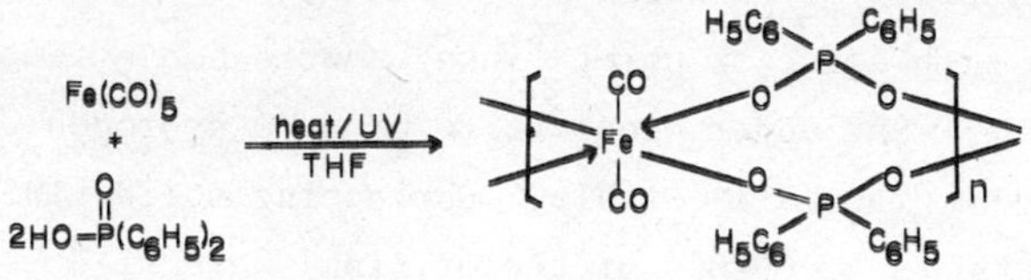

$$Fe(CO)_5 + 2HO{-}\overset{O}{\overset{\|}{P}}(C_6H_5)_2 \xrightarrow[THF]{heat/UV} \left[Fe(CO)_2 \left(O{-}P(C_6H_5)_2{-}O \right)_2 \right]_n$$

10.6 Recent Developments

10.6-1 Siloxanes

Synthetic and property studies have continued on several inorganic systems, but no significantly new types have appeared. Clear, viscous liquid N-phenylcyclosilazoxanes are possible and are stable nearly to 500°C, a higher stability than that of simple polysiloxanes [107]:

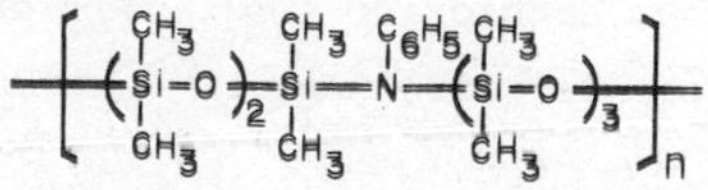

$$\left[\left(Si(CH_3)_2{-}O \right)_2 {-} Si(CH_3)_2 {-} N(C_6H_5) {-} \left(Si(CH_3)_2{-}O \right)_3 \right]_n$$

The increased stability is proposed as being due to the silazane function, which serves to disturb the regular backbone structure. This irregularity then prevents intramolecular depolymerizations and concommitant cyclization.

10.6-2 Carboranes

A series of papers has concentrated on D_2-m-carborane-siloxane synthesis [103] and properties [110] and trifluoropropyl modifications [111]. In a new, mild, very fast reaction of carborane disilanol with bisureidosilanes, molecular weights above 250,000 were obtained. The products showed useful properties after 300 hr at 315°C in air [103,109]. To obtain an elastomeric material, some of the m-carborane nuclei were replaced with p-carborane or p-phenylene, which disrupted the crystallinity [110].

Trifluoropropyl groups were placed on carborane siloxane backbones to improve the solvent resistance of these polymers [111]. The thermal stability of these new materials was superior to fluorosilicones, but solvent resistance was comparable in only a few instances. The fluorinated version was, however, much less thermally stable than commercial poly(carborane siloxane) (Ucarsil, Union Carbide).

Poly(phenylene carboranes) were synthesized at 20°C and contained carboranes both as backbone and as pendant functions [106]. These prepolymers (molecular weight 1100-4600) were heat-cured at 400°C for 4 hr to produce an insoluble, infusible mass. Although TGA/air data were given, they have little value for evaluating carborane systems, due to the weight gain realized in the oxidation of carboranes.

Carborane polymers with amide or ester linkages were subjected to controlled pyrolysis in air and under pressure to produce hydrogen gas and nondeformable, crosslinked, thermally stable, lubricating solid [105]. This cokelike residue amounts to 60-85% of the original weight, depending on structure.

10.6-3 Ferrocenes

In spite of the poor performance of ferrocene-containing polymers with regard to thermal stability, some research on these systems has continued with some interesting polymers resulting. A poly(titanocene ferrocene ether) was synthesized by the very rapid interfacial condensation of a ferrocene diol with cyclopenatadienyltitanium dichloride [96]. Yields are low to moderate, and thermal stabilities are quite poor.

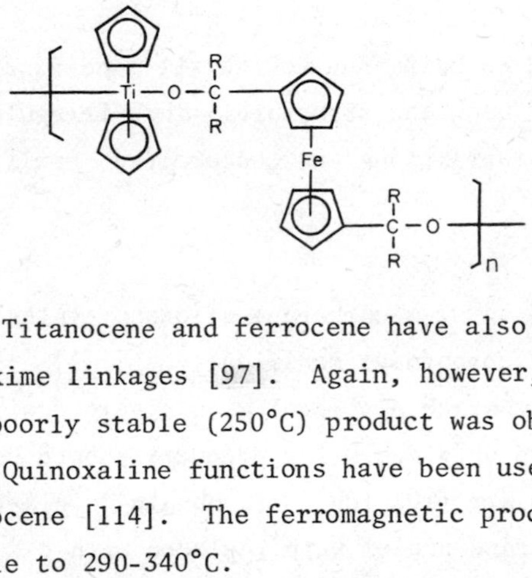

Titanocene and ferrocene have also been joined in a polymer backbone by oxime linkages [97]. Again, however, a soluble but low molecular weight and poorly stable (250°C) product was obtained.

Quinoxaline functions have been used as cofunctions in a backbone with ferrocene [114]. The ferromagnetic products are dark, poorly soluble, and stable to 290-340°C.

10.6-4 Phosphazenes

Still the most popular inorganic backbone other than siloxane is the phosphazene. The principal function of these investigations has been to relate the properties of the polymers to the composition of the pendant groups.

Numerous substituents, including aryloxy [101], arylamino [116], and fluoroalkoxy [95,104], have been investigated. The arylamino and aryloxy systems were prepared by nucleophilic substitutions onto the polydichlorphosphazene backbone. The others were obtained by ring-opening reactions on a properly substituted cyclic trimer.

The thermal degradation of polybis(p-isopropylphenoxy)phosphazene was studied to reveal a random chain scission process [102,113]. Depolymerization was found to be no serious problem.

Polyalkoxy- and fluoroalkoxyphosphazenes were shown to be moderately stable toward UV photolysis, with the former being better [108]. No-P=N- bond cleavage was observed, but mostly the C-O bond was broken. Residual chlorine on the backbone was detrimental to photostability.

The fluoroalkoxyphosphazene polymers are known now to be useable as seals from -60 to +200°C [115]. Also their flame-retardant character is valuable either in a foam configuration [112] or when the cyclic phosphazene function is pendant in a styrene copolymer [100].

10.6-5 Titanium Amines

Titanocene dichloride can condense with diamines [98] or tetraamine [99] to provide products of low to moderate stabilities. Of course, the latter system can be crosslinked if a greater than 1:1 ratio of reactants is used. Yields were low in the very rapid interfacial condensation. Weight losses by TGA begin at 100-200°C and proceed to 20-80% at 600°C.

References

1. H. R. Allcock, *Chem. Rev., 72,* 315 (1972).
2. H. R. Allcock, *Phosphorus-Nitrogen Compounds,* Academic Press, New York, 1972.
3. H. R. Allcock, *Science, 193,* 1214 (1976).
4. H. R. Allcock, R. W. Allen, and J. J. Meister, *Macromolecules, 9*(6), 950 (1976).
5. H. R. Allcock and G. Y. Moore, *Macromolecules, 8*(4), 377 (1975).
6. R. W. Allen and H. R. Allcock, *Macromolecules, 9*(6), 956 (1976).
7. L. H. Almond, *Organometallic Compounds, 27*(6), 156-166 (1975).
8. H. H. Anderson, *J. Amer. Chem. Soc., 73,* 5802 (1951).
9. C. Arnold, Jr., *J. Polym. Sci.: Polym. Chem. Ed., 13,* 517 (1975).
10. B. P. Block, S. H. Rose, C. W. Schaumann, E. S. Roth, and J. Simkin, *J. Amer. Chem. Soc., 84,* 3200 (1962).

11. B. P. Block, E. S. Roth, C. Schaumann, and L. Ocone, *Inorg. Chem., 1*, 860 (1962).

12. B. P. Block and C. W. Schaumann, U.S. Patent 3,245,953 (April 12, 1966).

13. B. P. Block, J. Simkin, and L. Ocone, *J. Amer. Chem. Soc., 84*, 1749 (1962).

14. C. E. Cain, A. Mashburn, Jr., and C. A. Hauser, *J. Org. Chem., 26*, 1030 (1961).

15. D. M. Carlton and P. E. Cassidy, unpublished results, Tracor, Inc., Austin, Texas.

16. C. E. Carraher, *Org. Coat. Plast. Preprints, 35*(2), 380 (1975).

17. C. E. Carraher, *Polym. Preprints, 17*(2), 365 (1976).

18. C. E. Carraher and T. Brandt, *Makromol. Chem., 126*, 66 (1969).

19. C. E. Carraher and D. Burger, *Makromol. Chem., 138*, 59 (1970).

20. C. E. Carraher and D. R. Burger, *Angew. Makromolec. Chem., 46*, 73 (1975).

21. C. E. Carraher and C. Kuregar, *Makromol. Chem., 133*, 219 (1970).

22. G. E. Coates and D. S. Golightly, *J. Chem. Soc.*, 2523 (1962).

23. T. M. Connelly and J. K. Gillham, *J. Appl. Polym. Sci., 20*(2), 473 (1976).

24. J. L. Cotter, G. J. Knight, J. M. Lancaster, and W. W. Wright, *J. Appl. Polym. Sci., 12*, 2481 (1968).

25. V. Crescenzi, V. Giancotti, and A. Ripamonti, *J. Amer. Chem. Soc., 87*, 391 (1965).

26. J. E. Curry and J. D. Byrd, *J. Appl. Polym. Sci., 9*, 295 (1965).

27. A. D. Delman, *J. Polym. Sci., A-1, 8*, 943 (1970).

28. A. D. Delman, H. N. Kovacs, and B. B. Simms, *J. Polym. Sci.: A-1, 6*, 2117 (1968).

29. N. J. Delollis, personal communication.

30. J. Ditter, E. B. Klusmann, J. D. Oakes, and R. E. Williams, *Inorg. Chem., 9*, 889 (1970).

31. G. H. Emblem, E. C. Oxyley, and A. S. Trow, *Brit. Polym. J., 2*, 83 (1970).

32. R. G. Gamper, P. T. Funke, and A. A. Volpe, *J. Polym. Sci., A-1, 9*, 2137 (1971).

33. R. E. Goldsberry, M. J. Adamson, and R. F. Reinisch, *J. Polym. Sci.: Polymer Chem. Ed., 11*(10), 2401 (1973).

34. G. L. Hagnauer and B. R. Laliberte, *J. Appl. Polym Sci., 20*(11), 3073 (1976).

35. H. J. Harwood and D. W. Grisley, *J. Amer. Chem. Soc., 82*, 423 (1960).

36. T. L. Heying, J. W. Agern, Jr., S. J. Clark, R. P. Alexander, S. Papetti, J. A. Reed, and S. J. Trotz, *Inorg. Chem., 2*, 1097 (1963).

37. M. Kajiwara, A. Sakamoto, and H. Saito, *Angew. Makromolec. Chem., 46*, 63 (1975).

38. P. T. Kan, C. T. Lenk, and R. L. Schaff, *J. Org. Chem.*, *26*, 4038 (1961).

39. R. E. Kesting, K. F. Jackson, E. B. Klusmann, and F. J. Gerhart, *J. Appl. Polym. Sci.*, *14*, 2525 (1970).

40. R. E. Kesting, K. F. Jackson, and S. M. Newman, *J. Appl. Polym. Sci.*, *15*, 1527 (1971).

41. K. O. Knollmueller, R. N. Scott, H. Kwasnik, and J. F. Sieckhaus, *J. Polym. Sci.*, *A-1*, *9*, 1071 (1971).

42. H. N. Kovacs, A. D. Delman, and B. B. Simms, *J. Polym. Sci.*, *A-1*, *6*, 2103 (1968).

43. H. Lederle, G. Ottmann, and E. Kober, *Inorg. Chem.*, *5*, 1818 (1966).

44. J. J. Lorkowski and R. Pannier, *J. Prakt. Chem.*, *311*(6), 958 (1969).

45. M. Maienthal, M. Hellmann, C. P. Haber, L. A. Lymo, S. Carpenter, and A. J. Carr, *J. Amer. Chem. Soc.*, *76*, 6392 (1954).

46. R. M. Minimmi and A. V. Tobolsky, *J. Appl. Polym. Sci.*, *16*(10), 2555 (1972).

47. J. E. Mulvaney, J. J. Bloomfield, and C. S. Marvel, *J. Polym. Sci.*, *62*, 59 (1962).

48. J. E. Mulvaney and C. S. Marvel, *J. Polym. Sci.*, *50*, 541 (1961).

49. T. Nakajima and C. S. Marvel, *J. Polym. Sci.*, *A-1*, *7*, 1295 (1969).

50. P. Nannelli, H. D. Gillman, H. G. Monsimer, and S. B. Advani, *J. Polym. Sci.: Polym. Chem. Ed.*, *12*, 2525 (1974).

51. P. Nannelli, H. D. Gillman, H. G. Monsimer, and S. B. Advani, *Makromol. Chem.*, *177*, 2607 (1976).

52. E. W. Neuse, *J. Polym. Sci.*, *A-1*, *6*, 1567 (1968).

53. E. W. Neuse, *Macromolecules*, *1*, 171 (1968).

54. E. W. Neuse, Encyclopedia of Polymer Science and Technology, Vol. 8, *Metallocene Polymers*, Interscience, New York, 1968, pp. 667-692.

55. E. W. Neuse and H. Rosenberg, *J. Macromol. Sci. Rev. Macromol. Chem.*, *C4*, 1 (1970).

56. E. W. Neuse and R. K. Crossland, *J. Organometal. Chem.*, *7*, 344 (1967).

57. E. W. Neuse, G. Horlbeck, H. Siesler, and K. Yannakow, *Polymer*, *17*(5), 423 (1976).

58. S. Papetti, B. B. Schaeffer, A. P. Grany, and T. L. Heying, *J. Polym. Sci.*, *A-1*, *4*, 1623 (1966).

59. W. J. Patterson, S. P. McManus, and C. U. Pittman, *J. Polym. Sci.: Polym. Chem. Ed.*, *12*, 837 (1974).

60. C. U. Pittman, *J. Paint Technol.*, *39*, 585 (1967).

61. C. U. Pittman, *Chem. Techn. (Berlin)*, July 1971, p. 416.

62. C. U. Pittman, W. J. Patterson, and S. P. McManus, *J. Polym. Sci.*, *A-1*, *9*, 3187 (1971).

63. C. U. Pittman, W. J. Patterson, and S. P. McManus, *J. Polym. Sci.: Polym. Chem. Ed.*, *14*, 1715 (1976).

64. L. Plummer and C. S. Marvel, *J. Polym. Sci.*, *A-2*, 2559 (1964).

65. J. E. Podall and T. L. Iapalucci, *J. Polym. Sci., B, 1,* 457 (1963).

66. I. C. Popoff and J. P. King, *J. Polym. Sci., B, 1,* 247 (1963).

67. K. A. Reynard, A. H. Gerber, and S. H. Rose, *Synthesis of Phosphonitrilic Elastomers and Plastics for Marine Applications,* Horizons, Inc., Cleveland, Ohio, Naval Ship Engineering Center, AMMRC CTR 72-29, December 1972 (AD 755188).

68. K. A. Reynard, R. W. Sicka, J. E. Thompson, and S. H. Rose, *Poly(aryloxyphosphazene) Foams, and Wire Coverings,* Horizons, Inc., Cleveland, Ohio, Naval Ship Engineering Center Contract No. N00024-73-5474, March 1975.

69. M. B. Roller and J. K. Gillham, *J. Appl. Polym. Sci., 16*(2), 3073 (1972).

70. M. B. Roller and J. K. Gillham, *J. Appl. Polym. Sci., 16*(2), 3095 (1972).

71. M. B. Roller and J. K. Gillham, *J. Appl. Polym. Sci., 16*(2), 3105 (1972).

72. S. H. Rose and B. P. Block, *J. Polym. Sci., A-1, 4,* 377 (1966).

73. S. H. Rose and B. P. Block, *J. Polym. Sci., A-1, 4,* 583 (1966).

74. S. H. Rose and B. P. Block, *J. Polym. Sci., A-1, 4,* 573 (1966).

75. H. Sakurai, R. Kohi, A. Hosomi, and M. Kumada, *Bull. Chem. Soc. Japan, 39,* 2050 (1966).

76. H. Sakurai, A. Hosomi, and M. Kumada, *Chem. Comm.,* 930 (1968).

77. A. J. Saraceno and B. P. Block, *J. Amer. Chem. Soc., 85,* 2018 (1963).

78. A. J. Saraceno and B. P. Block, *Inorg. Chem., 3,* 1699 (1964).

79. R. L. Schaaf, P. T. Kan, C. T. Lenk, and E. P. Deck, *J. Org. Chem., 25,* 1986 (1960).

80. F. Schmitt, Jr., *Experimental High Temperature Polymers for Protective Coatings,* AFML-TR-67-174 Part II, Sec. II, p. 2 (March 1969).

81. N. S. Schneider, C. R. Desper, and R. E. Singler, *J. Appl. Polym. Sci.,* (11), 3087 (1976).

82. H. A. J. Schoutissen, *J. Amer. Chem. Soc., 55,* 4545 (1933).

83. H. A. Schroeder, R. P. Alexander, T. B. Larchar, and O. G. Schaffling, Report on Contract No. bs-94278, Olin Mathieson Chemical Corp., July 31, 1967.

84. H. Schroeder, O. G. Schaffling, T. B. Larchar, F. F. Frulla, and T. L. Heying, *Rubber Chem. Technol., 39*(4), Part 2, 1184 (1966).

85. K. Shiina and M. Kumada, *J. Org. Chem., 23,* 139 (1958).

86. R. E. Singler, G. L. Hagnauer, N. S. Schneider, B. R. Laliberte, R. E. Sacher, and R. W. Matton, *J. Polym. Sci.: Polym. Chem. Ed., 12,* 433 (1974).

87. R. E. Singler, N. S. Schneider, and G. L. Hagnauer, *Polym. Eng. Sci., 15*(5), 321 (1975).

88. P. J. Slota, Jr., L. P. Freeman, and N. R. Fetter, *J. Polym. Sci., A-1, 6,* 1975 (1968).

89. W. R. Sorenson and T. W. Campbell, *Preparative Methods of Polymer Chemistry,* 2nd ed., John Wiley and Sons, New York, 1968, p. 187.

90. L. H. Sperling, S. L. Cooper, and A. V. Tobolsky, *J. Appl. Polym. Sci.*, *10*, 1725 (1966).

91. J. K. Stille, *Polym. Preprints*, *12*(1), 96 (1976).

92. S. S. Washburne and W. R. Peterson, Jr., *J. Organometal. Chem.*, *21*, 59 (1970).

93. J. E. White, R. E. Singler, and S. A. Leone, *Polym. Preprints*, *16*(2), 7 (1975).

94. S. Yajima, *Ind. Eng. Chem.*, *Prod. Res. Deve.*, *15*(3), 219 (1976).

95. H. R. Allcock, J. L. Schmitz, and K. M. Kosydar, *Macromolecules*, *11*(1), 179 (1978).

96. C. Carraher, Jr. and G. F. Burrish, *J. Macromol. Sci. Chem.*, *A10*(8), 1457 (1976).

97. C. E. Carraher, M. J. Christensen, and J. A. Schroeder, *J. Macromol. Sci. Chem.*, *A11*(11), 2021 (1977).

98. C. Carraher and P. Lessek, *Europ. Polym. J.*, *8*, 1339 (1972).

99. C. E. Carraher, R. Pfeiffer, and P. Fullenkamp, *J. Macromol. Sci. Chem.*, *A10*(7), 1221 (1976).

100. *Chem. Eng. News*, p. 15, Oct. 2, 1978.

101. R. L. Dieck and L. Goldfarb, *J. Polym. Sci.: Polym. Chem. Ed.*, *15*, 361 (1977).

102. L. Goldfarb, N. D. Hann, R. L. Dieck, and D. C. Messersmith, *J. Polym. Sci.: Polym. Chem. Ed.*, *16*, 1505 (1978).

103. E. Hedaya, J. H. Kawakami, P. W. Kopf, G. T. Kwiatkowski, D. W. McNeil, D. A. Owen, E. N. Peters, and R. W. Tulis, *J. Polym. Sci.: Polym. Chem. Ed.*, *15*, 2229 (1977).

104. M. Kajiwara, K. Nakashima, and H. Saito, *J. Macromol. Sci. Chem.*, *A11*(3), 449 (1977).

105. V. V. Korshak, I. A. Gribova, A. N. Chumaievskaya, and B. M. Mgeladze, *J. Appl. Polym. Sci.*, *23*, 1915 (1979).

106. V. V. Korshak, M. M. Teplyakov, Ts. L. Gelashvili, S. M. Komarov, V. N. Kalinin, and L. I. Zhakharkin, *J. Polym. Sci.: Polym. Lett. Ed.*, *17*, 115 (1979).

107. Z. Lasocki and M. Witekowa, *J. Macromol. Sci. Chem.*, *A11*(3), 457 (1977).

108. J. P. O'Brien, W. T. Ferrar, and H. R. Allcock, *Macromolecules*, *12*(1), 108 (1979).

109. E. N. Peters, E. Hedaya, J. H. Kawakami, G. T. Kwiatkowski, D. W. McNeil, and R. W. Tulis, *Rubber Chem. Technol.*, *48*, 14 (1975).

110. E. N. Peters, J. H. Kawakami, G. T. Kwiatkowski, E. Hedaya, B. L. Joesten, D. W. McNeil, and D. A. Owen, *J. Polym. Sci.: Polym. Phys. Ed.*, *15*, 723 (1977).

111. E. N. Peters, D. D. Stewart, J. J. Bohan, R. Moffit, C. D. Beard, G. T. Kwiatkowski, and E. Hedaya, *J. Polym. Sci.: Polym. Chem. Ed.*, *15*,

112. J. L. Thompson and K. A. Reynard, *J. Appl. Polym. Sci.*, *21*, 2575 (1977).

113. J. K. Valaitis and G. S. Kyker, *J. Appl. Polym. Sci.*, *23*, 765 (1979).
114. M. Vernois and H. L. Williams, *J. Appl. Polym. Sci.*, *23*, 1601 (1979).
115. J. C. Vicic and K. A. Reynard, *J. Appl. Polym. Sci.*, *21*, 3185 (1977).
116. J. E. White and R. E. Singler, *J. Polym. Sci.: Polym. Chem. Ed.*, *15*, 1169 (1977).

Epilogue

Many people who have been involved in the field of thermally stable materials from the beginning have been conjecturing recently as to the future of this technical area. Their comments at first appear to be pessimistic, as perhaps they are for the one who seeks to synthesize a new organic backbone with outstanding stability. Many people have stated that we have reached the limit for organic polymers which are to be used in air for extended periods up to 300°C. Various modifications of bonding, functional groups, and pendant groups have been made with consequent improvement in processing and solubility but little or no increase in stability.

What remains to be done is more, however, than what has been done. Most of these new goals pertain to use of the polymers, and not all of them being in thermally stable applications. More applications are becoming known for polymers with controlled degradation. New composites and adhesives will find larger use due to their moduli, environmental resistance, and strength-to-weight ratios. Some inroads have been made into reverse-osmosis membranes, corrosion protective coatings, electrical or photoconductive polymers, fire-resistant fibers or building materials, and elastomers which retain their properties in geothermal fluids, all of which will require more work and will realize a larger market than they do presently. Several other research needs can be listed, some of which have been touched upon in the past few years:

1. Crosslinking without the formation of byproducts
2. Liquid crystal development
3. Environmental effects on specialty polymers and composites
4. Degradation mechanisms
5. Battery separators, and
6. New processing techniques

Two areas of research are yet to be extensively exploited by scientists who develop thermally stable polymers: inorganic backbones and natural products (cellulose, lignin, etc.). The former area has shown to provide some easily synthesized, soluble polymers with impressive ranges of operation temperatures; the latter is being recognized now by those people working in biomass programs for food and fuel production. It might behoove the high-temperature polymer chemists and engineers to look more seriously toward synthesizing and processing these types of polymers.

Appendix A

Structure-Nomenclature Index for Functional Groups*

PART 1: According to Structure

Acyclic

Structure	Name
–CH=N–	imine
–CH=N–N=CH=	azine or azomethine
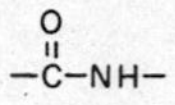 –C(=O)–NH–	amide
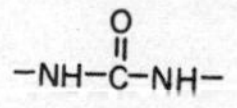 –NH–C(=O)–NH–	urea
 –S(=O)(=O)–	sulfone
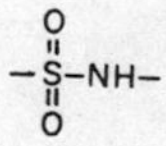 –S(=O)(=O)–NH–	sulfonamide
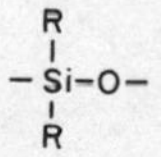 –Si(R)(R)–O–	siloxane
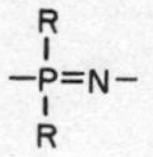 –P(R)(R)=N–	phosphazene
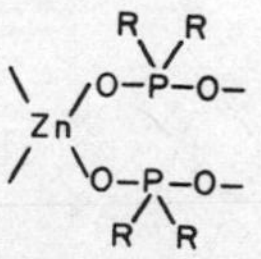 Zn(–O–P(R)(R)–O–)₂	zinc phosphinate

---

*Some structures are shown with bonds at points at which they are usually incorporated into backbones.

Cyclic

Five-member ring *with nitrogen only*

One N

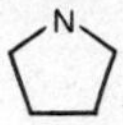

pyrrolidine

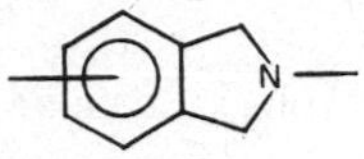

trihydrobenzopyrrole or isoindoline

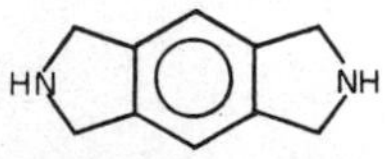

diisoindoline or hexahydrobenzodipyrrole

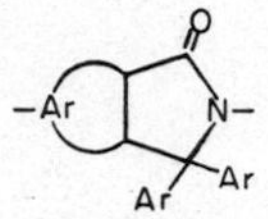

imidine

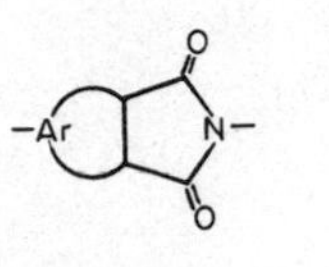

imide

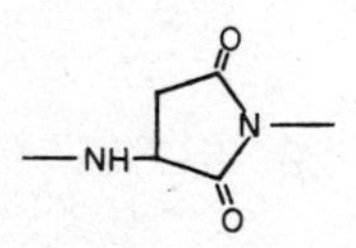

aspartimide

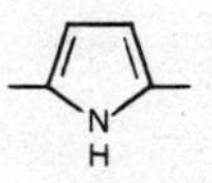

pyrrole

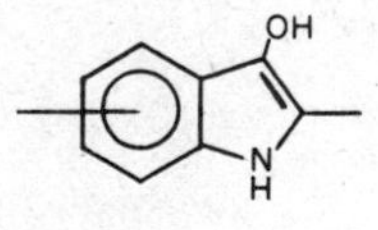

indoxyl

Two N

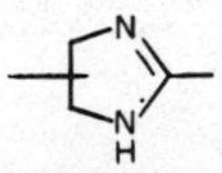

imidazole

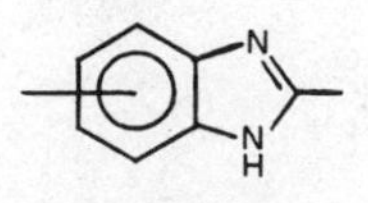

benzimidazole

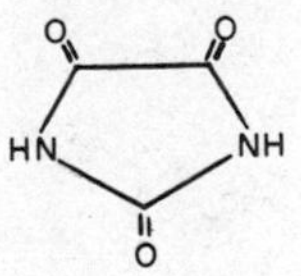

parabanic acid or
1,3-imidazolidine-2,4,5-trione

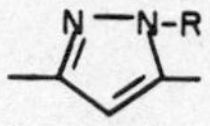

pyrrazole

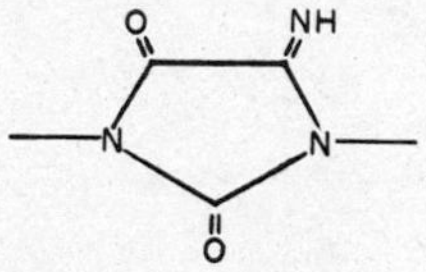

5-imino-1,3-imidazolidinedione

Three N

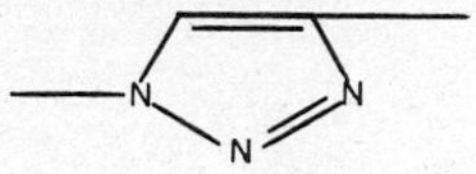

1,2,3-triazole

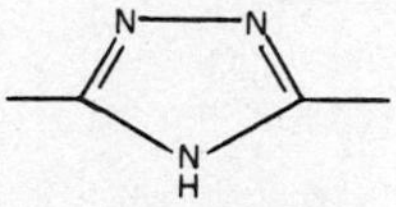

1,2,4-triazole

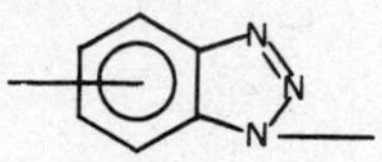

benzotriazole

Four N

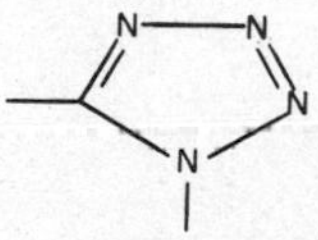

tetrazole

With nitrogen and oxygen

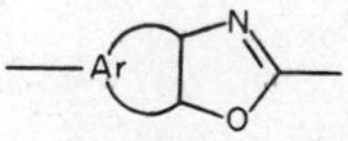

oxazole

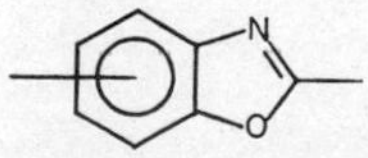

benzoxazole or indolone

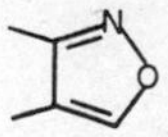

isoxazole

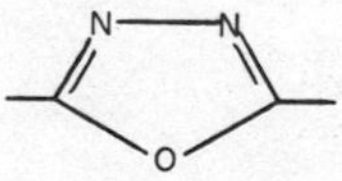

1,3,4-oxadiazole

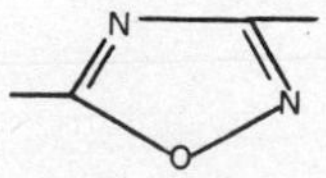

1,2,4-oxadiazole

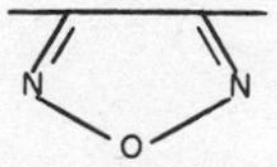

1,2,5-oxadiazole

With nitrogen and sulfur

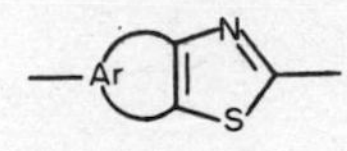

thiazole

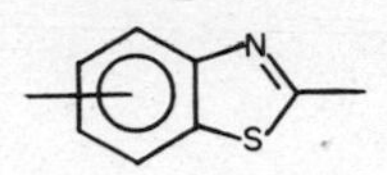

benzothiazole

arylene sulfimide

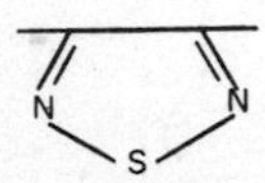

1,2,5-thiadiazole

With sulfur only

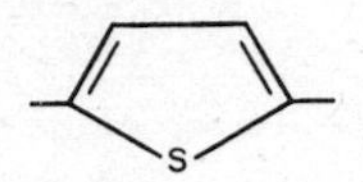

thiophene

Six-member rings with nitrogen only

One N

quinoline

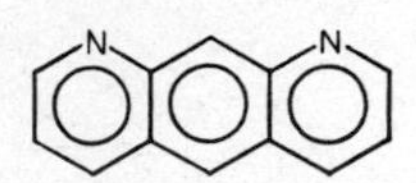

anthrazoline

isoanthrazoline

1,7-phenanthroline

piperazine

Two N

pyrazine

quinoxaline

quinoxalone

quinazolone

quinazolinedione

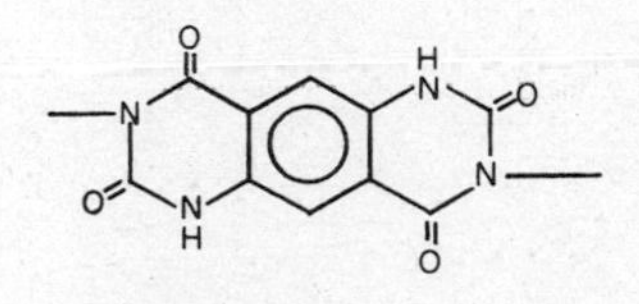

pyrimidoquinazolinetetraone

Three N

s-triazine

as-triazine

Four N

1,2,4,5-tetrazine

1,2,3,4-tetrazine

1,2,4,5-dihydrotetrazine

With nitrogen and oxygen

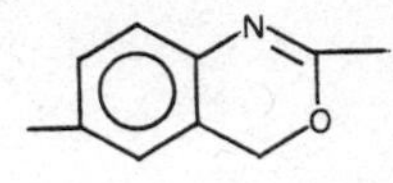

benzoxazine

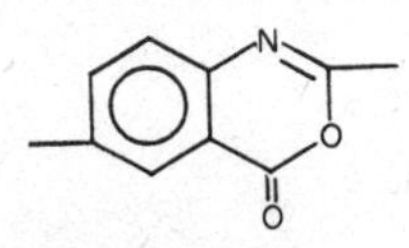

benzoxazinone

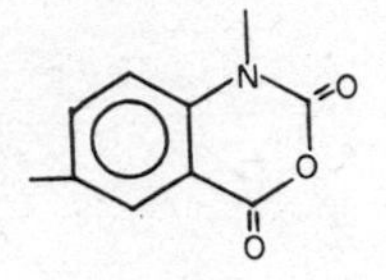

benzoxazinedione

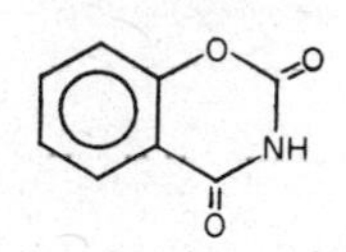

benzoxazadione

With nitrogen and sulfur

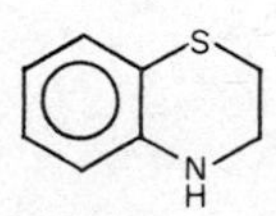

dihydrobenzothiazine

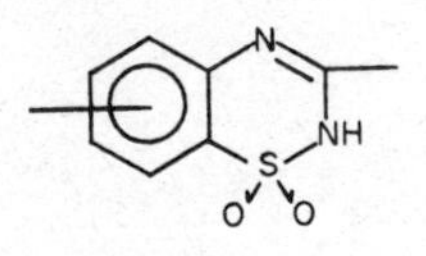

benzothiadiazine dioxide

Fused heterocyclic rings

Two fused five-member rings *with nitrogen only*

Two N

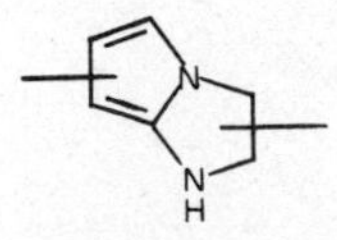

pyrroloimidazole

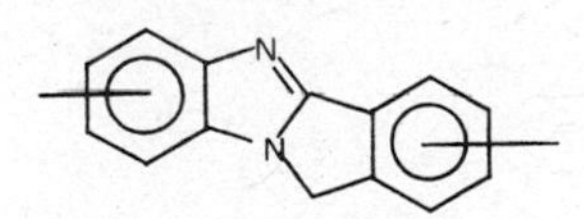

benzylenebenzimidazole

imidazopyrrolone or pyrrone or
benzoylene benzimidazole

Three N

aroylene-s-triazole

With nitrogen and sulfur

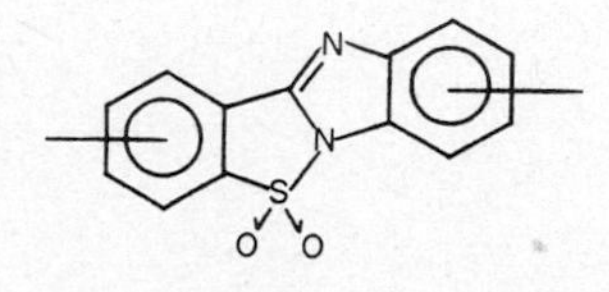

benzothiazone

thiazolothiazole

Fused five- and six-member rings

Two N

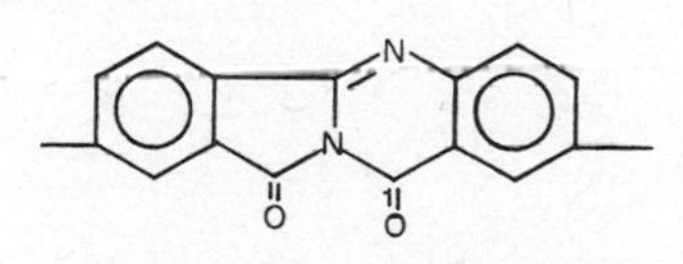

isoindoloquinazolinedione

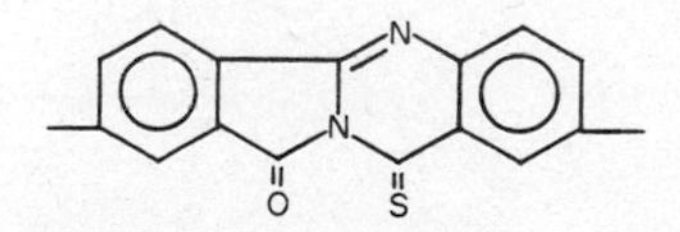

isoindolothioquinazolinedione
or thioquinazolopyrrolone

Three N

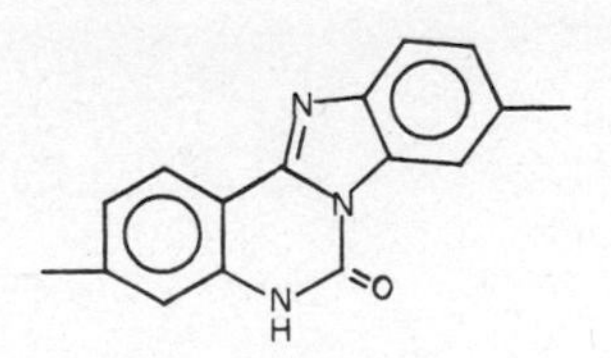

benzimidazoquinazolone

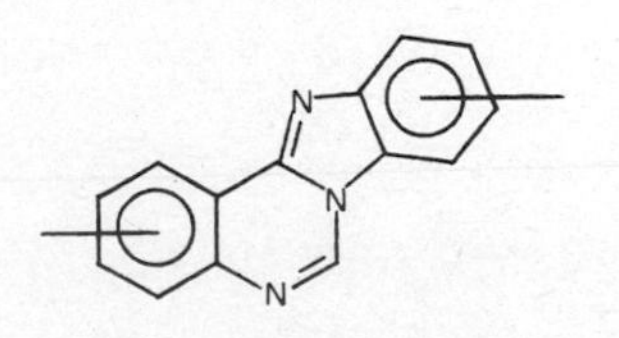

benzimidazoquinazoline

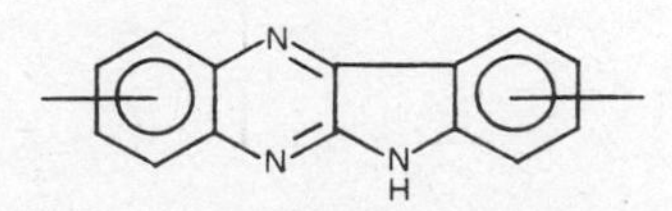

indoloquinoxaline or
benzopyrroloquinoxaline

Four N

triazoloquinaoline or quinazolotriazole

Nitrogen and sulfur

oxoisoindolobenzothiadiazine dioxide

Two fused six-member rings

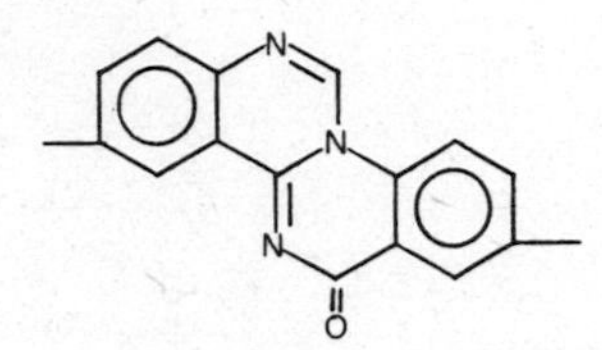

quinazolinoquinazolone

Miscellaneous groups

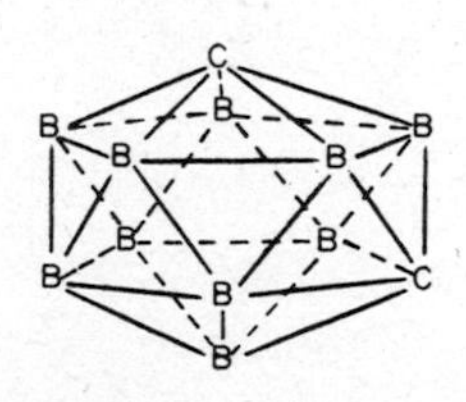

carborane

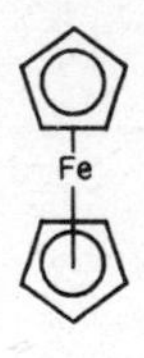

ferrocene

PART 2: According to Nomenclature

amide —C(=O)—NH—

anthrazoline

aroylene-s-triazole

arylene sulfimide

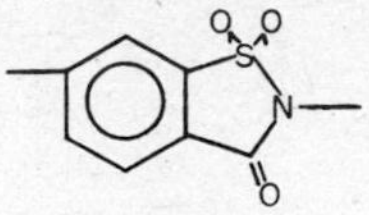

aspartimide

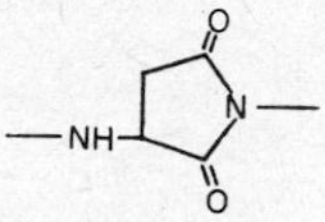

azine —CH=N–N=CH=

benzimidazole

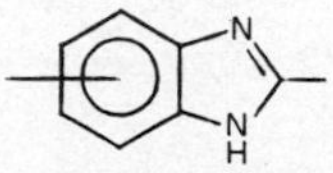

benzimidazoquinazoline

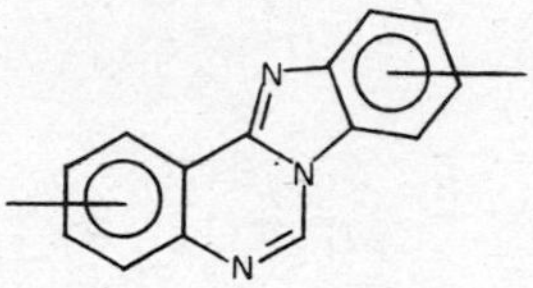

benzimidazoquinazolone

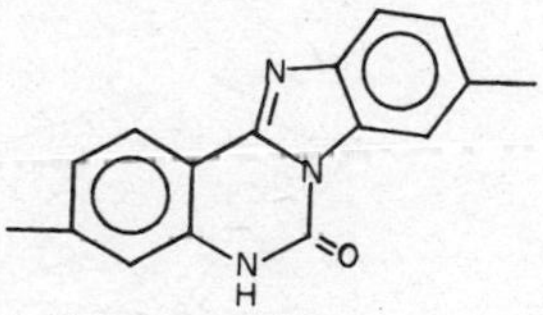

benzothiadiazine dioxide

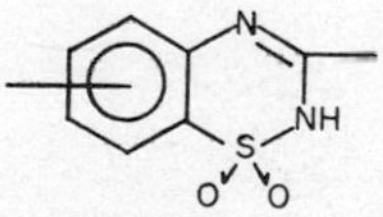

benzothiazole

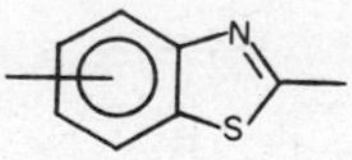

benzothiazone

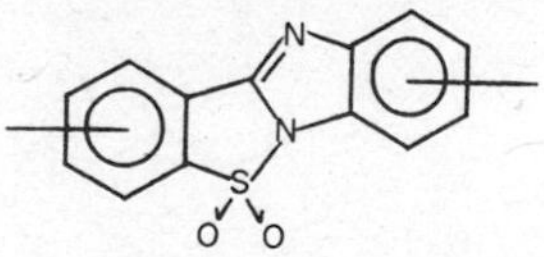

benzotriazole

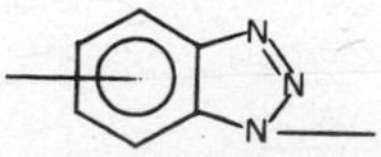

benzoxazadione

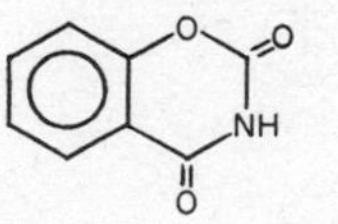

benzoxazine

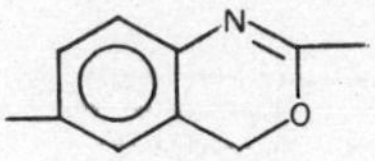

Benzoxazinedione

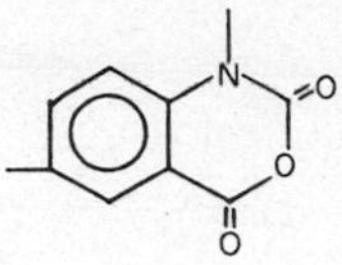

benzoxazinone

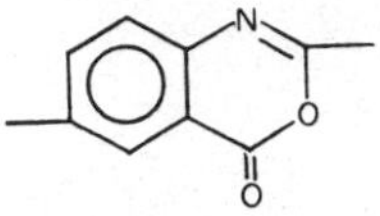

benzoxazole or indolone

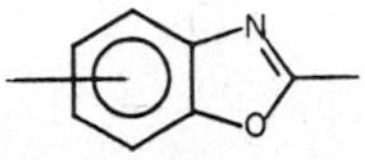

benzylenebenzimidazole

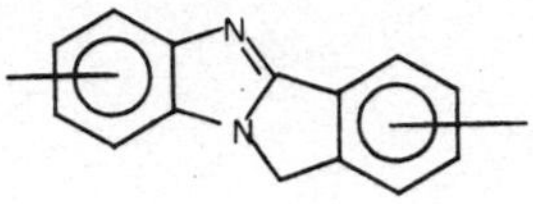

carborane

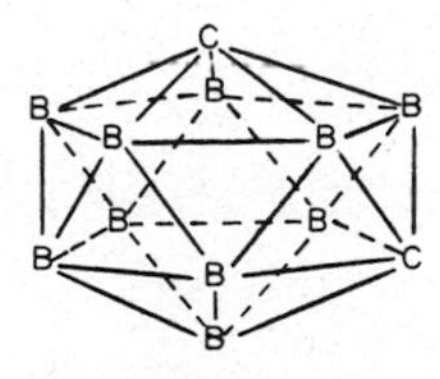

dihydrobenzothiazine

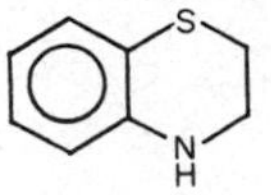

1,2,4,5-dihydrotetrazine

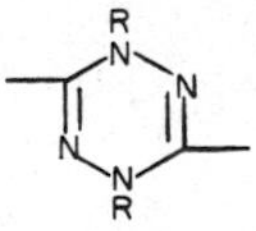

diisoindoline or
hexahydrobenzodipyrrole

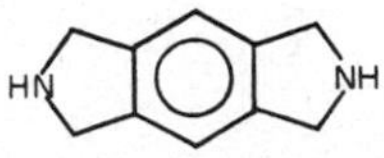

ferrocene

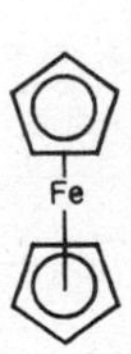

imide

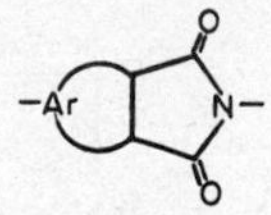

imidine

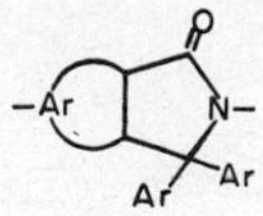

imine $-CH=N-$

imidazole

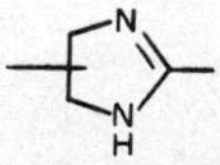

5-imino-1,3-imidazolindinedione

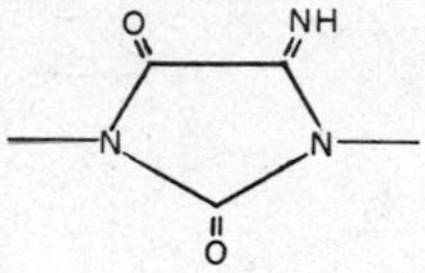

1,3-imidazolidine-2,4,5-trione

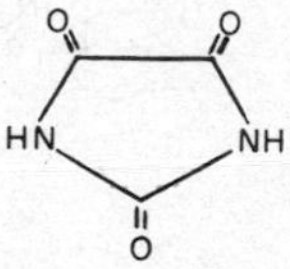

imidazopyrrolone or pyrrone or
benzoylene-benzimidazole

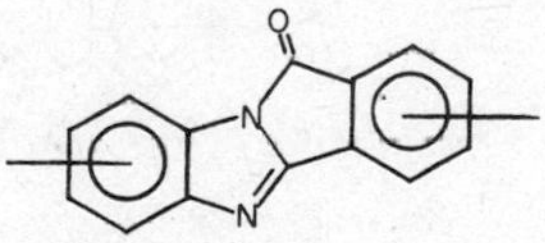

indoxyl

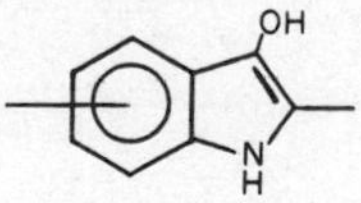

isoanthrazoline

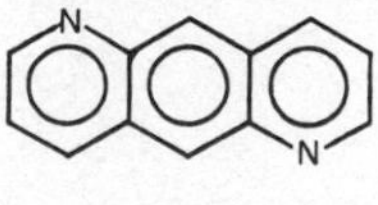

indoloquinoxaline or
benzopyrroloquinoxaline

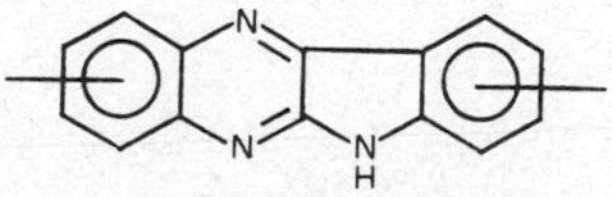

isoindoloquinoline dione

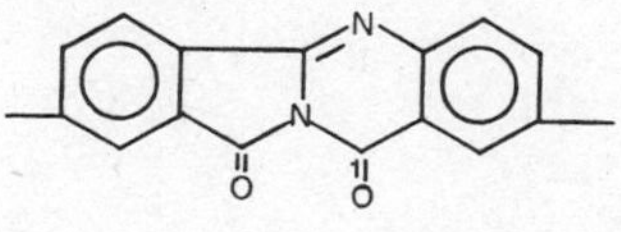

isoindolothioquinazoline dione
or thioquinazolopyrrolone

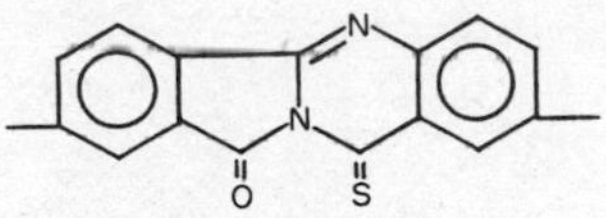

isoxazole

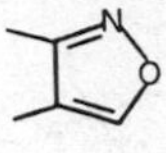

1,2,4-oxadiazole

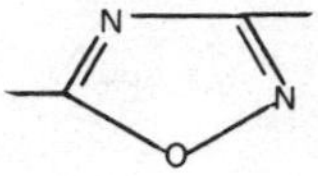

1,2,5-oxadiazole

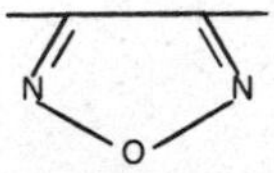

1,3,4-oxadiazole

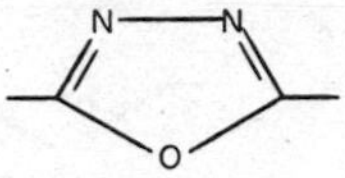

oxazole

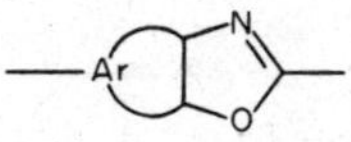

oxoisoindolobenzothiadiazine
dioxide

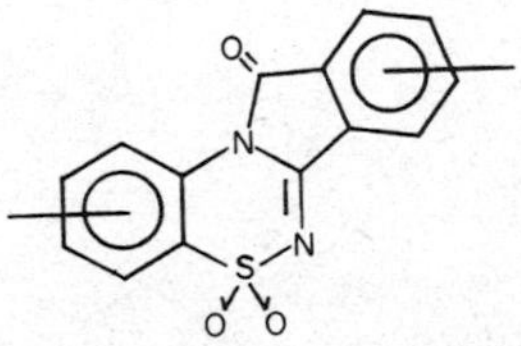

parabanic acid

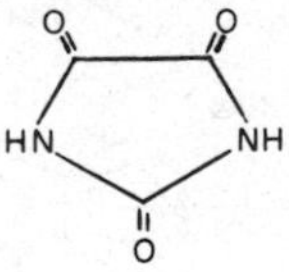

1,7-phenanthroline

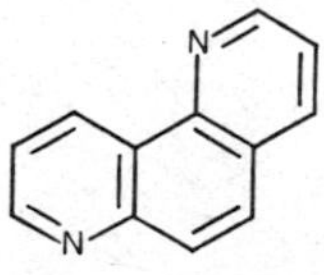

phosphazene

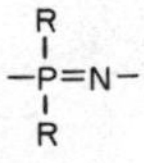

piperazine

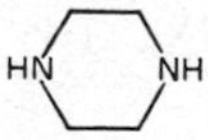

pyrazine

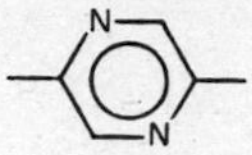

pyrimidoquinazolinetetraone

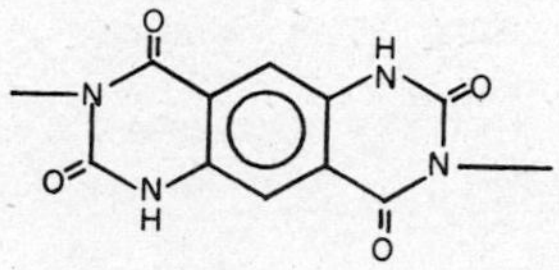

pyrrazole

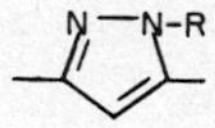

pyrrole

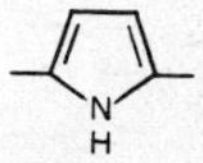

pyrrolidine

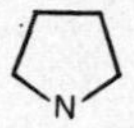

pyrroloimidazole

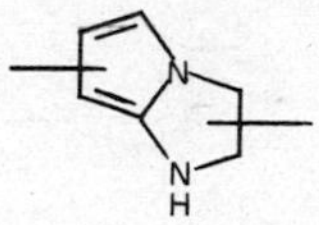

quinazolinedione

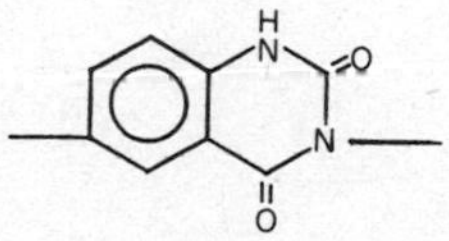

quinazolinoquinazolone

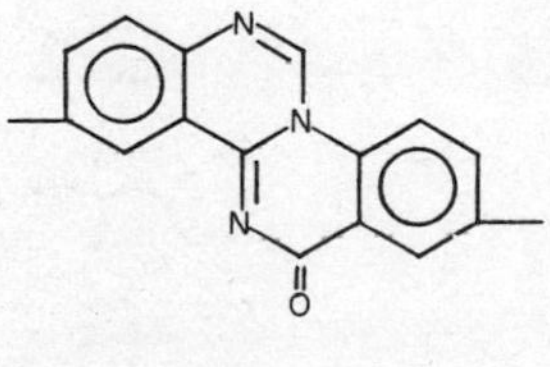

quinazolone

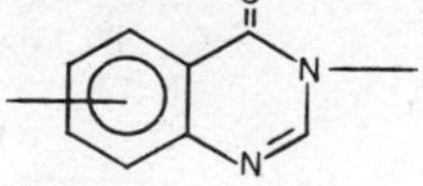

quinoline

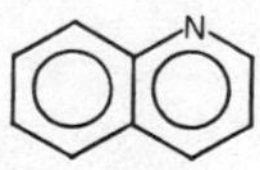

quinoxaline

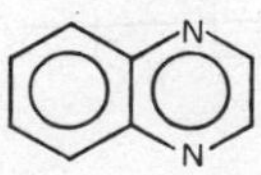

quinoxalone

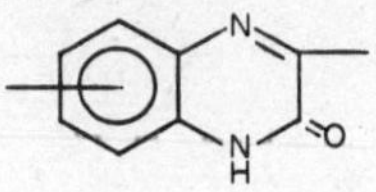

siloxane

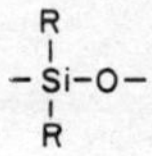

sulfone

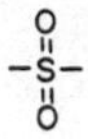

sulfonamide

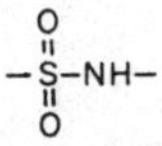

1,2,3,4-tetrazine

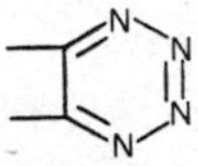

1,2,4,5-tetrazine

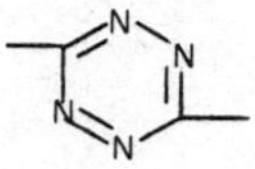

tetrazole

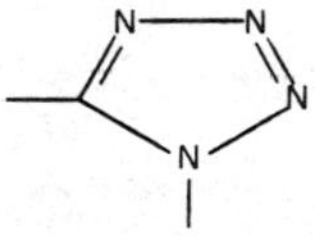

1,2,5-thiadiazole

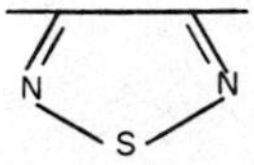

thiazole

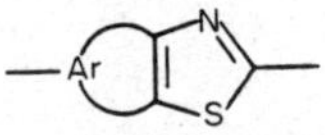

thiazolothiazole

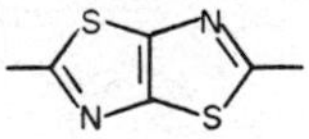

thiophene

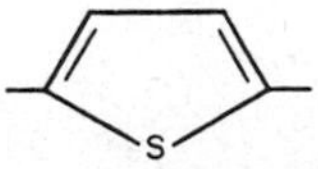

s-triazine

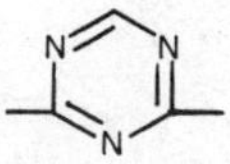

as-triazine

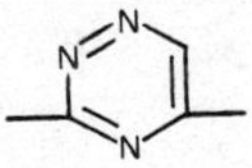

1,2,3-triazole

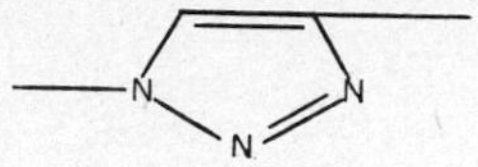

1,2,4-triazole

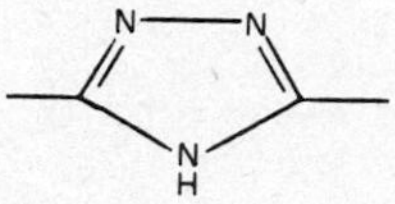

triazoloquinaoline or
quinazolotriazole

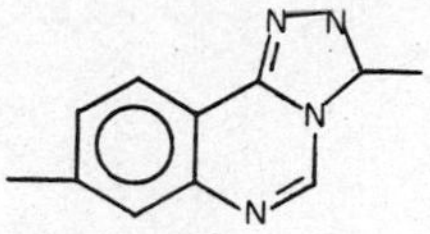

trihydrobenzopyrrole or
isoindoline

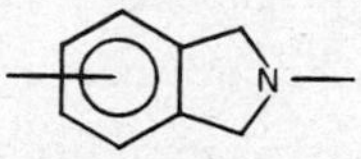

urea

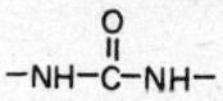

zinc phosphinate

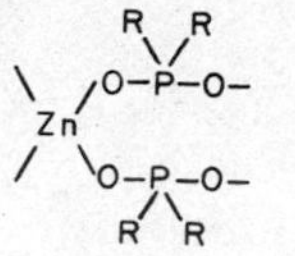

Appendix B

Abbreviations*

BBB poly(bisbenzimidazobenzophenanthroline)
BBL poly(benzimidazobenzophenanthroline) ladder polymers
Bisphenol A 2,2-bis(4-hydroxyphenyl)propane
BTDA 3,4,3',4'-benzophenonetetracarboxylic cianhydride
DAB 3,3'-diaminobenzidine
DMAc N,N-dimethylacetamide
DMF dimethylformamide
DMSO dimethylsulfoxide
DSC differential scanning calorimetry
DTA differential thermal analysis
DTG differential thermogravimetric analysis
EGA effluent gas analysis
EGD evolved gas detection (determination)
EPR ethylene-propylene rubber
HDT heat deflection (or distortion) temperature
HMPA hexamethylphosphoramide
ICTA International Committee for Thermal Analysis
MSA methane sulfonic acid
NMP N-methyl-2-pyrrolidone
PBI polybenzimidazole
PI polyimides
PMDA pyromellitic dianhydride
PMT polymer melt temperature
PPA polyphosphoric acid
PPQ phenylated polyquinoxaline
PQ polyquinoxaline

*Excluding commercial product designations.

PQT	polyquinazolotriazole
SiB	silicone-m-carborane cofunctions
TAB	1,2,4,5-tetraaminobenzene
TBA	torsional braid analysis
TEA	thermal evolution analysis
Tg	glass transition temperature
TGA	thermogravimetric analysis
THF	tetrahydrofuran
Tm	crystalline transition temperature
TMA	thermal mechanical analysis

Appendix C

Commercial Materials

Name	Type	Company
200 P	Poly(arylether sulfone) Tg 230°C	ICI
360	(See Astrel 360)	
720 P	Poly(arylether sulfone) Tg 285°C	ICI
AFLAS	Alternating copolymer of tetrafluoroethylene and propylene	Asahi Glass Co.
Amoco A-I	Poly(amide imide)	American Oil Co.
Arenka	High-modulus fiber of aromatic polyamide	Akzo Co.
Astrel 360	Polyphenylenesulfone	3M
Bakelite Polysulfone	(see: P-1700)	
BK 692	Polyamide	duPont
Conex	Aromatic polyamide [poly(m-phenylene isopthtalamide)]	Teijin
D-202	Poly(siloxane carborane)	Olin-Matheson
Dexsil (now Ucarsil)	Poly(siloxane-m-carborane)	Olin Chemicals, Olin-Matheson
ECD-006	Copolymer of tetrafluoro-ethylene and perfluoro-methyl vinyl ether	duPont
Ekkcel	Polyester	Carborundum
Ekonol	Aromatic polyester	Carborundum
Exten	High-modulus fiber of aromatic polyamide	Goodyear
Fenilon	Aromatic polyamide [poly(m-phenylene isophthalamide)]	USSR
Fiber B (now Kevlar)	High-modulus fiber of aromatic polyamide	duPont Textile Fibers Dept.

Fluorel	Polyfluorocarbon	3M
H-Film (now Kapton)	Polyimide	duPont
Halon	Alternating copolymer of ethylene and chlorotrifluoroethylene (useful to 325°F)	duPont
Imidite 850	Polybenzimidazole-coated, glass fabric adhesive	Whittaker Corp.
Imidite 1850	Polybenzimidazole-coated glass cloth laminating stock	Whittaker Corp.
Kalrez (formerly ECD-006)	Terpolymer of tetrafluoroethylene, perfluoromethyl vinyl ether, and an acrylic acid or derivative thereof (as a site for ionic crosslinking with divalent metal ions)	duPont
Kapton (formerly H-Film)	Polyimide	duPont
Kel F	Polytrifluorochloroethylene	duPont
Kerimid 601	Polyimide, addition cured via maleimide end cap, laminating grade prepolymer	Rhodia, Inc.
Kevlar 49 (formerly Fiber B)	Aromatic polyamide fiber for composites	duPont Textile Fibers Dept.
Kinel 5502 or 5504	Polyimide, addition cured via maleic anhydride termination, 65% E-glass fiber (use temperature 230°C)	Rhodia (Rhone-Poulenc)
KS	Aromatic polyamide resin (see Vespel)	duPont
Kynar	Polyvinylidenefluoride [HDT 204°C (300°F) at 450 KPa (66 psi)]	Pennwalt
Meldin	Polyimide	Dixon
Modmor (no longer available)	Graphite fiber	Morganite
Nolimid A-380	Polyimide, addition-cured, titanium adhesive	Rhodia, Inc.
Nomex	Aromatic polyamide [poly(m-phenylene isophthalamide)]	duPont Textile Fibers Dept.
Noryl	Polyphenylene oxide (compounded with polystyrene)	General Electric
NR-150 A or B	Polyimide	duPont
P-13N	Polyimide, addition-cured via Nadic anhydride end cap	TRW

P-105A	Polyimide, addition cured via Nadic anhydride end cap	Ciba-Geigy
P-1700 or Bakelite Polysulfone	Bisphenol A/dichlorodi-phenylsulfone copolymer	Union Carbide
PBI	Polybenzimidazole	Celanese
PI-2080	Thermoplastic polyimide for compression molding	Upjohn
PI-4701	Polyimide adhesives	duPont
PMR	Polyimide, norbornene- or itaconic acid-terminated	Riggs Engineering Gulf Oil Chemicals Ferro Corp. Fiberite US Polymeric Hexcel
PNF	Fluorinated polyphosphazine	Firestone
PPQ-O	Polyphenylquinoxaline	Whittier Corp.
PPQ-401	Poly(phenylquinoxaline) for laminating or compression molding	Whittaker Corp.
PRD-49	High-modulus fiber of aromatic polyamide	duPont
Parylene-C	Chlorinated polyxylylene	Union Carbide
Parylene-N	Polyxylylene	Union Carbide
Penton	Poly[3,3-bis(chloromethyl) oxetane]	Hercules
Phenylone	(see Fenilon)	
Polymer SP	Polyimide resin (see Vespel)	duPont
Pyralin	Polyimide	duPont
Pyre-ML	Polyimide wire coating, enamel, varnish, or glass composite	duPont Plastics Dept.
QX-13	Polyimide from diacetyl-methylenedianiline and BTDA	ICI
RC #5081	Polyimide	duPont
Ryton	Polyphenylenesulfide	Phillips Petroleum
Skybond 703 (formerly Skygard)	Polyimide	Monsanto
Skygard (now Skybond)	Poly(amide imide) or polyamide or polyimide	Monsanto
SP	Polyimide resin (see Vespel)	duPont
Technoflon	Copolymer of $CH_2 = CF_2$ and $CF_2 = CHCF_3$	Montedison

Teflon FEP	Copolymer of tetrafluoroethylene and hexafluoropropylene	duPont
Tefzel	Copolymer of ethylene and tetrafluoroethylene (HDT 220°F at 66 psi)	duPont
Thermid 600	Polyimide from BTDA (acetylene-terminated)	Gulf Oil Chemicals
Torlon	Poly(amide imide) unfilled injection molding grade (Tg 270-280°C)	American Oil Co.
Ucarsil (formerly Dexsil)	Poly(siloxane-m-carborane) F_1 and F_2: $-CH_3$ and $-CH_2CH_2CF_3$ groups F_3: $-CH_2CH_2CF_3$ groups Me_2: $-CH_3$ groups	Union Carbide
UDEL	Poly(arylethersulfone) from bisphenol A (Tg 190°C)	Union Carbide
Vespel	Formed parts from SP (polyimide) or KS (aramid) resins	duPont
Viton	Copolymer of vinylidene fluoride and hexafluoropropylene	duPont
X-500	Aromatic poly(amide hydrazide)	Monsanto

Bibliography

Allcock, H. R. *Heteroatom Ring Systems and Polymers.* Academic Press, New York, 1967.

Alfrey, T., Jr. Structure-Property Relationships in Polymers, in *Applied Polymer Science* (J. K. Craver and R. W. Tess, Eds.). ACS Division of Organic Coatings and Plastics Chemistry, 1975.

Arnold, C. Sandia Corporation, Albuquerque, New Mexico, Report SC-M-720559, September 1973.

Billmeyer, F. W. *Textbook of Polymer Science*, 2nd Ed. Interscience, New York, 1971.

Block, B. P. *Inorganic Polymers* (F. G. A. Stone and W. A. G. Graham, Eds.). Academic Press, London, 1962.

Braunsteiner, E. E. and H. F. Mark. Aromatic Polymers, *J. Polym. Sci., D, Macromol. Rev., 9,* 83 (1974).

Burns, R. L. Recent Developments in Thermally Resistant Polymers, *J. Oil Colour Chem. Assoc., 53*(1), 52-68 (1970).

Campbell, T. W. *Condensation Monomers* (J. K. Stille, Ed.). Interscience, New York, 1972.

Cassidy, P. E., and N. C. Fawcett. Thermally Stable Polymers: Polyoxadiazoles, Polythiadiazoles, Polythiazoles and Polyoxadiazole-n-oxides, *J. Macromol. Sci.--Rev. Macromol. Chem., C17*(2), 209-266 (1979).

Conley, R. T., Ed. *Thermal Stability of Polymers*, Vols. 1 and 2. Marcel Dekker, New York, 1970.

Cotter, R. J., and M. Matzner. *Ring-Forming Polymerization, Part A, Carbocyclic and Metallorganic Rings.* Academic Press, New York, 1969.

Cotter, R. J., and M. Matzner. *Ring-Forming Polymerizations, Part B, Heterocyclic Rings*, Vols. 1 and 2. Academic Press, New York, 1972.

D'Alelio, G. F., and J. A. Parker. *Ablative Plastics.* Marcel Dekker, New York, 1971.

Frazer, A. H. *High Temperature Resistant Polymers* (Vol. 17 of Polymer Reviews, H. F. Mark and E. H. Immergut, Eds.). Interscience, New York, 1968.

Gerber, A. H., and E. F. McInerney. Survey of Inorganic Polymers, Horizons Research, Inc., Cleveland. Document NASA CR-159563, HRI-396, under NASA contract NAS3-21369, June 1979.

Golub, M. A., and J. A. Parker. Polymeric Materials for Unusual Service Conditions, *Applied Polymer Symposia No. 22* (1973).

Ham, G. E., Ed. Symposium on Polymer Characterization by Thermal Methods of Analysis, *J. Macromol. Sci. Chem.*, *A8*(1), 1-239 (1974).

Jones, J. I. High Temperature Resistant Organic Polymers, *Chem. Brit.*, *6* (6), 251-259 (1970).

Ke, B. *Thermal Analysis of High Polymers*. Interscience, New York, 1964.

Keattch, C. J., and D. Dollimore. *An Introduction to Thermogravimetry*, 2nd Ed. Sadtler Research Laboratories, Philadelphia, 1975.

Korshak, V. V. *Heat Resistant Polymers*. Academy of Sciences of the USSR, 1969. (Translation by the Israel Program for Scientific Translations, Jerusalem, 1971.)

Mark, H., and N. Gaylord, Eds. *Encyclopedia of Polymer Science and Technology, 11*. Interscience, New York, 1969.

Hasegawa, M. Polyhydrazides and Polyoxadiazoles, pp. 169-187.

Levine, H. H. Polybenzimidazoles, pp. 188-230.

Bell, V. L. Polymidazopyrrolones, pp. 240-246.

Sroog, C. E. Polyimides, pp. 247-272.

Kovacic, P. Polyphenylenes, pp. 380-388.

Berenbaum, M. B. Polysulfides, pp. 425-446.

Johnson, R. N. Polysulfones, pp. 447-463.

Stille, J. K. Polyquinoxalines, pp. 389-395.

Marvel, C. S. Polymer Needs and New Synthetic Approaches, *Polym. Preprints*, *17*(1), 239 (1976).

Marvel, C. S. Needs and New Synthetic Approaches, *Polym. Preprints*, *17*(1), 239 (1976).

Marvel, C. S. Thermally Stable Polymers with Aromatic Recurring Units, *J. Macromol. Sci.*, *A-1*(1), 8-12 (1967).

Morgan, P. W. *Condensation Polymers: By Interfacial and Solution Methods*. Interscience, New York, 1965.

Nielson, L. E. *Mechanical Properties of Polymers*. Reinhold, New York, 1962.

Preston, J. Aromatic Polyamide Fibers. *Encyclopedia of Polymer Science and Technology*, Supplement Vol. II (1977), pp. 84-112.

Preston, J., and J. Economy. High Temperature and Flame Resistant Fibers, *Applied Polymer Symposia, 21* (1973).

Ray, N. H. *Inorganic Polymers*. Academic Press, New York, 1978.

Sandler, S. R., and W. Karo. *Polymer Synthesis*, Vol. 1 (Vol. 29 of *Organic Chemistry*, A. T. Blomquist and H. Wasserman, Eds.). Academic Press, New York, 1974, pp. 215-266.

Segal, C. L., Ed. *High Temperature Polymers*. Marcel Dekker, New York, 1967.

Seymour, R. B. *Modern Plastics Technology*. Reston, Reston, Virginia, 1975.

Slade, P. E., Jr., and L. T. Jenkins, Eds. *Thermal Characterization Techniques*. Marcel Dekker, New York, 1970.

Stille, J. K. High Temperature Polymers: Evolution and Future, *Polym. Preprints, 17*(1), 96 (1976).

Wendlandt, W. W. *Thermal Methods of Analysis*. Interscience, New York, 1974.

Willard, H. H., L. L. Merritt, and J. A. Dean. *Instrumental Methods of Analysis*, 5th Ed. D. Van Nostrand, New York, 1974.

Wrasidlo, W. *Advances in Polymer Science*, Vol. 13 of Thermal Analysis of Polymers, Springer-Verlag, New York, New York, 1974.

Author Index

Subject Index